Emerging Technology
Management Trends in Environment and Sustainability

About the Conference

The International Conference (EMTES 2022) is oriented to include the themes like Water Quality Management, Advanced Water Treatment, Advanced Wastewater Treatment, Assessment and Control of Air Pollution, Solid and Hazardous Waste Management, Prevention of Groundwater Contamination, Wetland Management/Phyto-remediation, Case studies in Industrial Pollution Control, Liquid waste management, recent advancement in engineering, technology and management for optimization of environmental issues, application of IOT and IT in remedial measure of Environment and sustainability, Health issues and safety.

Emerging Technology and Management Trends in Environment and Sustainability

Proceedings of the International Conference, EMTES-2022

Edited by
Dr. Sushovan Sarkar (Corresponding Editor)
Professor and HOD(Civil),
Budge Budge Institute of Technology
Dr. Shubhangi Gupta
Executive Director, Budge Budge Institute of Technology
Dr. Ashok Kumar Shaw
Professor and HOD (BSH) and Dean (R &D)

CRC Press is an imprint of the
Taylor & Francis Group, an **informa** business

First edition published 2023
by CRC Press
4 Park Square, Milton Park, Abingdon, Oxon, OX14 4RN

and by CRC Press
6000 Broken Sound Parkway NW, Suite 300, Boca Raton, FL 33487-2742

British Library Cataloguing-in-Publication Data
A catalogue record for this book is available from the British Library

Library of Congress Cataloging-in-Publication Data
[Insert LoC Data here when available]

ISBN: 978-1-032-41095-1 (pbk)
ISBN: 978-1-003-35623-3 (ebk)

DOI: 10.4324/9781003356233

Typeset in: Times LT Std
Typeset by: Aditiinfosystems

Table of Contents

List of Figures

List of Tables

Foreword

It is proud to feel that our country is marching ahead keeping pace with global development. However, it is known that environmental pollution is inevitable and essentially associated with civilization, urbanization, industrialisation and globalization as well. Today, it has become a challenging problem to the scientific community, administrators, policy makers, engineers and technologists, social workers and reformers, stake holders and over and above the beneficiaries of environmental services such as water supply, management of wastewater and solid waste, air quality control, health issue and safety etc. to mitigate the problem arising out human activities in a large scale and to an indiscriminate extent. Especially, in the sectors of water supply, management of water, wastewater and solid waste, industrial pollution control , air pollution monitoring and such other areas, our Governments including Civic Bodies and Local Governments are passing through a tough time to provide the aforesaid environmental facilities to the community, primarily for the region of poor economic edifice of the developing countries like ours.

In the above backdrop the Department of Civil Engineering of BBIT has felt an urgent need to hold this International Conference with special emphasis on Environment, Sustainability, Health issue and safety to take an opportunity to share a scientific platform for addressing the key issues relating to environmental conservation.

For this Conference, I record my sincere gratitude and thanks to entire Trustee of BBIT, all patrons, program chairs, co-convener, organising secretaries, treasurers, media coordinator of conference , our technical partners Nihon University (Japan), The Institution of Engineers (India), The Computer Society of India and finally our publisher Taylor and Francis for their whole hearted cooperation and contribution for making this International Conference EMTES 2022 a grand success.

I hope that the participants /delegates will bear with the organizer for inconvenience, if any during the conference period.

Thanking you

Professor (Dr) Sushovan Sarkar
Convener, EMTES 2022

Preface

Developmental activities and industrialization due to rapid population explosion are two major concerns for environmental exploitation around the world. For balancing the relationship among economic development, operational and environment compliance, effective resource utilization is the greatest challenge leading to innovative approaches to combat these problems. In view of that, technological thrust is necessary to evolve a sustainable approach for abatement of water, air, soil, noise pollution and for the contaminants which create threat even in low concentration over a limited area. To comprehend the research initiatives in the field of sustainable technologies, the present international conference EMTES 2022 is a unique international platform for researchers and academicians to share their views and ideas.

The objective of this conference is to provide a technical get-together for the academicians, engineers, scientists, personnel from research laboratories and industry for reciprocating the advances/recent findings of their research through oral and poster presentation of the technical papers on various themes. This conference will certainly provide an opportunity to learn the recent trends in sustainable Environmental Engineering for mutual benefit of the Human beings and Society.

The world's vision and commitment to sustainable development are growing, and it has been changing over time. The development of engineering, technology and science is to build the foundations. The cause and consequences of climate change are simultaneous. Further, the conferences on the environment impact the lives of the commoners to let them know the issues. Sustainable development is commonly practiced everywhere. The conferences are about the environmental concerns organized at different places. The conference's primary goal is to make people aware of the climate and environment. The Environment conferences enhance the world's population in preserving the human environment. The knowledge of the climatic change affects how natural and anthropogenic the causes are! The other objective is to analyze the possible future climate variability. Moreover, the implication of human society is vast.

The causes we make with pollution are immensely affecting the environment. It aims at developing a conference framework for the protocol on the protection of the atmosphere. Through this conference, awareness is created. The main message of this international conference on the environment helps to bring the necessary changes. The conscious impact will enable the needs of the commoners.

The International Conference (EMTES 2022) is oriented to include the themes like Water Quality Management, Advanced Water Treatment, Advanced Wastewater Treatment, Assessment and Control of Air Pollution, Solid and Hazardous Waste Management, Prevention of Groundwater Contamination, Wetland Management/ Phyto-remediation, Case studies in Industrial Pollution Control, Liquid waste management, recent advancement in engineering, technology and management for optimization of environmental issues, application of IOT and IT in remedial measure of Environment and sustainability.

In EMTES 2022, we received more than 100 numbers abstracts and 50 numbers full papers. Only 21 full papers have been accepted for the proceedings. These papers and deliberations in EMTES 2022 have explored all the technology and management applied in environment and sustainability. These volumes of the proceedings will be useful for research scholars, scientists, faculty members, industry experts, practitioners and the government.

EMTES 2022 express gratitude to Taylor and Francis, NIHON UNIVERSITY, Japan, Institution of Engineers (India) and Computer Society of India for their technical collaboration in this International Conference EMTES 2022.

Professor (Dr) Sushovan Sarkar
Convener, EMTES 2022

Introduction

Welcome to the International Conference on "Emerging Technology & Management Trends in Environment and Sustainability-EMTES 2022 at Budge Budge Institute of Technology., Budge Budge, India. This publication is a collection of full length papers of twenty one (21) delegates from various places all around the country. The main theme of the conference is on the areas like water and wastewater Quality Management, assessment and Control of air pollution, solid and hazardous waste management, prevention of groundwater contamination, wetland management/ phyto-remediation, case studies in Industrial Pollution control, recent advancement in engineering, technology and management for optimization of environmental issues including health and safety, application of IOT and IT in remedial measure of Environment and sustainability.

The conference is being organized by the department of Civil Engineering. Budge Budge Institute of Technology. Budge Budge Institute of Technology (A NAAC and NBA accredited institute) is one of the leading Engineering Colleges in Kolkata under Gupta Trustee Educational Initiatives, one of the premier Education Service Providers in Eastern India offering top quality Technical and Management Education to aspiring students. The Institution is located at Budge Budge, 24 Parganas (S) on a land area of 20 acres, well connected by bus and Eastern Railway. The college was established in 2009and has already churned out 9 batches of B.Tech & MBA and 2 batches of M.Tech. All are placed in Industry & Academics in their respective fields.

The basic aim of the conference is to expose together engineers, scientists, research scholars, industry persons from India and Overseas to discuss about the updated technology and management practices on Environment and Sustainability, present challenges to the world. As we all know that there is a vital need for the researchers and practicing engineers to keep up-to-date of the latest trends and developments in technology and management practices on the areas affecting environment and sustainability including health and safety. Increasing of demands for drinking water, energy, fresh and clean air, safe waste disposal and transportation require environmental protection and infrastructural development. Therefore, there is a serious need of intra-disciplinary, cross- disciplinary and multi-disciplinary collaboration on projects and also in research and development to deal with these demanding problems.

The goal of EMTES-2022 is basically to fulfil the above mentioned perspectives. This international conference is going to provide an opportunity, a platform, and a forum for the students, faculty members, scientists, researchers from Institutes, Industries and research organisations to keep well-informed about the current trends and developments on Environment, sustainability health and safety issues.

Professor(Dr) Sushovan Sarkar
Convener, EMTES 2022

Details of Programme Committee

Chief Patron:

Mr. Jagannath Gupta, Chairman, BBIT

Patrons:

Mrs Urmila Devi Gupta, Trustee

Mr Krishna Kumar Gupta, Vice Chairman, BBIT

Dr Shubhangi Gupta,Executive Director, BBIT

Dr Balram Gupta, CE and MD, BBIT

Ms Visha Gupta, Trustee

Ms Arti Shaw, Trustee

Ms Renu Gupta, Trustee

Ms Shubhadra Gupta, Trustee

Dr Bhabes Bhattacharya, Director General Academics, BBIT

Programme Chair

Dr Sandeep Malik, Registrar and Dean CSE, IT and ECE

Convener

Dr Sushovan Sarkar, Professor and HOD(Civil)

Co-convener

Dr Ashok Kumar Shaw, Professor and HOD(BSH) & Dean (R&D)

Organizing secretary

Dr Narentra Nath Jana, Dean Academics

Jt. Organizing secretary

Dr Siladitya Bandopadhaya, Dean, Student Affairs

Jt. Organizing secretary

Dr Moumita Poddar, Dean, MBA

Treasurer

Mr Parmanand Pandit, Senior Manager (Finance)

Mr Sanjay Gupta (Head - Operation)

Media Coordinator

Mr Amit Gupta

International Advisory Committee

Dr Joe Otsuki, Professor, NIHON UNIVERSITY, JAPAN

Dr Iacopo Carnacina, Liverpool John Moores University, UK

Dr Oscar Castillo, TijuanaInstitute Technology, Mexico

Dr Hossein Hashemi, Lund University, Sweden

Dr Dac-Nhuong Le, Haiphong University, Vietnam

Dr Gonzalo Carrasco, Nanyang Technological University, Singapore

National Advisory Committee

Dr Kalyan Rudra, Chairman, WBPCB

Dr Saikat Maitra, Vice Chancellor, MAKAUT

Dr Dhrubajyoti Chattopadhaya, Vice Chancellor, Sister Nivedita University

Dr Chiranjib Bhattacharya, Pro VC, Jadavpur University

Dr Somnath Mukherjee, Professor, Department of Civil Engineering, Jadavpur University

Dr Sunil Kumar, Senior Principal Scientist and Head, Waste Reprocessing Division CSIR-NEERI, Nagpur

Dr Papiya Mandal, Principal Scientist, CSIR-NEERI, Delhi

Dr Makarand M.Ghangrekar, Professor, IIT Kharagpur

Dr Aniruddha Nag, Former Chairman, CSI

Mr Jagdish Chand Singhal, Chairman Environmental Engineering Division Board, The Institution of Engineers India

CHAPTER 1

Climate Change and its Importance for Assessment of Environmental Sustainability

Debasmita Baidya[1]
Student, Department of Civil Engineering, Jadavpur University, Kolkata

Somnath Mukherjee[2]
Professor, Environmental Engineering Division,
Department of Civil Engineering, Jadavpur University, Kolkata

Abstract

Climate change along with environmental sustainability is a widely discussed topic buzzing around the world. The theme slogan of "Save Earth" for the year 2022 also has direct relevance to this topic. Climate change refers to long-term shifts in temperature and weather patterns. These shifts may be natural, such as thorough variations in the solar and hydrological cycle. But through the years, due to rapid industrialisation, urbanisation and globalizing efforts, ascending trend of deforestation imparted by various anthropogenic activities has acted as the main driver of climate change. It is reported in the literature that an amalgamation of increased use of fossil fuels for various daily life purposes and deforestation for land use becomes the main cause of climate change. The last decade (2011–2020) was the warmest decade on record. According to the World Health Organisation, 24% of deaths can be traced back to avoidable environmental factors, which are majorly driven by the adverse effects of climate change. The Climate Change Performance Index (CCPI) is an independent monitoring tool for tracking the climate protection performance of 60 countries and the European Union (EU). In order to save Earth from these key issues, climate change performance and its quantity evaluation are extremely important. It is exhibited in the literature. The climate change performance is assessed in four (4) categories viz. Greenhouse Gas Emissions (40% of overall score), Renewable Energy (20% of overall score), Energy Use (20% of overall score), and Climate Policy (20% of overall score). A country's performance in each of the categories as 1–3 is defined by its performance regarding four different equally weighted indicators, reflecting four different confined of the category: "Current Level", "Past Trend (5-year trend)", "2°C-Compatibility of the Current Level" and the "2°C-Compatibility of 2030 Target". India has consecutively ranked tenth twice in the overall CCPI rating. The present paper discusses basic aspects of CCPI and its evaluation through different categorical analyses and scores based on information available in the literature.

Keywords: assessment, CCPI, climate change, sustainability, global scenario

[1]debasmita.baidya11299@gmail.com
[2]somnath.mukherjee@jadavpuruniversity.in, mukherjeesomnath19@gmail.com

Introduction

The ability of humans and ecological populations depends heavily on the weather and climate. The latter is greatly surmounted by a weather pattern's long-term influence. The commitment to "Save Earth" for World Environment Day 2022 has a strong connection with the climate and how it has changed the world since its inception.

Long-term changes in temperature and weather are referred to as climate change. These changes could have arisen naturally, perhaps as a result of oscillations in the solar cycle (UN General Assembly, United Nations Framework Convention on Climate Change). But since the 1800s, fast industrialisation, widespread urbanisation and overcompensation have been observed as the primary drivers of climate change. These factors are burning fossil fuels, building skyscrapers and emitting other harmful gases from stacks.

The burning of fossil fuels emits greenhouse gases (GHG) that act like a blanket wrapped around the earth, trapping the sun's heat and raising temperatures under unfavourable meteorological conditions.

Major contributors to climate change include GHGs, such as carbon dioxide with ascending concentration and methane. As individual factors continue to expand, more and more of these gases are released into the atmosphere. This rapid industrialisation is then followed by urbanisation, which mostly involves deforestation to make room for additional industrial complexes with more houses and apartments for human dwelling. The biggest contributor to climate change is an amalgamation of rising fossil-fuel consumption for varied daily needs and deforestation for land usage. The world is currently 1.1°C warmer than it was in the late 1800s despite rising emissions. According to the Climate and Energy Eco-Justice Collaborative, the most recent 10 years (2011–2020) were the warmest ever recorded (Eco-Justice Collaborative).

Human beings have long attributed climate change's impacts to rising temperatures, but because the world is a system, a change in one area inevitably affects all other aspects of it. Intense droughts, water scarcity, destructive fires, increasing sea levels, flooding, melting polar ice, catastrophic storms and dwindling biodiversity are just a few of the repercussions of climate change that are currently occurring (Eco-Justice Collaborative). In addition to upsetting the ecological balance, climate change has a direct impact on human health, food production, housing, safety and employment. Long-lasting droughts are putting people in danger of starvation, while climatic effects, such as sea-level rise and saltwater intrusion, have already forced entire villages to evacuate. If real action is not taken now to stop climate change, the number of "climate refugees" is only predicted to increase in the future. The World Health Organisation estimates that 24% of deaths are attributable to avoidable environmental causes, the majority of which are caused by the negative consequences of climate change (Eco-Justice Collaborative).

The problem of environmental sustainability has gained attention since the last decade with the onset of climate change. In order to support health and welfare both today and in the future, it is necessary to safeguard global ecosystems and conserve natural resources (Sphera). It can be simply defined as "filling today's requirements without compromising the ability of future generations to satisfy their needs", according to the US Environmental Protection Agency. Depending on the geographical, economic, social and environmental circumstances, there are many different environmental requirements for sustainability. Every country, as well as the governments of each state and province, has different criteria for the quality of the air, water, soil, wildlife habitats and carbon emissions. These standards are upheld through monetary penalties and legal action. However, studies have shown that in order to achieve environmental sustainability, a more extensive worldwide set of legislation or a stronger commitment from businesses themselves will be required.

The concept of sustainability covers a wide range of environmental ideas and business obligations. Both the average person and major industries must realise that their efforts to adopt environmental sustainability do not have to conflict with society's ability to function or with their desire to make a profit. As the definition of development as it has been used up to this point must shift to create room for bettering global living standards in the near future, renewable energy and resources should be the new future. A growing number of nations have pledged to achieve net zero emissions by 2050, but to keep global warming to 1.5°C, nearly half of those reductions must be implemented by 2030 (United Nations, Key Findings). Between 2020 and 2030, fossil-fuel production must decrease by around 6% annually (United Nations, Key Findings).

This current paper aims to provide an introduction to one of the recent methods of assessment of climate change, namely the Climate Change Performance Index (CCPI). In the coming sections, the assessment procedure and results of the CCPI 2022 and India's performance in comparison to the global scenario are discussed which in turn provide valuable hints for the improvement of climate change protection practises.

Recent Methods of Assessment—Need for the CCPI

Climate change and its resulting effect on sustainability are translated in a term known as the CCPI. The CCPI is a term for global index on the climate change issue. In accordance with the Paris Agreement, global warming must be kept to a maximum of 1.5°C to lessen the severity of its effects. The only way to minimise GHG emissions, which cause climate change, is by firm action. The CCPI plays a key role in providing information on the Paris Agreement's implementation phase as an independent monitoring instrument (Burch et al., 2021). The CCPI has been analysing national climate protection efforts since 2005. It establishes transparency in climate policy, enables comparison of climate protection initiatives and enables us to track advancements and setbacks. 60 nations and the European Union, which together account for more than 90% of global GHG emissions, are evaluated by the CCPI. Using standardised criteria, the CCPI uses as a tool in four broad diversions with 14 indicators: GHG Emissions (40% of the overall score), Renewable Energy (20%), Energy Use (20%) and Climate Policy (20%). The distinctive climate policy part of the CCPI assesses how well each nation is doing in terms of framing policies into place that advance the objectives of the Paris Agreement. India is currently ranked 10th on the CCPI as of 2021. Although Denmark topped the index in 2021, no nation performs well enough across all CCPI index categories to receive an overall very high grade, leaving the top three spots in the overall ranking (Burch et al., 2021). The Portuguese government has made use of the CCPI to share successes and provide explanations for disappointments. It has been a yearly indicator used as an evaluation tool of the development of Portugal's climate change policies by stakeholders and the media (Burch et al., 2021). In a global setting where nations are already implementing newer and better policies to perform better in the next rankings, this demonstrates the credibility of the CCPI.

Assessment Procedure of the CCPI

Climate change performance is assessed in four distinct categories:

1. GHG Emissions (40% of overall score);
2. Renewable Energy (20% of overall score);
3. Energy Use (20% of overall score);
4. Climate Policy (20% of overall score).

A country's performance in each of the categories 1–3 is defined by its performance regarding four different equally weighted indicators, reflecting four different confines of the category: "Current Level", "Past Trend (5-year trend)", "2°C-Compatibility of the Current Level" and the "2°C-Compatibility of 2030 Target". These 12 indicators are complemented by two indicators under the category "Climate Policy", measuring the country's performance regarding its national climate policy framework and implementation as well as regarding international climate diplomacy (Burch et al., 2021). Figure 1.1 exhibits an overview of the composition and weightage factor of the four categories and 14 indicators defining a country's overall score in the CCPI (Burch et al., 2021).

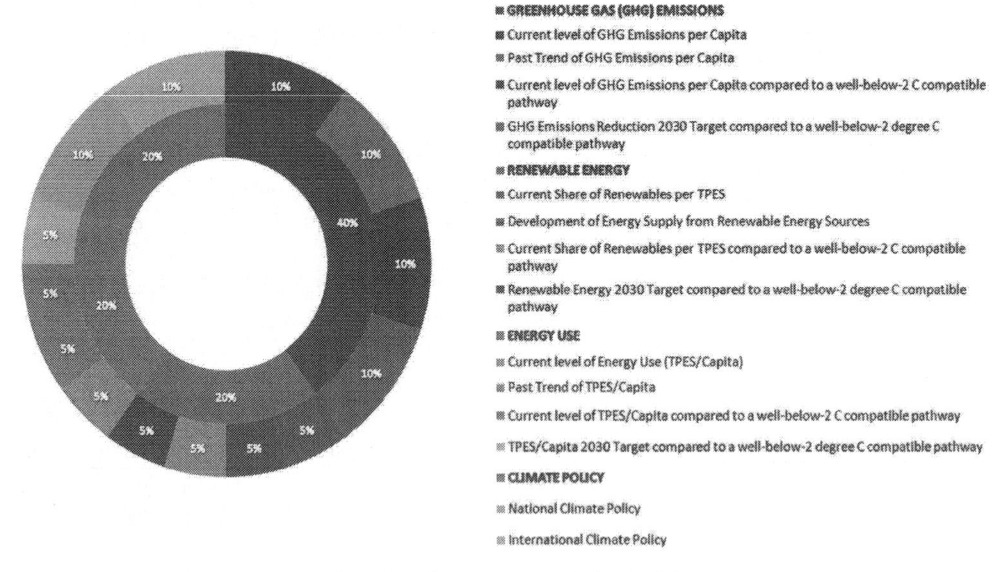

Fig. 1.1 Components of the CCPI

Source: (Burch et al., 2021)

Climate policy indicator and scoring value indicate to the serious contribution to save Earth by the countries evaluated by the CCPI. The index rewards policies that aim for climate protection, both at the national level and in the environment of international climate diplomacy. Whether or not countries are stimulating and seeking a better performance can be derived from their scores in the "Climate Policy" indicators. Whether these policies are effectively enforced, can be read – with a time lag of a few years – in the country's improving scores in the indicators "Renewable Energy" and "Energy Use", and incipient positive developments in the indicator "GHG Emissions". Following this sense, the index takes into account the progress in the three areas eventually showing their effect on a country's GHG emissions performance with a weighting of 20% each:

- An effective climate policy
- An expansion of renewable energy
- Advancements in energy effectiveness and therefore control over domestic energy use

With this weighting system, the CCPI is a sufficiently flexible tool to capture current changes in climate policy and successes in the effort to reduce GHG emissions. This category has the largest weight in the index (40%), as it is necessary to reduce GHG emissions to prevent serious climate change. The CCPI measures both trends in emissions and levels within this category, giving a complete view of a country's performance without favouring simply those that are cutting emissions from a genuinely high position or those that still have low levels but are increasing them significantly. A balanced assessment of a country's performance is ensured by the combination of looking at emissions from several angles and, as of 2017, also taking into account a country's performance in connection to its unique well-below-2°C pathway. For each of these pointers, the countries admit a standing between "very high" and "very low". The indicator-specific limits for the rating can be set up in the section fastening on the very indicator. In the following sections, the categories for assessment are further discussed.

GHG Emissions (40% of Overall Score)

The GHG emission, caused by anthropogenic activities apart from other natural activities, such as a volcano, is the driving factor that increases the greenhouse effect, which leads to climate change. It being the largest contributing factor to climate change, is perceived as the most significant measure of the success of climate policies. It contributes 40% to the overall score of a country. The data used for evaluation is taken from Paris Reality Check (Paris Reality Check: PRIMAP-hist) dataset provided by the Potsdam Institute for Climate Impact Research (PIK). Owing to the diversity in the countries evaluated in the CCPI, more than one perspective of GHG emissions level and their development is taken for proper evaluation. The four GHG emissions level indicators are shown in Fig. 1.2 (Burch et al., 2021).

Current Level of GHG Emissions Per Capita

The level of current per capita GHG emissions changes only in a longer-term perspective which makes it an indicator of the starting point of the country being evaluated. It takes into account each country's development situation which in turn addresses the equity issue. A "very high" ranking is awarded for a maximum of 2.5 CO_2te/Capita, a "high" or "medium" rating is awarded to emissions within the 5.5 to 8 CO_2te/capita range while a "very low" rating represents emissions more than 11 CO_2te/capita (Burch et al., 2021).

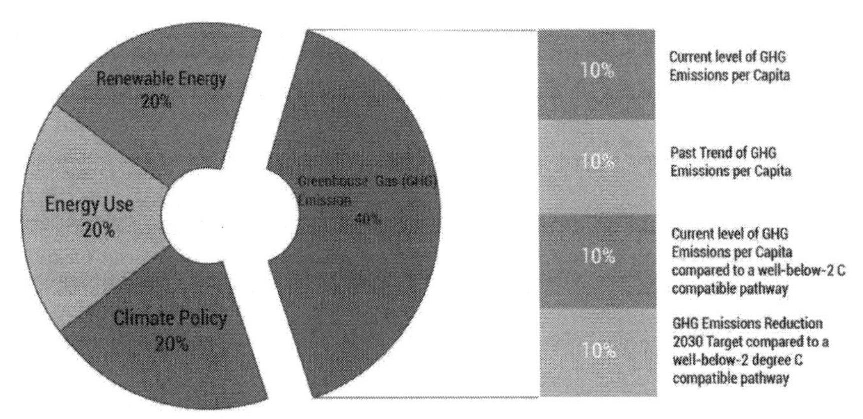

Fig. 1.2 Weighting of GHG Emissions Indicators

Source: (Burch et al., 2021)

Past Trend of GHG Emissions Per Capita

The trend over five years of GHG emissions per capita is considered for evaluation of this category because measuring recent development in emissions relates directly to the successful implementation of a country's climate policy. A "very high" rating is awarded to the countries with a 20% decrease in emissions, while a "high" rating is assigned for a 7% decrease in emissions. In case of an increase in emissions, a country receives a "low" rating, and if the increase is more than 5% in the past 5 years, then the country is assigned a "very low" rating (Burch et al., 2021).

Current Level of GHG Emissions Per Capita Compared to a Well Below-2°C Compatible Pathway

The criterion for a well below-2°C compatible pathway is based on reaching GHG neutrality on a global scale within the 2nd half of the century, which is almost synonymous with the long-term target according to the Paris Agreement. For estimating this indicator, the distance of a country's current level of per capita emissions to its well below-2°C compatible pathway is calculated, as illustrated in Fig. 1.3. In case the country's per capita emissions level undercuts its pathway, a "high" rating is awarded. For cases where the difference is above 3 CO_2te/capita, a "very high" rating is assigned. For differences of up to 2 and 4 CO_2te/capita, a "medium" or "low" rating is awarded, respectively, and for any value above this, a "very low" rating is awarded.

GHG Emissions Reduction 2030 Target Compared to a Well-Below-2°C Compatible Pathway

The CCPI also assesses a nation's 2030 mitigation target, or its efforts to reduce emissions by that year. The distance between this aim and the country's pathway, which was established using the shared but differentiated convergence technique, is measured in order to do this. The absolute difference is expressed as tCO_2e/capita.

An objective that is lower than the nation's pathway can result in a "high" or "very high" grade. A nation will obtain the best rating of "very high" if the difference is even higher than

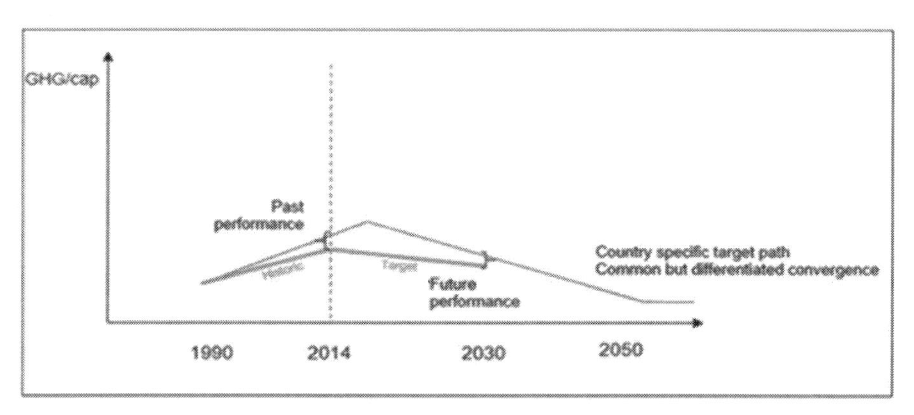

Fig. 1.3 The actual pathway of GHG emissions (green) vs. well-below-2°C target pathway (orange) is shown for an over-performing country

Source: (Burch et al., 2021)

–1 CO_2te/capita. A difference of up to 1 CO_2te/capita will result in the country receiving the "high" classification. Targets that deviate from the pathway by up to 2 CO_2te/capita are rated as having a "mid" performance and those that deviate by up to 4 CO_2te/Capita as having a "poor" performance. The country is rated as "extremely low" if there is no target or if it even exceeds 4.

Renewable Energy (20% of Overall Score)

Renewable energy is defined as energy collected from renewable sources that are naturally replenished on a human timescale. With CO_2 concentrations crossing 400 parts per million in the atmosphere, a rapid increase in renewable energy production and simultaneous phasing out of fossil fuel-derived energy is required for achieving the goal of limiting global warming well below 2°C. It is equally important to increase global energy efficiency, which consequently

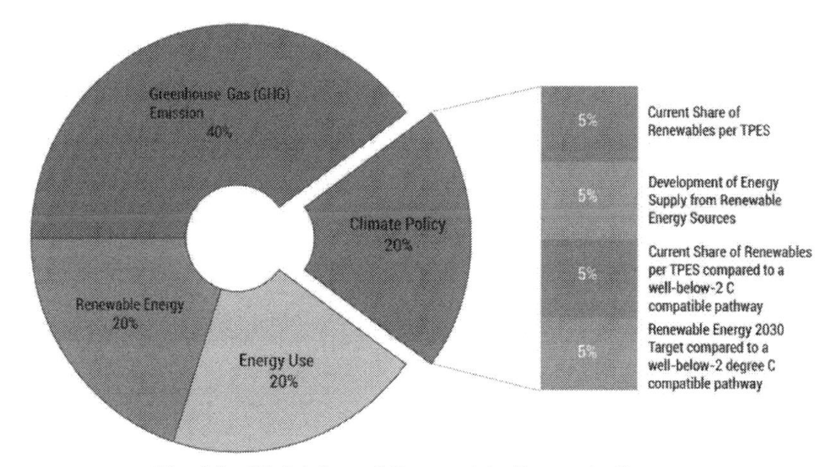

Fig. 1.4 Weighting of Renewable Energy Indicators

Source: (Burch et al., 2021)

will lead to a reduction in global energy use. This category, hence, evaluates the targeted increase in the share of renewable energy for emissions reduction and contributes to 20% of the overall rating of a country. It comprises four indicators, which are shown in Fig. 1.4 (Burch et al., 2021).

Current Share of Renewable Energy Sources per Total Primary Energy Supply (TPES)

In total, 5% of the overall ranking is based on the share of renewable energy sources per TPES, which includes hydropower. A country receives a "very high" ranking for a minimum share of 35%, a "high" ranking for a share of 20%, a "medium" ranking for a share of 10% and less than 5% represents a "very low" rating.

Past Trend of Energy Supply from Renewable Energy Sources per TPES

The recent development of energy supply from renewable sources is considered for evaluation of this indicator, which accounts for 5% of the CCPI score. A "very high" rating is awarded for an increase in the share of renewable by 75% in the past 5 years, "high" for 30%, "medium" for 15%, "low" for 5%, and if the share of renewable energy sources decreases, then a "very low" rating is assigned to that particular country.

Current Share of Renewables per TPES Compared to a Well-Below-2°C Compatible Pathway

According to the Paris Agreement, GHG emissions should reach net zero by the middle of this century. To achieve the target, the transition to renewable energy resources is indispensable. This sets the benchmark for a well-below-2°C compatible pathway within the "Renewable Energy" category to be a share of 100% renewable energy by 2050 (including hydropower). The global benchmark to target performance can be visualised as shown in Fig. 1.5. A "very high" rating is awarded if a country overreaches its pathway, a "high" rating for undercutting up to 10%, a "medium" rating for a difference up to 15%, a "low" rating for undercutting by 17.5% and anything above results in a "very low" rating (Burch et al., 2021).

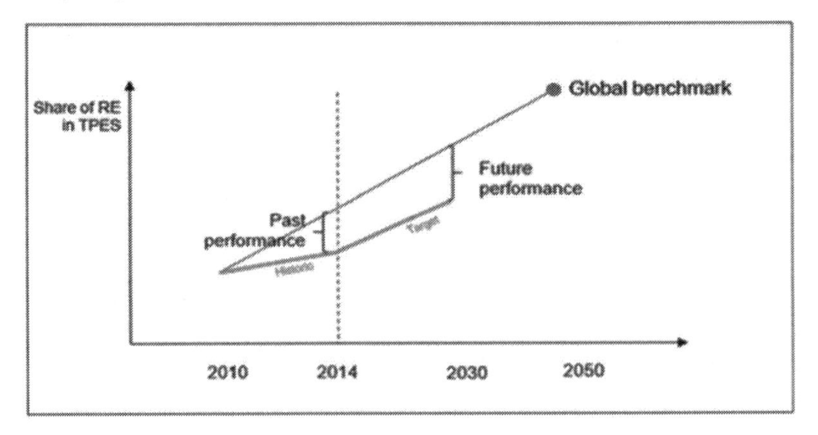

Fig. 1.5 Renewable Energy Pathway

Source: (Burch et al., 2021)

Renewable Energy 2030 Target Compared to a Well-Below-2°C Compatible Pathway

The CCPI assesses the gap between a nation's 2030 renewable energy target and its preferred path from 2010 to 100% renewable energy in 2050, which takes into account hydropower (using a linear pathway for methodological reasons). Since there are no common guidelines for achieving such targets, comparing renewable energy targets is a significant difficulty because different countries present their renewable energy targets in different ways. Some nations only have national targets, whereas others have targets for sub-national states. Instead of focusing on the percentage of renewable energy in the TPES, others describe their goals in terms of installed capacity.

Energy Use (20% of Overall Score)

As previously mentioned, an extensive increase in energy efficiency is instrumental in achieving GHG neutrality by mid-century as per the Paris Agreement. For this category, the per capita energy usage and measures progress for each country are evaluated. To provide a more balanced evaluation, as in the case of the previous two categories, the energy use category is represented by four indicators which are shown in Fig. 1.6 (Burch et al., 2021).

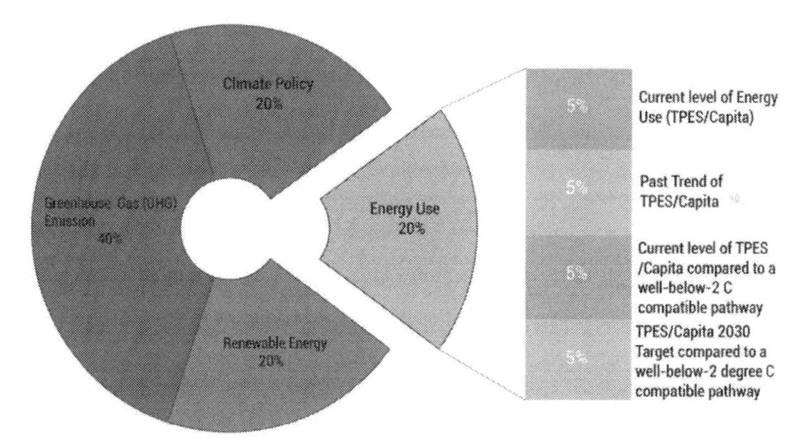

Fig. 1.6 Weighting of Energy Use Indicators

Source: (Burch et al., 2021)

Current Level of Energy Use Measured as TPES per Capita

The current level of energy use measured as TPES per capita is used as an indicator to account for the countries which from a very low level are still increasing their energy use per capita. A "very high" ranking is awarded for a maximum of 60 unitTPES/capita, a "high" ranking for 90 unitTPES/capita, a "medium" ranking for unitTPES/capita, and a "very low" rating for more than 160 unitTPES/capita (Burch et al., 2021).

Past Trend of Energy Use Measured as TPES per Capita

Similar to the previous two categories, the recent developments in the effort to increase energy efficiency in per capita energy use in the past 5 years are measured in this category. A "very high" rating is awarded for more than a 15% decrease in energy use and a "high" rating for a

5% decrease. A "low" or "very low" rating is assigned in cases where energy use has increased by more than 10% over the last 5 years (Burch et al., 2021).

Current Level of TPES per Capita Compared to a Well-Below-2°C Compatible Pathway

The CCPI chosen benchmark for a well-below-2°C compatible pathway in 2050 is the same as the energy use per capita as the current global average. The current global average is 80 gigajoules per capita in TPES. A linear pathway from 1990 to the described benchmark 2050 is calculated, after which the distance of each country's level is measured to this calculated pathway. An example of the same has been visualized in Fig. 1.7. A "high" or "very high" rating in case of a 30% difference is awarded when a country undercuts its pathway. For 10% or 30% difference, "medium" and "low" ratings are assigned respectively and any value above this is deemed as a "very low" rating (Burch et al., 2021).

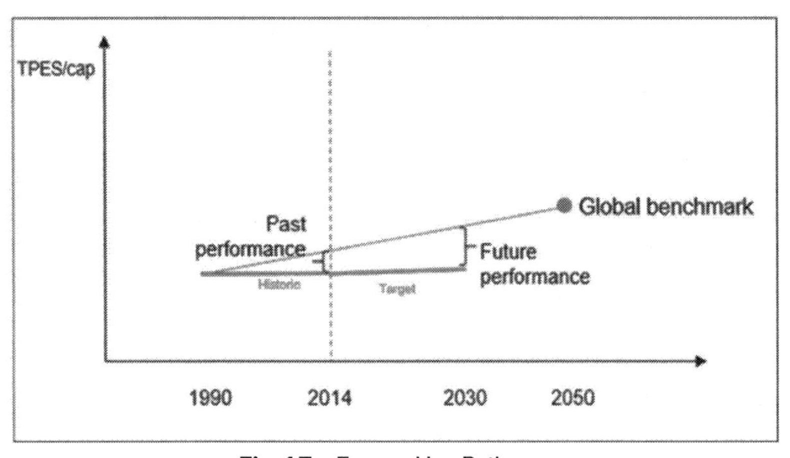

Fig. 1.7 Energy Use Pathway

Source: (Burch et al., 2021)

Energy Use TPES per Capita 2030 Target Compared to a Well Below-2°C Compatible Pathway

This statistic represents the separation between the nation's 2030 energy targets along its pathway and the 2050 benchmark. Instead of using a relative scale, this distance is measured in absolute terms. The result is extrapolated or interpolated as necessary to achieve the figure for 2030 if the targets are for any year other than 2030. The IEA energy balances database includes all TPES data. A target undermining the nation's pathway can result in a "very high" or "high" rating. A country will be rated as having the second-best pathway if the difference is only up to 30%. Targets that deviate from the pathway by up to 10% are classified as performing "medium"; those that deviate by up to 40% or more are classified as performing "low" or "very low" (Burch et al., 2021).

Climate Policy (20% of Overall Score)

In the CCPI, the "Climate Policy" category results are based on the measures taken by the governments of the participating countries in countering the effects of GHG emissions. The data for this category is rated on basis of a country's performance. This rating is issued by climate and energy policy experts from various non-governmental organisations, universities, and think tanks within the countries, which are evaluated. The importance of this category lies in the fact that only through the implementation of successful climate change mitigation policies can the overall ratings in all the other categories be perfected in future scores within the CCPI. For this category, grades above 4.5 in the survey receive a "very high" rating, grades above 3.5 receive a "high" rating, any grade about 2.5 receives a "medium" rating and any grade below 1.5 receives a "very low" rating. The two indicators under this category are the National Climate Policy and International Climate Policy, as demonstrated in Fig. 1.8 (Burch et al., 2021).

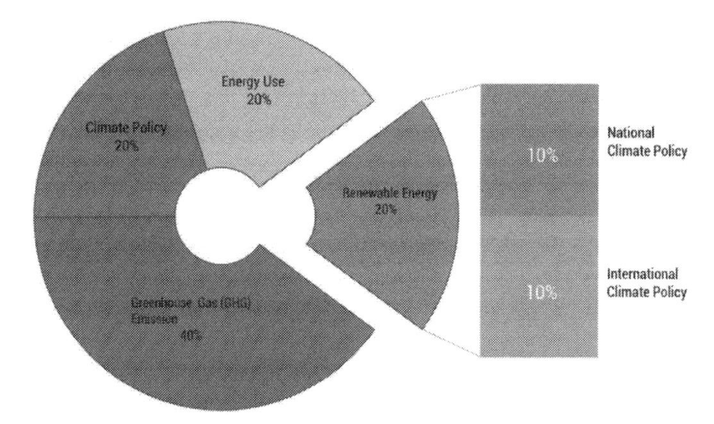

Fig. 1.8 Weighting of Climate Policy Indicators

Source: (Burch et al., 2021)

National Climate Policy

The annual climate policy performance questionnaire for the indicator "National Climate Policy" encapsulates specific policies for the promotion of renewable energies, an increase in energy efficiency, and other actions to reduce GHG emissions in the manufacturing and construction sectors, the transportation and other residential sectors, as well as other sectors. Furthermore, the effectiveness of the current climate policy is assessed in relation to national peat land protection, support for and protection of biodiversity within forest ecosystems, and a reduction in deforestation and forest degradation. Experts assess the quality and degree of execution of the specific policy framework within each of these policy areas. The ambition level and well-below 2°C compatibilities of each country's Nationally Determined Contributions (NDCs) are also evaluated by experts in accordance with the Paris Agreement as well as their potential to reach specified goals.

International Climate Policy

The CCPI assesses how well nations perform at United Nations Framework Convention on Climate Change meetings as well as other international meetings and multilateral accords. The specialists are tasked with evaluating how their nation has performed recently in the international forum.

Calculations and Results

The overall index assigns a value to each country between 0 and 100, with higher values denoting more "climate-friendly" behaviour. The following formula is used to generate the final CCPI ranking from the weighted average of the scores attained in each of the different indicators (Puertas et al., 2021).

$$I = \sum_{i=1}^{n} w_i X_i$$

where,

I: Climate Change Performance Index

X_i: normalised indicator

w_i: weighting of X_i

$$\sum_{i=1}^{n} w_i = 1 \quad \text{and} \quad 0 \leq w_i \leq 1$$

where, i: 1, …, n: number of partial indicators (currently 14) and finally,

$$\text{Score} = 100\left(\frac{\text{Actual value} - \text{Minimum value}}{\text{Maximum value} - \text{Minimum value}}\right)$$

The attained score, not the ranking itself, is the only place where variations in countries' climate protection measures may be readily recognised. The highest-ranking nation, Denmark, was not at the top of all indicators, let alone had it score 100 points, as can be seen when examining the CCPI 2022 in greater detail. This illustration demonstrates that a nation's shortcomings and strong points can only be identified within the many categories and indicators (Burch et al., 2021).

Global Scenario and India's Position

The key results of the CCPI published for the year of 2022 rate the countries in four categories and score them aggregately on a rating of "Very High", "High", "Medium", "Low" and "Very Low" (Burch et al., 2021). For the year 2022, as in all other years, no countries are ranked in the top three overall positions because no country performs well enough in all index categories to achieve an overall "very high" rating in the CCPI. This indicates the fact that no country among the 64 evaluated countries is doing enough to prevent dangerous climate change. Denmark is the highest ranked country in CCPI 2022, yet it does not perform well enough to achieve an overall very high rating. The G20 countries, which are responsible for 75% of the world's GHG emissions, had four countries in the high-performing category in CCPI 2022 – United Kingdom (7th), India (10th), Germany (13th) and France (20th). Saudi Arabia is the lowest ranked G20 country, ranked 63rd among the evaluated 64 nations. The last position on the CCPI 2022 is taken up by Kazakhstan (Burch et al., 2021).

India secured consecutively the number 10th position on the CCPI for last two years. It was ranked "high" in the categories of GHG emissions, energy use and climate policy categories, while achieving a "medium" rating in the renewable energy category. Given the fact that today, India is the fifth largest economy in the world; it has fared well in the CCPI in comparison to its other economically high-achieving counterparts. India's commendable performance can be credited to its focus on the implementation and achievements of the NDCs and the improvement in achieving the renewable energy usage targets. India is on its course to meet its 2030 emissions target, which is compatible with a well-below-2°C scenario (Burch et al., 2021). According to India's NDC, it targets to derive 40% renewable energy by replacing fossil fuels by 2030. It also framed a mission to achieve a 33%–35% reduction in energy intensity by 2030 and India is on schedule to meet both of these goals too. However, with no clear coal phase-out goals in view, there are still some downsides to investigate and rectify. (Burch et al., 2021) Given the fact that not enough has been done into introducing more solar power sources, it can be understood why only in the renewable energy category, India did not achieve a high rating. Notwithstanding the fact that an overall high rating was achieved by the country, experts in the field urge for an explicit net zero target for 2050 and focus more on the renewable energy sector for aiding in said targets. As long as equity and social development are lagging, implementation of climate policy changes will not be adequate to lower the climate change risk. The overall ratings and indicator-wise ratings for India are exhibited in Table 1.1 for a better understanding of India's position.

Table 1.1 Indicator-wise ranking of India in the CCPI 2022

India's Rank	Category	Score	Overall Rating	Indicators Under Each Category	
				Indicator	**Rating**
9	GHG Emissions	31.42	High	GHG per capita – current level	Very High
				GHG per capita – current trend	Very Low
				GHG per capita – compared to a well-below-2°C benchmark	Very High
				GHG 2030 Target – compared to a well-below-2°C benchmark	Very High
24	Renewable Energy	9.10	Medium	Share of RE in Energy Use (TPES)** – current level (including hydro)	Medium
				RE current trend (excluding hydro)	High
				Share of RE in Energy Use (TPES) (including hydro) – compared to a well-below-2°C benchmark	Low
				RE 2030 Target (including hydro) – compared to a well-below-2°C benchmark	Medium
14	Energy Use	14.68	High	Energy Use (TPES)** per capita – current level	Very High
				Energy Use (TPES) per capita – current trend	Very Low
				Energy Use (TPES) per capita – compared to a well-below-2°C benchmark	Very High
				Energy Use 2030 Target – compared to a well-below-2°C benchmark	High
16	Climate Policy	14.00	High	National Climate Policy Performance	Medium
				International Climate Policy Performance	Medium

Conclusion

The recent issue of climate change and its associated effect on sustainability triggers a great impact on environmental scientists and engineers in Indian and global content. To address such events, the CCPI has been recognised as a tool to measure the status of environmental quality with respect to climate change in different countries. Different components need to be examined on the basis of GHG emissions, renewable energy, energy use and climate policy. Critical analysis is very much necessary to monitor and assess the level of environmental quality. As per the standard criterion, India is positioned tenth since the last two years. There is a scope for improvement of the situation by changing the fuel alternatives with possible changes towards using more renewable sources of energy to reduce emissions and greenhouse effect. It is also suggested that more mathematical tools, models and monitored data would be required for the application of CCPI analysis in Indian climatic conditions.

Acknowledgment

I would like to thank Professor Dr. Somnath Mukherjee for enlightening me about the theme and guiding me in every step of the way while working on this topic. I would also like to thank all members of the Environmental Engineering Division, Department of Civil Engineering, Jadavpur University, Kolkata - 700032, India for their support and assistance.

References

1. Burck, J., Uhlich, T., Höhne, N., et al. (2021). The Climate Change Performance Index 2022, Background and Methodology, Germanwatch, Bonn, Germany.
2. Burck, J., Uhlich, T., Höhne, N., et al. (2021). The Climate Change Performance Index 2022, Results, Germanwatch, Bonn, Germany.
3. Eco-Justice Collaborative, Building a Just and Sustainable World. https://ecojusticecollaborative.org/climate-and-energy/(accessed August 28, 2022).
4. Paris Reality Check: PRIMAP-hist. https://www.pik-potsdam.de/paris-reality-check/primap-hist/ (accessed August 28, 2022).
5. Puertas, R. & Marti, L. (2021). International ranking of climate change action: An analysis using the indicators from the Climate Change Performance Index. Renewable and Sustainable Energy Reviews 148(2): 111316.
6. Sphera, What Is Environmental Sustainability?. https://sphera.com/glossary/what-is-environmental-sustainability/(accessed August 28, 2022)
7. UN General Assembly, United Nations Framework Convention on Climate Change: resolution / adopted by the General Assembly, 20 January 1994. https://www.refworld.org/docid/3b00f2770.html (accessed August 28, 2022).
8. United Nations, Key Findings. https://www.un.org/en/climatechange/science/key-findings (accessed August 28, 2022).

CHAPTER 2

Development of Cost Functions for Construction of Wastewater Treatment Plant with State-of-the-Art Technology

Bhaskar Sengupta[1]
Ph. D Research Scholar, Environmental Engineering Division, Department of Civil Engineering, Jadavpur University, Kolkata, India

Somnath Mukherjee[2]
Professor, Environmental Engineering Division, Department of Civil Engineering, Jadavpur University, Kolkata, India

Abstract

The availability of many conventional and updated technological options for biological treatment of municipal and industrial wastewater often leads to the selection of inappropriate, uneconomical and non-site specific schemes as a whole. Several decisive criteria, for example, capital project cost, operation expenditures and cost pertaining to the acquisition of required land, are to be explored and considered in the appropriate selection of biological reactor mechanism. The sustainability of chosen technology is also needed to be exercised during the process to discard confusing apprehension of both clients and designing agencies for developing countries like India.

The necessity for exhibiting a sensible and sustainable decision quickly for designing and evaluation of project costs for funding wastewater treatment plants is recognised worldwide. A handful number of models are cited in the literature, which proclaim to prepare project cost estimates of the wastewater treatment plant. Some models are useful and sensitive to practicing engineers involved in accurately projecting the cost of construction by considering an optional alternative solution.

The objective of this paper is to discuss the importance of cost functions for economic analysis and predicting the most economic treatment scheme among various kinds of state-of-the-art treatment systems containing conventional and space-saving advanced-level bioreactor systems.

The cost function is a useful economical technique to justify and compare costs for system alternatives of wastewater treatment. It targets better engineering decisions and suitable selection of the appropriate treatment based on parametric criteria, requirement of space, and capital and annual cost of operation, as well as maintenance, particularly energy for the effective

[1]senguptabhaskar1962@yahoo.co.in & senguptabhaskar1962@gmail.com
[2]somnath.mukherjee@jadavpuruniversity.in & mukherjeesomnath19@gmail.com

life of the plant. The paper also highlights the methodology and criteria for the development of the aforesaid decision-making tools in order to attain the desired goal with reference to various kinds of established treatment technologies.

Keywords: appropriate reactor, biological treatment, cost function, methodology, wastewater

Introduction

Extensive urban developments, extension of cities, vibrant industrialisation, use of improvised materials and other miscellaneous reasons have resulted in the discharge of wastewater, which is a significant cause of water pollution. Globally, around 80% of wastewater is discharged in untreated form into several surface water streams [32]. This practice introduces a negative impact on the biodiversity in the aquatic environment and affects the universal food chains.

India is also progressing through rapid developments and in this context, there are several uphill challenges before the local governments and urban planners. Scientific and economic management of treating waste within the constraint of the scarcity of land is one of them.

Several programmes have been formulated by the Central Government of India, such as Namami Ganga Projects, to protect the urban environment from the kind of pollution addressed above. The main concern of the projects is to protect the quality of the water source.

An outline of the Indian inventory of municipal Wastewater Treatment Plants (WWTPs) – 2021 is furnished as follows [5]:

(a) In total, 1631 WWTPs inclusive of those proposed for construction are there and have a total capacity of 36,668 MLD (million litres per day). This encompasses 35 States/UTs. Among the above, 1093 WWTPs are under operation, 274 are under construction and 162 WWTPs are proposed for construction and the balance WWTPs are defunct.
(b) Among operational WWTPs, only 578 WWTPs with an overall capacity of around 12,200 MLD conform to the regulations stipulated by the Pollution Control Boards.
(c) Wastewater generation from urban communities is around 72,368 MLD.

A budget of ₹1,41,678 crores or 18,400 million USD for the period of 2021–2026 has been laid for sanitation of the urban sector [4] and therefore numbers of WWTPs are anticipated to be constructed in the next few years.

In this connection, urban planners often need a ready set of information on the cost of WWTPs for various capacities and technologies as well as the requirement of land area and energy for operation.

In the past decade, few new technologies have been used for the treatment of wastewater. Moving Bed Bio Reactor (MBBR) and Sequential Batch Reactor (SBR) are among these options for secondary treatment of wastewater. Membrane Bio-Reactor (MBR) is another option and may be designated as one of the advanced processes.

Facts of use of different technologies in WWTPs within India are furnished in Table 2.1.

Table 2.1 Different technologies in WWTPs

Sl. No.	Technology	Capacity in MLD	Number of WWTPs
1.	Activated sludge process (ASP)	9486	321
2.	Extended aeration (EA)	474	30
3.	SBR	10638	490
4.	MBBR	2032	201
5.	Fluidised aerobic bio-reactor (FAB)	242	21
6.	Up-flow anaerobic sludge blanket (UASB)	3562	76
7.	Waste stabilisation pond (WSP)	789	67
8.	Oxidation pond (OP)	460	61
9.	Any other	8497	364

Source: Ref. [5]

In India, SBR, activated sludge process (ASP) and MBBR have been adopted extensively by concerned local authorities.

The densities of the population in the last few years have increased in several urban local communities. Therefore, the issue of the ready availability of vacant land required for the construction of WWTPs has appeared as a major constraint. Wherever vacant lands are available, the local population often raises objections to the construction of WWTPs in the vicinity. Therefore, constructions of WWTPs need to be planned nowadays in such ways so that the requirements of areas are lesser than those used for WWTPs in earlier days. This will help in the easier acquisition of land.

Figure 2.1 demonstrates the requirement of plan area for use of various wastewater treatment technologies. It may be noted that the requirement of the area ranges from 0.2 to 1.0 hectares per MLD for WWTP based on the technology adopted except for WSP.

The Fig. 2.1 represents that area requirements in respect of various technologies for wastewater treatment, such as MBBR and SBR, are lower compared to that for ASP. Many urban local bodies in India have started to deploy MBBR and SBR as well as MBR to reduce the requirements for the acquisition of land areas, optimise operation and maintenance (O&M) costs and improve the quality of treated wastewater.

The concern or challenging task is to determine the selection of the most suitable technology among MBR, MBBR and SBR for a particular wastewater flow rate that would be overall cost-wise most effective, with due consideration of space required.

Therefore, it is necessary to explore a methodology for rapid cost estimation (forecast level estimation) with adequate accuracy for WWTP with each MBR, MBBR and SBR technologies.

Newness/Target Objective with Citation Reference

It is experienced that setting up the target objective for selection of the most suitable technology among MBR, MBBR and SBR through an optimization process is very much essential.

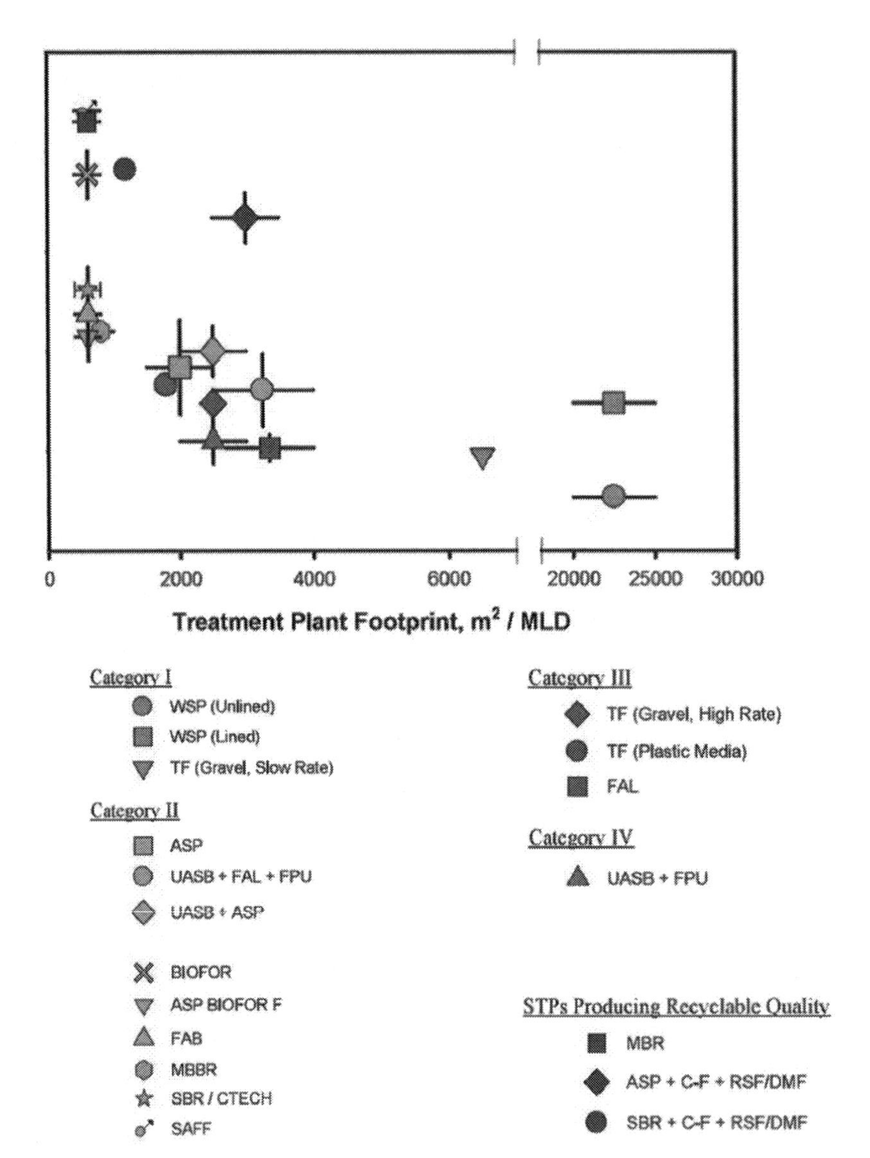

Fig. 2.1 Requirement of plan area for use of various wastewater treatment technologies

Source: [6]

As per the theory of micro-economy cost (*C*) may be defined as a function of output (O) and input prices (P) viz. $C(O, P)$. However, in practice, engineers often include other system-specific variable (X)s related to technical parameters and economic issues. The cost function, therefore, is designated as $C(O, P, X)$ [28].

In wastewater treatment, these variables are capacity, quality of wastewater with reference to regulatory norms for discharge, cost of scheduled items and so on.

As tools for selection or cost optimization of WWTP, cost functions may be used for each of the above technologies. A cost function is an expression used to predict how expenses will change at different output levels.

From the literature review, it has been noted that in the history of the last 50 years, several studies are being carried out in connection with the development of cost functions and cost indices for the construction and operation of WWTPs. The intent behind such efforts was the economic analysis of ubiquitous alternative treatment technologies.

Few studies carried out are addressed below for ready reference.

Shah and Reid (1970) developed cost functions for WWTPs based on the feedback survey of over 563 plants located in 48 states in the United States.

USEPA (1976) analysed available WWTP construction bid data and developed two cost curves (one for green field secondary plants and the other for upgradation of plants from primary to secondary).

Qasim et al. (1992) published generalised equations to assess the cost in different categories for the construction, operation and maintenance of treatment plants.

Vanrolleghem et al. (1996) discussed the importance of formulation of the cost function for the design and operation of WWTPs.

Tsagarakis et al. (2003) developed functions to assess requirements of land, cost of construction and O&M cost for activated sludge systems based on a data survey carried out in Greece.

Friedler & Pisanty (2006) have developed cost functions by analysis of the costs of 55 WWTPs constructed in Israel.

Sato et al. (2007) developed cost functions for capital cost, operation, maintenance and land requirements for UASB and WSP based on existing available data collected from various reports on the Indian environment.

Based on cost data collected from a number of municipal WWTPs, Wichitra Singhirunnusorn and Michael K. Stenstrom (2010) proposed a cost function to assess ASPs, oxidation ditches (ODs), ALs and WSPs for land, construction, operation and maintenance of WWTPs in Thailand.

Hernandez-Sancho et al. (2011) published a set of cost functions for different technologies based on collected data from 341 WWTPs in Spain.

Pannirselvam and Navaneetha Gopalakrishnan (2015) compiled the cost records for 30 WWTPs under operation which were built with conventional activated sludge technology. Cost data were adjusted to the base year of 2014 by use of the authentic construction cost indices. Cost functions were developed by regression analysis.

Koul and John (2015) attempted life cycle cost analysis for UASB, SBR and MBBR based on collected data from different plants.

By use of CapdetWorks, Jafarinejad (2016) estimated the costs for use of conventional ASP, EA-activated sludge and SBR technologies for a WWTP in Tehran and made a comparative analysis.

Gautam et al. (2017) made comparative scrutiny among MBR, MBBR, SBR, EA and submerged aerobic fixed film process based few small sizes WWTPs in India and derived cost functions for capital, electro-mechanical, electricity, operation and maintenance from collected data. The study concluded that SBR as the most economic technology and MBR as the most expensive one.

Acampa et al. (2019) made a study based on the estimation of construction costs of an urban WWTP with a capacity of less than 50,000 population equivalent. Cost estimation was made by two methods: synthetic cost estimation for civil works and multiple linear regression for electromechanical components. It was revealed that higher economic benefit will prevail with the increase in the size of WWTP for a population equivalent of 5000 to 10,000, while a further increase is less beneficial.

The literature available in the context of the selection of appropriate technology for biological treatment of wastewater on the basis of engineering economics and other issues applicable in India are quite limited.

No other study related to cost functions for WWTP apart from that addressed above, especially with space-saving technologies (MBR, MBBR and SBR) related to Indian conditions has been found reported.

It has been noted that in most of the studies addressed above, the cost functions for land, construction, operation and maintenance of WWTPs have been derived from region-specific historic cost data collected.

In a few instances, cost comparisons among a few conventional technologies have been made by use of CapdetWorks software on a specific inflow rate basis.

Further, the technologies taken into account are conventional ASPs, ODs, ALs, UASBs and WSPs. Technologies such as MBR, MBBR and SBR being new technologies might not have been taken into consideration.

An initiative and approach to explore a tool for cost estimation with due consideration of the cost of land acquisition, especially for space-saving technologies (MBR, MBBR and SBR) appears to be very much appropriate and meaningful for the installation of WWTP. Research in this new domain will add value to the set of tools available for the selection of technology to be used in a WWTP.

This paper discusses the objectives with reference to the target or goal. The development of a tool for rapid life cycle cost estimation in connection with the construction and operation of a WWTP with the use of one of the space-saving technologies (MBR, MBBR and SBR).

Approach Methodology

Major components for wastewater treatment generally include three stages – primary, secondary (biological) treatment and biological sludge treatment as well as the disposal of settled innocuous solids.

The primary treatment of wastewater using grit chambers and sludge treatment systems are more or less similar irrespective of the kind of technologies under concern in this study and the choice of different alternatives for such treatment is also very much limited. The cost of such physical treatment and biological sludge treatment for a definite inflow rate of wastewater is more or less the same while the cost of secondary treatment will vary based upon the selection of the bio-reactor technology. Therefore, the secondary treatment technology along with case-specific requirements for pre-treatment and post-treatment will dictate the level of life-cycle cost for the WWTP.

With the intent to compare and explore the cost economy for the selection of technology among MBR, MBBR and SBR, the approach to develop the cost functions is not envisaged to include the design and cost estimations for physical treatment and biological sludge treatment for the reasons stated above.

As the obvious forecasted cost of WWTP as to be projected by the developed cost function will not represent the overall cost of the entire project. The actual estimated cost of a project is always site-specific and should be worked out based on all components to be included in the project.

A novel and unique approach has been introduced in this study to develop the cost functions.

Design of equipment and accessories for biological treatment systems and technology-specific pre-treatment as well as post-treatment requisites are envisaged to be carried out by use of design algorithms developed in Microsoft Excel with reference to a specific design inlet flow rate and inlet characteristics of waste as well as the quality of treated wastewater.

The philosophy for selection of the number of treatment batteries and the number of biological reactors as well as other equipment therein vis-a-vis their relative orientations for a specific design flow rate will be adopted as per the CAPDET – USEPA [34].

The above approach will determine the number as well as the sizes of equipment required and enable further estimation of the cost of civil, mechanical and electrical items to find out the capital expenditure by use of applicable estimation algorithms available in CAPDET – USEPA [34].

With due consideration of the same inlet characteristics of wastewater and the quality of treated wastewater, this approach for design and estimation will be repeated to create a set of cost data at different inflow rates over a wide range (from 0.5 MLD to 150 MLD). Based on this data set, cost functions may be determined by the regression technique. Cost of operation, maintenance and requirement of land need also to be included in this exercise to arrive at the overall cost function for WWTP.

Under this kind of approach, it is indeed a comprehensive exercise that is needed to be carried out but cost functions envisaged to be derived will provide a tool for quick and accurate cost estimation for WWTP with any one of MBR, MBBR and SBR technologies. The output of this exercise will also enable to avoid the use of collected cost data, and deluge the traditional practice or preferred approach for the selection of technology with reference to biological treatment of wastewater.

The technical features as envisaged for the development of the model are described hereinafter:

Capacity Groups

The costs of WWTP generally decrease with an increase in capacity due to the economy of scale.

Therefore, a single cost function for all the capacity ranges is not considered appropriate.

WWTPs have been categorised capacity-wise in three groups viz. small (0.5–5.0 MLD), medium (5.050.0 MLD) and large (50.0–150.0 MLD).

Design Inlet Quality of Wastewater

The characteristics of raw wastewater depend on the rate of water supply and pollution load per capita.

The concentrations of impurities in raw wastewater as considered for design purposes, are furnished below. This is based on water supply @ 135 L/cap /day.

Table 2.2 Design inlet quality of wastewater

Parameters	Value	Unit
Biological oxygen demand (BOD)	250.00	g/m^3
Chemical oxygen demand (COD)	425.00	g/m^3
Volatile suspended solids (VSS)	262.50	g/m^3
Total suspended solids (TSS)	375.00	g/m^3
Ammonia nitrogen (NH_4-N)	32.50	g/m^3
Organic nitrogen (ON)	17.50	g/m^3
Nitrate nitrogen (NO_3-N)	5.00	g/m^3
Total kjeldahl nitrogen (TKN)	50.00	g/m^3
Total phosphorous (TP)	7.10	g/m^3

Design Treated Quality of Wastewater

The design treated water quality as adopted for the target objectives is presented in Table 2.3.

Table 2.3 Design treated water quality of wastewater

Parameters	Value	Unit
Biological oxygen demand (BOD) – design limit	10.00	g/m^3
Total suspended solids (TSS)	10.00	g/m^3

The above are in line with the stipulations documented by the Ministry of Environment & Forests (MoEF) of the Government of India (GOI)

Treatment Processes

MBR

MBR technology comprises ASP and membrane separation process. In general, low-pressure membranes are used. Membranes are kept submerged in the reactor itself or otherwise in a separate chamber to promote the separation of solids from the liquid.

Primary sedimentation tank, final sedimentation tank and disinfection facilities are not required to be installed in this process.

A schematic for MBR process cycle is reproduced in Fig. 2.2.

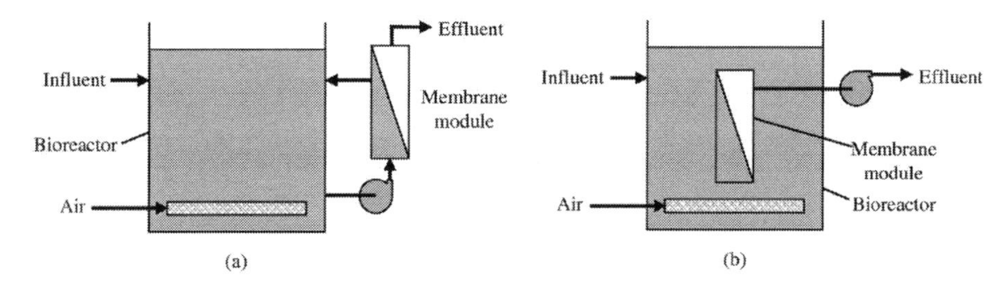

Fig. 2.2 Schematic for MBR process cycle

Source: https://www.politesi.polimi.it/bitstream/10589/94842/5/2014-10-Zaerpour.pdf

MBBR

In this process, a tank similar to that in an ASP is provided for biological reactions.

Carrier media made of polymeric material is used in MBBR technology to attach the microorganisms over the surface itself. The carrier media is kept in suspension by the supply of air in case of aerobic process or by mechanical agitation in the case of the anoxic or anaerobic process. A sieve is provided at the exit of MBBR to retain the carrier media within the system.

This technology includes a primary clarifier at upstream of MBBR. The secondary clarifier is also required to be provided. However, there is no requirement to recycle activated sludge at the inlet of MBBR, since an adequate population of microorganisms is maintained within MBBR by virtue of the presence of carrier media and biofilms attached to the same.

A schematic for MBBR process cycle is reproduced in Fig. 2.3.

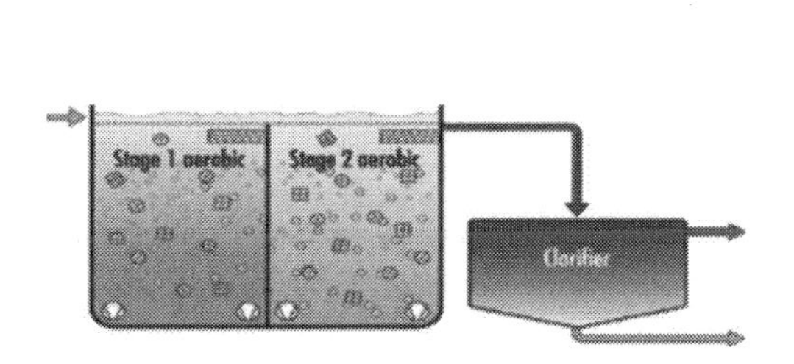

Fig. 2.3 Schematic for MBBR process cycle

Source: https://www.lenntech.com/processes/mbbr.htm

Fig. 2.4 Typical view of MBBR media

Source: https://dynamixinc.com/anoxic-and-aerobic-mbbr-mixing/

A typical view of MBBR media is furnished in Fig. 2.4.

SBR

SBR is an improvement of ASP. In the case of ASP, primary clarifier, an aeration tank and then a secondary clarifier is required for the treatment of wastewater, whereas in the SBR, the aeration and settling are carried out in a sequential manner within a single tank.

Primary clarifiers are not required to be provided. As a minimum two SBR basins are needed for parallel operation such that one is in the aeration phase and the other in the settling phase for subsequent decantation of the supernatant.

A schematic for SBR process cycle is reproduced in Fig. 2.5.

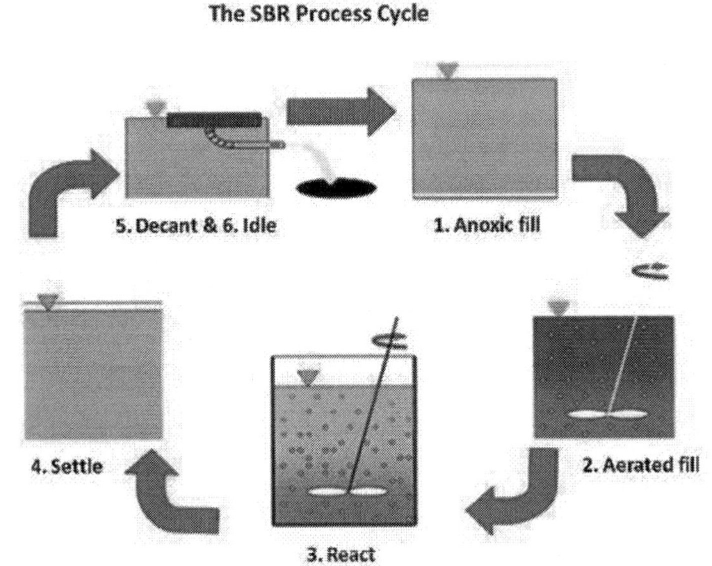

Fig. 2.5 Schematic for SBR process cycle

Source: https://kyocp.wordpress.com/2016/04/08/sequencing-batch-reactor-process-gaining-popularity/

Design Parameters

The design parameters as adopted for biological treatment in WWTPs are furnished in Table 2.4.

Table 2.4 Design parameters as adopted for biological treatment in WWTPs

Parameters	MBR		MBBR		SBR	
	Value	Unit	Value	Unit	Value	Unit
Peak factor	2.25		2.25		2.25	
Lean Factor	0.45		0.45		0.45	
Thickener overflow return as fraction of plant flow	0.15		0.15		0.15	

Parameters	**MBR**		**MBBR**		**SBR**	
	Value	**Unit**	**Value**	**Unit**	**Value**	**Unit**
BOD in thickener overflow return	500.00	g/m^3	500.00	g/m^3	500.00	g/m^3
Centrate from sludge dewatering as fraction of plant flow	0.0060		0.0060		0.0060	
BOD in centrate from sludge dewatering return	380.00	g/m^3	380.00	g/m^3	380.00	g/m^3
BOD_u/VSS	1.42	g BOD/g VSS	1.42	g BOD/g VSS	1.42	g BOD/g VSS
BOD_5/BOD_u	0.67		0.67		0.67	
Kinetic parameters for BOD removal						
Reference temperature for kinetic parameters	20.00	°C	20.00	°C	20.00	°C
Half Velocity Constant	20.00	g bs COD/m^3	20.00	g bs COD/m^3	20.00	g bs COD/m^3
Maximum specific bacterial growth rate	6.00	(g VSS/g VSS)/d	6.00	(g VSS/g VSS)/d	6.00	(g VSS/g VSS)/d
Endogenous Decay co-efficient	0.06	(g VSS/g VSS)/d	0.06	(g VSS/g VSS)/d	0.06	(g VSS/g VSS)/d
True Yield co-efficient	0.3125	g VSS/g b COD	0.3125	g VSS/g b COD	0.3125	g VSS/g b COD
	0.5000	g VSS/g BOD	0.5000	g VSS/g BOD	0.5000	g VSS/g BOD
Fraction of biomass that remains as cell debris	0.15		0.15		0.15	
θ values						
Temperature activity co-efficient for K_s	1.00		1.00		1.00	
Temperature activity co-efficient for μ_m	1.07		1.07		1.07	
Temperature activity co-efficient for k_d	1.04		1.04		1.04	
Design temperature of reactor basin	12.00	°C	12.00	°C	12.00	°C
Design MLSS	8000.00	g/m^3	5000.00	g/m^3	4000.00	g/m^3
Ratio of VSS to TSS	0.70		0.70		0.70	
Design MLVSS	5600.00	g/m^3	3500.00	g/m^3	2800.00	g/m^3

Parameters	MBR		MBBR		SBR	
	Value	Unit	Value	Unit	Value	Unit
Percentage of clean water oxygen transfer efficiency (for fine bubble ceramic diffusers)	35.00	%	35.00	%	35.00	%
Elevation at site	9.00	M	9.00	m	9.00	m
Atmospheric pressure at elevation of site	95.60	kPa	95.60	kPa	95.60	kPa
Effective liquid depth in reactor basin	4.07	M	4.07	m	4.07	m
Point of air release for ceramic diffusers from bottom of reactor basin	0.50	M	0.50	m	0.50	m
Standard temperature	20.00	°C	20.00	°C	20.00	°C
Concentration of dissolved oxygen at standard temperature and pressure of 1,01,325 N/m^2	9.08	g/m^3	9.08	g/m^3	9.08	g/m^3
Aeration α factor for BOD removal	0.50		0.50		0.50	
Salinity and surface tension correction factor for both conditions, i.e., BOD removal	0.95		0.95		0.95	
Diffuser fouling factor	0.90		0.90		0.90	
Percentage (by weight) of oxygen in air	23.20	%	23.20	%	23.20	%
Density of air	1.20	kg/m^3	1.20	kg/m^3	1.20	kg/m^3
Oxygen transfer efficiency	8.00	%	8.00	%	8.00	%
Factor of safety	2.00		2.00		2.00	
Oxygen consumption	1.42	mg/mg of cell	1.42	mg/mg of cell	1.42	mg/mg of cell

Components of WWTPs

The major components as envisaged for biological treatment in WWTP are as follows:

MBR

(a) Reaction basins and accessories
(b) Membrane chambers and accessories

(c) Reactor basin waste transfer pumps and pump-house
(d) Internal recirculation pumps and pump-house
(e) Mixed liquor recirculation pumps and pump-house
(f) Blowers and blower building

Primary clarifiers and secondary clarifiers are not applicable for WWTPs with MBR.

MBBR

(a) Primary clarifiers
(b) Primary clarifier sludge sump
(c) Primary clarifier sludge transfer pumps and pump-house
(d) Reaction basins and accessories (in two stages)
(e) Reactor basin waste transfer pumps and pump-house
(f) Secondary clarifiers
(g) Secondary clarifier sludge sump
(h) Secondary clarifier sludge transfer pumps and pump-house
(i) Blowers and blower building

SBR

(a) Reaction basins and accessories
(b) Reactor basin waste transfer pumps and pump-house
(c) Blowers and blower building

Primary clarifiers and secondary clarifiers are not applicable for WWTPs with SBR.

Design, Detailing and Cost Estimation

Design and estimations are envisaged to be carried out by means of a model developed in Microsoft Excel Spread Sheets.

A model has been developed to facilitate the following tasks:

(a) Process design and sizing of the major components of the WWTP as per the design criteria.
(b) Estimation of the bill of quantities for the designed components by use of algorithms suggested for basins with diffused aeration, pumping and blowers as illustrated in CAPDET – USEPA [34].
(c) Estimation of cost for each of the major components for biological treatment in the WWTP by use of the Schedule of Rates (latest publication) of Public Works Department (PWD), Government of India, for scheduled components of civil works (for cost of earthwork, RC wall in-place, RC slab in-place and handrails in-place) and market rates (rate data collected from vendors) for non-scheduled mechanical as well as electrical equipment with applicable accessories including contingencies at @ 10% to account for the minor cost items, such as liquid piping system, control equipment, painting, site cleaning and preparation.
(d) Determination of Total Bare Construction Cost (CAPEX), levelised cost based on energy requirement, operation and maintenance for 25 years of the life of WWTP (OPEX), cost of land for installation of the complete system (based on space requirements with due consideration of adequate access for operation and maintenance of major equipment).

The inputs to the model will be the design capacity of WWTP in MLD, concentrations of BOD_5, SS of raw wastewater and the permissible effluent BOD_5 and SS. The output from the model will be the overall cost of major components for biological treatment in the WWTP inclusive of CAPEX, OPEX and required land with 2021 considered as the base year.

Development of Capital Cost Functions

Data sets comprised different groups in terms of capacities [at the interval of 0.5 MLD for a small group, the interval of 5 MLD for a medium group, and the interval of 10 MLD for a large group] and respective integrated costs inclusive of capital cost, operation, as well as maintenance cost and cost of land, are envisaged to be determined as addressed above by means of model algorithms for detail design and estimation developed in Microsoft Excel.

Regression analysis of data generated from the model are envisaged to be carried out in Microsoft EXCEL based on trend lines with five different equations (viz., exponential, linear, logarithmic, polynomial and power) to determine the relationship between capacity and respective overall cost for the treatment envisaged. The equation that will correspond to the maximum coefficient of determination shall be selected as the cost function for each case.

The cost function derived for each of the three capacity groups will be validated based on the comparison between the predicted value by the use of the cost function and the respective estimated cost. The accuracy of predicted cost shall be analysed based on Mean Absolute Percentage Error– a widely used measure of accuracy. MAPE is defined as an average of absolute percentage errors. It is calculated as follows:

$$\text{MAPE} = \frac{\sum \frac{|A - F|}{A} \times 100}{N},$$

where

A = Estimated cost,

F = Forecasted cost by use of cost function as determined and

N = Number of elements in the data set.

The significance of MAPE value is interpreted in Table 2.5 to express the accuracy of prediction:

Table 2.5 Interpretation of MAPE

MAPE	Interpretation
< 10	Accurate forecasting result
10–20	Good forecasting result
20–50	Reasonable forecasting result
> 50	Inaccurate forecasting result

Conclusion

The exercise for the development of cost functions for rapid life cycle cost estimation in connection with the construction and operation of a WWTP with the use of one of the space-saving technologies (MBR, MBBR and SBR) is under progress. It is anticipated that accurate cost forecasts can be made by cost functions for small, medium and large capacities for the selection and use of state-of-the-art technologies, such as MBR, MBBR and SBR.

Acknowledgment

I would like to thank my research advisor, Professor Dr. Somnath Mukherjee, for enlightening me first on the theme and for valuable guidance for the research since 2019.

I would also like to thank all members of the Environmental Engineering Division, Department of Civil Engineering, Jadavpur University, Kolkata - 700032, India for their support and assistance.

References

1. Acampa, G., Giustra, M. G., and Parisi, C. M. (2019). Water Treatment Emergency: Cost Evaluation Tools' MDPI – 07 May.
2. Arif, A. U. A., Sorour, M. T., and Aly, S. A. (2020). Cost analysis of activated sludge and membrane bioreactor WWTPs using CapdetWorks simulation program: Case study of Tikrit WWTP (middle Iraq). Alexandria Engineering Journal: Hosted by ELSEVIER.
3. Balmer, P. and Mattson, B. (1994). Wastewater treatment plant operation costs. Wat. Sci. Tech. vol. 30, no. 4, pp. 7–15.
4. Budget, Government of India. (2021-2022).
5. Central Pollution Control Board – Government of India. (2021). National inventory of Sewage Treatment Plant.
6. Central Public Health and Environmental Engineering Organization in collaboration with JICA. (2013). Manual on Sewerage and Sewage Treatment Systems.
7. Doherty, E. (2017). Development of new benchmarking systems for wastewater treatment facilities. A thesis submitted to the College of Engineering and Informatics, National University of Ireland, Galway, in partial fulfilment of the requirements for the Degree of Doctor of Philosophy. Dysert, L. R. (2008). An Introduction to Parametric Estimating. AACE International Transactions
8. Dysert, L. R. (2008). An Introduction to Parametric Estimating. AACE International Transactions
9. Friedier, E. and Pisanty, E. (2006). Effects of design flow and treatment level on construction and operation costs of municipal wastewater treatment plants and their implications on policy making. Water Res. vol.40, no.2, pp. 3751–3758.
10. Gautam, S., Ahmed, S., Dhingra, A., and Fatima, Z. (2017). Cost-Effective Treatment Technology for Small Size Sewage Treatment Plants in India. Journal of Scientific & Industrial Research - Vol. 76, April 2017, pp. 249–254.
11. Gillot, S., De Clercq, B., Defour, D., Simoens, F., Gernaey, K., and Vanrolleghem, P.A. (1999). Optimization of wastewater treatment plant design and operation using simulation and cost analysis. 72nd annual conference WEFTEC, New Orleans, USA.
12. Gratziou, M.K., Tsalkatidou, M., and Kotsovinos, N.E. (2006). Economic evaluation of small capacity sewage processing units. Global Nest Jl. vol.8, no.1, pp. 52–60.

13. Gumerman, R.C., Culp, R. L., & Hanesan, S.P. (1978). Estimating cost for water treatment as a function of size and treatment efficiency. USEPA, (EPA-600/2-78-182).
14. Hernandez, S.F., Senante, M.M., and Garrido, R.S. (2011). Cost Modeling in Wastewater Treatment Processes. Desalination', vol.268, pp. 1–5.
15. Jafarinejad, S. (2016). Cost estimation and economical evaluation of three configurations of activated sludge process for a wastewater treatment plant (WWTP) using simulation. Applied Water Science I Springer – July.
16. Koul, A. and John, S. (2015). A Life Cycle Cost Approach for Evaluation of Sewage Treatment Plants. International Journal of Innovative Research in Advanced Engineering (IJIRAE) ISSN: 2349-2163 Issue 7, Volume 2, July 2015.
17. Martins, R., Fortunato, A., and Fernando Coelho, F. (2006). ESTUDOS DO GEMF.
18. McNamara, G. (2018). Economic and Environmental Cost Assessment of Wastewater Treatment Systems: A Life Cycle Perspective. A Thesis Submitted in Fulfilment of the Requirements for the Degree of Doctor of Philosophy (PhD) - School of Mechanical and Manufacturing Engineering - Dublin City University.
19. Metcalf & Eddy, Inc. (Fourth Edition). Wastewater Engineering – Treatment and Reuse.
20. Miranda, J. P. R., Ubaque, C. A. G., and Londoño, J. C. P. (2015). Analysis of the investment costs in municipal wastewater treatment plants in Cundinamarca. Universidad Nacional de Colombia. DYNA 82 (192), pp. 230-238. August, Medellín.
21. Nogueira, R., Ferreira, I., Janknecht, P., Rodriguez, J. J., Oliveria, P., and Brito, A. G. (2007). Energy-saving wastewater treatment systems: formulation of cost functions. Water Sci. Tech., vol. 56, no.3, pp. 85–92.
22. Panaitescu, M., Panaitescu, F. V., and Anton, I. A. (2015). Theoretical and experimental researches on the operating costs of a wastewater treatment plant. IOP Conf. Series: Materials Science and Engineering 95 (2015) 012131.
23. Papadopoulos, B., Konstantinos, P., Tsagarakis, and Yannopoulos, A. (2007). Cost and land functions for wastewater treatment projects: Typical simple linear regression versus Fuzzy linear regression. ASCE. J. Env. Eng. vol.133, pp. 581.
24. Prado, O. J., Gaabriel, D., and Lafuente, J. (2009). Economical assessment of the design, construction and operation of open-bed biofilters for waste gas treatment. J. Environ. Manag. vol. 90, no.8, pp. 2515–2523.
25. Public Works Department (PWD), Government of India. (2021). Schedule of Rates.
26. Qasim, S.R., Lim, S.W., Motley, E.M., and Heung, K. G. (1992). Estimating costs for treatment plant construction. J. Am. Wat. Work. Ass. vol.84, pp. 56–62.
27. Sato, N., Okubo, T., Onodera, T., Agrawal, L. K., Ohashi, A., and Harada, H. (2007). Economic evaluation of sewage treatment processes in India. J. Env. Manage. vol. 84, pp. 447–460.
28. Sekandari, A. W. (2019). Cost Comparison Analysis of Wastewater Treatment Plants. International Journal of Science Technology & EngineeringI Volume 6 I Issue 1.
29. Senante, M. M., Sancho, F. H., and Garrido, R. S. (2012). Economic feasibility study for new technological alternatives in wastewater treatment processes: a review. Water. Sci. Technol., vol. 65, no.5, pp. 896–906.
30. Singhirunnusorn, W and Stenstrom, M. K. (2010). A Critical Analysis of Economic Factors for Diverse Wastewater Treatment Processes: Case Studies in Thailand. Sustain. Environ. Res., 20(4), 263–268
31. Tsagarakis, K. P., Mara, D. D., and Angelakis, A. N. (2003). Application of Cost Criteria for Selection of Municipal Wastewater Treatment Systems. Water Air and Soil Pollution - January Issue.

32. United Nations Environment Programme. (2015). Economic Valuation of Wastewater - The cost of action and the cost of no action.
33. USEPA. (1976). An analysis of construction cost experiences for wastewater treatment plants. EPA 430-9-76-002.
34. USEPA. (1982). Process Design and Cost Estimating Algorithms for The Computer Assisted Procedure for Design and Evaluation of Wastewater Treatment Systems (CAPDET).
35. USEPA. (1998). Detailed Costing Document for the Centralized Waste Treatment Industry', EPA 821- R-98-016.
36. Vanrolleghem, P. A., Jeppsson, U., Carstensen, J., Carlsson, B., and Olsson, G. (1996). Integration of wastewater treatment plant design and operation – a systematic approach using cost functions. Wat. Sci. Tech., vol. 34, pp. 159–171.
37. Wen, C. G. and Lee, C. S. 1999. Development of cost functions for wastewater treatment system with fuzzy regression. Fuzzy Sets and Systems, vol.106, pp. 143–153.
38. Yengejeh, R. J. Z., Davideh, K., and Baqeri, A. (2014). Cost/Benefit Evaluation of Wastewater Treatment Plant Types (SBR, MLE, Oxidation Ditch), Case Study: Khouzestan, Iran. Bulletin of Environment, Pharmacology and Life Sciences - Vol 4 [1] December: 55–60.

CHAPTER 3

Flood Frequency Analysis with Focus on Rivers Mundeswari, Ajoy and Churni

Gargi De[1]
MS Student, Department of Civil and Environmental Engineering, Syracuse University, New York, USA

Manik De[2]
Ex-Director, River Research Institute, West Bengal, India

Abstract

For the design and economic appraisal of a variety of engineering works, such as the construction of hydraulic structures, bridges and flood estimates are required. The hydrological data available are mostly samples of limited sizes. The basic tools of mathematical statistics possess enormous possibilities and empower us to extract the requisite information from the available data. One of the most important issues is the identification of the best method by using some statistical criteria for fitting models on hydrological variables of different rivers. The main objective of this paper is to address this particular issue by means of flood frequency analysis.

In analytical or statistical approach, the observed data are used to fit suitable probability distribution. In this study, the distributions used are lognormal distribution (LN), Pearson Type III distribution (PT III), Log Pearson Type III distribution (LPT III), General Extreme Value distribution (GEV) and Gumbel distribution. In this paper, annual maximum discharge data (in cumec) and gauge data (R.L. in metre) of the river Mundeswari (site Harinkhola), annual maximum water level data (in feet) of river Ajoy (Darbardanga) and annual maximum discharge data (in cumec) of river Churni (site Hanskhali) of West Bengal, have been taken into consideration.

It is very difficult to identify a particular distribution that can be used to estimate design flood, because a number of distributions may equally fit the observed data on the basis of well-established statistical criteria (Chi-Square test). Hence, the need for appropriate statistical criteria, which can differentiate quantitatively and identify the best-fitted distribution for a particular data series. In this paper, data of the rivers mentioned above, have been considered and an explicit study has been carried out to find out the procedure for selecting the best-fitted distribution for different complex situations.

Keywords: flood frequency analysis, return period, design flood, probability density function, extreme events, cumulative density function (CDF)

[1]gargi09@hotmail.com; [2]de_manik@yahoo.co.in

Introduction

The study of the characteristics of flow variables is one of the most important aspects of hydrological investigations. The continuous measurement and recording of flows of different rivers yield basic information on their flow variability over time. For analysis purpose, the basic data are discretized to suit the objective of the specific investigation required to be made. Flood estimates are required for the design and economic appraisal of a variety of engineering works.

To estimate flood magnitudes of desired frequencies, we can use flood frequency analysis. Generally, flood frequency analysis is carried out by two approaches viz. analytical approach and graphical approach. In graphical approach, the sample data are plotted on various probability papers using an appropriate plotting position formula, and a straight line is fitted to the data. This plot is used to estimate T years return period flood Q_T. In analytical or statistical approach, the observed data are used to fit suitable probability distribution.

The physical characteristics of different rivers vary from one to other. Hence, it is necessary to conduct an explicit study for each river. It is also required to study whether the different hydrological variables, such as discharge and gauge of a river, follow the same distribution pattern.

Methodology

The nature and extent of the data available determine the estimation procedure to be used to estimate the return period flood. Now, if we have a continuous discharge or water level record for N years, then it is possible to get the yearly maximum instantaneous discharge/or water level data for N years. In this way, N values $x_1, x_2, \ldots x_N$ may be derived from the whole series. If these maximum discharge/water level data are independently distributed and a probability density function $f(x, \theta_1, \theta_2, \ldots)$ can be fitted to the data series then there will be a value X_T such that

$$\int_{X_T}^{\infty} f(x, \hat{\theta}_1, \hat{\theta}_2, \ldots) dx = \frac{1}{T} \tag{1}$$

i.e.,

$$F(X_T) = 1 - \frac{1}{T} \tag{2}$$

wherein F (.) is the cumulative density function (CDF) of X.

The value X_T is such that the probability of peak discharge/gauge in a particular year that exceeds X_T is $1/T$. Hence, it is the estimate of the T-year flood by definition. $\hat{\theta}_1, \hat{\theta}_2, \ldots$ are the estimates of the parameters $\hat{\theta}_1, \hat{\theta}_2, \ldots$

Generally, two different types of distribution functions are available: (i) The distribution can be expressed in the inverse form $X_T = \phi(F)$. Here, X_T is calculated directly by replacing F with its value from Eq. (2). Examples of this type are the Gumbel distribution and general extreme value distribution. (ii) These types of distributions cannot be expressed directly in the inverse form $X_T = \phi(F)$. In this case, numerical methods are used to calculate X_T corresponding to a given value of F. The general form used to calculate X_T is $X_T = \mu + \sigma \cdot K_T$ (3) which was proposed by Chow (1964), where μ is the population mean, σ is the population standard deviation and K_T is the frequency factor (which depends upon the probability distribution used for flood frequency analysis and return period T).

Kite (1978) has given a detailed discussion with mathematical formulae to estimate different parameters of concerned distributions.

K_T values of Pearson Type III distribution are given in the tables in Bulletin 17B Hydrological Sub Committee (1981), the values can be calculated by the formula:

$$k_T = \frac{2}{C_s}\left[\left\{\left(k_n - \frac{C_s}{6}\right)\frac{C_s}{6} + 1\right\}^3 - 1\right],$$

where k_n is standard normal deviation corresponding to the probability of non-exceedance $\left(1 - \frac{1}{T}\right)$, given in the report: Bulletin 17B Hydrological Sub Committee (1981).

We have to test the randomness and stationarity of the data in order to judge whether the probability distribution function can be fitted to the data series.

The data to be used for flood frequency analysis should be (1) relevant (2) adequate and (3) accurate.

There are certain assumptions to fit the distribution to the data series: (1) the events are random, (2) the series is stationary and (3) the sample is representative of the population, or, in other words, the population parameters can be estimated by the sample.

To test the randomness of the sequence of discharges/gauges we have used:

1. Turning point test
2. Anderson's Correlogram test

Test of Randomness

Turning Point Test

In a sequence of data if a peak or a trough occurred, then the peak or the trough is defined as a turning point. Suppose $\{y_t\}$, $t = 1,2, \ldots N$ be a sequence of data then a peak is defined as the occurrence of a value y_t satisfying $y_{t-1} < y_t > y_{t+1}$ and a trough is defined by the occurrence of a value y_t satisfying $y_{t-1} > y_t < y_{t+1}$

The total number of turning points is approximately normally distributed with mean $2(N-2)/3$ and variance $(16N-29)/90$.

Given the series $\{y_t\}$, $t = 1, 2, \ldots N$, if too many or too few turning points exist, then the series is not random.

Here, the test statistic is $Z = \frac{p - E(p)}{\sqrt{\text{var}(p)}}$, where p is the total number of turning points.

$$E(p) = \frac{2(N-2)}{3} \quad \text{and} \quad \text{Var}(p) = \frac{16N - 29}{90}$$

Critical region ω: $|Z| > Z_{1-\frac{\alpha}{2}}$

Anderson's Correlogram Test

The kth-order autocorrelation, denoted by r_k is defined as follows:

$$r_k = \frac{\frac{1}{N-k}\sum_{1}^{N-k}\left\{\left(y_t - \frac{1}{N-k}\sum_{1}^{N-k} y_t\right)\left(y_{t+1} - \frac{1}{N-k}\sum_{1}^{N-k} y_{t+1}\right)\right\}}{\left[\frac{1}{N-k}\sum_{1}^{N-k}\left(y_t - \frac{1}{N-k}\sum_{1}^{N-k} y_t\right)^2 \cdot \frac{1}{N-k}\sum_{1}^{N-k}\left(y_{t+1} - \frac{1}{N-k}\sum_{1}^{N-k} y_{t+1}\right)^2\right]^{\frac{1}{2}}} \tag{4}$$

Our hypothesis that the series is random is H_0: $\rho_k = 0$, where ρ_k is population autocorrelation of order k.

Under H_0, r_k sample, kth-order autocorrelation can be approximated as Normal Distribution with $E(r_k) = -\frac{1}{N-k}$, $V(r_k) = \frac{N-k-1}{(N-k)^2}$

Test statistics $Z = \frac{r_k - E(r_k)}{\sqrt{\text{var}(r_k)}}$

Critical region ω: $|Z| > Z_{1-\frac{\alpha}{2}}$

Test of Stationarity

Kendal's rank correlation test is used to test the stationarity of the data.

The test statistic is, $Z = \frac{\tau}{\sqrt{Var(\tau)}}$ where, $\tau = \frac{4p}{N(N-1)} - 1$, $\text{Var}(\tau) = \frac{2(2N+5)}{9N(N-1)}$

where p is the total score for which observations exceed the previous values.

Critical region ω: $|Z| > Z_{1-\frac{\alpha}{2}}$

If the tests discussed above show no evidence of serial correlations and the discharge/water level data series may be considered to be a random series, then our task is to decide which probability distribution will fit the data series. The choice is very wide, and several distributions may fit the data equally well. Since tests of goodness of fit possess little power, the decision as to which probability distribution is to be selected becomes subjective, if the number of observations is small.

Goodness of Fit

The validity of the probability distribution function to be fitted on the available data set is tested analytically by Chi-Square Test and D-Index method.

Chi-Square Test

Here, our null hypothesis is H_0: The assumed distribution fits the given data.

The test statistics is
$$\chi^2 = \sum_{j=1}^{m} \frac{(O_j - E_j)^2}{E_j} = \sum_{j=1}^{m} \frac{O_j^2}{E_j} - N \tag{5}$$

Where O_j = observed frequency

E_j = expected frequency

m = number of classes

N = total number of observations

P = number of parameters

Critical region is defined as ω: $\chi^2 > \chi^2_{1-\alpha,m-p-1}$ (6)

That is, accept H_0 if $\chi^2_{cal} < \chi^2_{1-\alpha,m-p-1}$

D-Index Test

The D-index is defined as:

$$\text{D-index} = \frac{1}{\overline{X}} \sum_{i=1}^{6} \left| X_{i,observed} - X_{i,computed} \right| \tag{7}$$

where $\overline{X}$ is the mean of the observed series. X_i's are the largest six observations.

$X_{i,\ computed}$ is estimated using frequency factor [Eq. (3)] corresponding to the value $P(X \geq X_{i,\ observed})$.

The probability of exceedance is estimated by Weibull plotting position formula

$$P(X \geq X_{(m)}) = \frac{m}{N+1} \tag{8}$$

where N is the total number of observations and m is the rank of hydrological observed data arranged in descending order of magnitude. Here, our aim is to predict the design flood with the help of the most suitable distribution. Hence, it is necessary to judge the accuracy of the prediction of the estimate.

The measure of the accuracy of predictions is the standard error of the estimate.

Standard Error of Estimation

Let, X_T = T-year return period flood

$\overline{X}$ = Sample mean

S_X = Sample standard deviation

$\overline{Y}$ = Sample mean of log-transformed series

S_Y = Sample standard deviation of log-transformed series

K_T = Frequency factor corresponding to $P(X \geq X_T) = \frac{1}{T}$

C_S = Coefficient of skewness of original series

Y_T = T-year return period flood of log-transformed series

The T-year return period flood is written as $X_T = \bar{X} + K_T S_X$ or $X_T = \exp(\bar{Y} + K_T S_Y)$

The standard error of estimate [SE(X_T)] of different distributions is given as follows:

Lognormal Distribution:

$$\mathrm{SE}(X_T) = \frac{X_T}{2}\left\{e^{Se(Y_T)} - e^{-Se(Y_T)}\right\} \tag{9}$$

where
$$\mathrm{Se}\,(Y_T)\ \frac{S_Y}{\sqrt{N}}\left[1+0.5K_T^2\right]^{0.5}, \tag{10}$$

Pearson Type III Distribution:

$$\mathrm{SE}(X_T) = \frac{s_x}{\sqrt{N}}\left[1+K_T C_S + \frac{K_T^2}{2}\left(3\frac{C_S^2}{4}+1\right)+3K_T V_T\left(C_S+\frac{C_S^3}{4}\right)+3V_T^2\left(2+3C_S^2+\frac{5C_S^4}{8}\right)\right]^{\frac{1}{2}} \tag{11}$$

where
$$V_T = \frac{y_t^2-1}{6}+\frac{4(y_t^3-6y_t)}{6^3}C_S-\frac{3(y_t^2-1)}{6^3}C_S^2+\frac{4y_t}{6^4}C_S^3-\frac{10}{6^6}C_S^4 \tag{12}$$

y_t = Standard normal deviation corresponding to return period T.

Log Pearson Type III Distribution:

The standard error of estimate of X_T for LPT III distribution is computed from

$$\mathrm{SE}(X_T) = \frac{X_T}{2}\left[e^{SE'(Y_T)} - e^{-SE'(Y_T)}\right] \tag{13}$$

where $SE'(Y_T)$ satisfies Eq. (10)

Gumbel Distribution:

If parameters are estimated by MOM, the standard error is given by

$$\mathrm{SE}(X_T) = \frac{\alpha}{\sqrt{N}}\left(1.17+0.196Y_T+1.099Y_T^2\right)^{0.5} \tag{14}$$

where, $\alpha = \frac{\sigma\sqrt{6}}{\pi}$

Analysis and Results

All the data series of different rivers are tested, and it is found that all the data series can be treated as random sequence and the trend is absent, hence, a probability distribution can be fitted to these data series. The mean, standard deviation, skewness and kurtosis of discharge and gauge data of different rivers are given in Table 3.1.

Table 3.1 Sample mean, standard deviation, skewness and kurtosis

	N		Mean	SD	Skew	Kurt
Mundeswari(Q)	26	Original	2222.446	919.5944	–0.7787	–0.2766
		log transformed	7.56143	0.64594	–1.69282	2.10451
Mundeswari(h)	24	Original	12.613	1.4800	–1.1732	0.4281
		log transformed	2.527	0.12692	–1.35758	1.0039
Ajoy(h)	25	Original	292.836	4.5143	0.5146	0.2391
		log transformed	5.6795	0.01536	0.47322	0.17797
Churni(Q)	20	Original	227.356	68.9413	0.48737	0.6309
		log transformed	5.38025	0.32159	-0.68997	1.75549

Goodness of fit:

Table 3.2 Calculation of χ^2 value for the lognormal distribution, river Mundeswari (Q)

Class	Prob limits	Reduced limit		Limits	Obs(Oj)	Exp(Ej)
1	0–1/6	–∞	–0.967	–∞ ≤ x < 1029.4768	4	4.3333
2	1/6–2/6	–0.967	–0.43	1029.4768 ≤ x < 1455.3910	1	4.3333
3	2/6–3/6	–0.43	0	1455.3910 ≤ x < 1922.5929	2	4.3333
4	3/6–4/6	0	0.43	2539.7734 ≤ x < 2539.7734	7	4.3333
5	4/6–5/6	0.43	0.967	3590.5260 ≤ x < 3590.5260	12	4.3333
6	5/6–1	0.967	1	3590.5260 ≤ x < ∞	0	4.3333

$$\chi^2 = \sum_{i=1}^{m} \frac{O_i^2}{E_i} - N = 23.385$$

From χ^2 table, $\chi^2_{0.95,3} = 7.81$. The calculated value of χ^2 obtained from Table 3.2 is therefore not significant and it may be inferred that there is reason to believe that log normal distribution does not fit the given data (discharge, Q).

Table 3.3 Calculation of χ^2 value for lognormal distribution, river Mundeswari(h)

Class	Prob limits	Reduced limit		Limits	Obs(Oj)	Exp(Ej)
	0–1/6	–∞	-0.967	–∞ ≤ x < 11.075	4	4
2	1/6–2/6	–0.967	-0.43	11.075 ≤ x < 11.856	3	4
3	2/6–3/6	–0.43	0	11.856 ≤ x < 12.521	1	4
4	3/6–4/6	0	0.43	12.521 ≤ x < 13.224	4	4
5	4/6–5/6	0.43	0.967	13.224 ≤ x < 14.156	10	4
6	5/61	0.967	1	14.156 ≤ x < ∞	2	4

$$\chi^2 = \sum_{i=1}^{m} \frac{O_i^2}{E_i} - N = 12.5$$

From χ^2 table, $\chi^2_{0.95,3} = 7.81$. The calculated value of χ^2 obtained from Table 3.3 is therefore significant and it may be inferred that there is reason to believe that log normal distribution does not fit the given data (gauge, h).

Table 3.4 The calculation of χ^2 value to fit Pearson type III distribution river Mundeswari (Q), $C_s = -0.7787$.

Class	Prob limits	Reduced limit		Limits	Obs (Oj)	Exp (Ej)
1	0–1/6	−∞	−0.8937	$-\infty \le x < 1365.8630$	5	4.3333
2	1/6–2/6	−0.8937	−0.2454	$1365.8630 \le x < 1934.4821$	2	4.3333
3	2/6–3/6	−0.2454	0.1882	$1934.4821 \le x < 2339.7980$	3	4.3333
4	3/6–4/6	0.1882	0.5521	$2339.7980 \le x < 2701.5191$	6	4.3333
5	4/6–5/6	0.5521	0.9183	$2701.5191 \le x < 3094.2850$	7	4.3333
6	5/6–1	0.9183	∞	$3094.2850 \le x < \infty$	3	4.3333

$$\chi^2 = \sum_{i=1}^{m} \frac{O_i^2}{E_i} - N = 4.462$$

From χ^2 table, $\chi^2_{0.95,2} = 5.99$. The calculated value of χ^2 obtained from Table 3.4 is therefore not significant and it can be inferred that there is reason to believe that Pearson type III distribution fits the given data well.

Table 3.5 Calculation of χ^2 value for the Pearson type III distribution, river Mundeswari(h), Cs = −1.1732

Class	Prob limits	Reduced limit		Limits	Obs(Oj)	Exp (Ej)
1	0–1/6	−∞	−0.8937	$-\infty \le x < 11.2903$	4	4
2	1/6–2/6	−0.8937	−0.2454	$11.2903 \le x < 12.2498$	4	4
3	2/6–3/6	−0.2454	0.1882	$12.2498 \le x < 12.8915$	2	4
4	3/6–4/6	0.1882	0.5521	$12.8915 \le x < 13.4301$	4	4
5	4/6–5/6	0.5521	0.9183	$13.4301 \le x < 13.9721$	8	4
6	5/6–1	0.9183	∞	$13.9721 \le x < \infty$	2	4

$$\chi^2 = \sum_{i=1}^{m} \frac{O_i^2}{E_i} - N = 6$$

From χ^2 table, $\chi^2_{0.95,2} = 5.99$. The calculated value of χ^2 obtained from Table 3.5 is therefore not significant and we can say that there is no reason to believe that Pearson type III distribution fits the given data well.

Similarly, χ^2 values of all distributions for all the rivers are calculated and given in Table 3.6.

Table 3.6 Calculated χ^2 values

	Log normal	PT III	LPT III	Gumbel	GEV
DF (m-p-1)	3	2	2	3	2
Mundeswari (Q)	23.39	4.46	10.92	8.15	18.40
Mundeswari (*h*)	12.50	6.00	6.00	20.50	9.50
Ajoy (*h*)	0.68	1.64	1.64	5.48	20.36
Churni (Q)	0.40	0.40	1.60	0.40	2.20
$\chi^2_{0.95,df}$	7.81	5.99	5.99	7.81	5.99
$\chi^2_{0.99,df}$	11.30	9.21	9.21	11.30	9.21

From Table 3.6, it can be inferred that at 5% level of significance, no distribution fits the gauge data of Mundeswari river but there is reason to believe that both PT III and LPT II distributions fit the given data very well at 1% level of significance.

Table 3.7 Goodness fit using χ^2 test

	Mundeswari (Q)*	Mundeswari (*h*)**	Ajoy (*h*)*	Churni (Q)*
Lognormal	Does not fit well	Does not fit well	Fits well	Fits well
PT III	Fits well	Fits well	Fits well	Fits well
LPT III	Does not fit well	Fits well	Fits well	Fits well
Gumbel	Does not fit well	Does not fit well	Fits well	Fits well
GEV	Does not fit well	Does not fit well	Does not fit well	Fits well

* 5% level of significance, ** 1% level of significance

From Table 3.7, it is clear that different distributions equally fit the data set for a particular site of a river. Hence, there is a need to select some other test criterion to choose the best-fitted distribution. For this purpose, the D-Index test is used, which is based on the highest six values.

Table 3.8 D-Index values

	Mundeswari (Q)	Mundeswari (*h*)	Ajoy (*h*)	Churni (Q)
Lognormal	2.7206	0.2832	0.0248	0.3041
PT III	0.1731	0.0936	0.0299	0.4567
LPT III	1.9275	0.1269	0.0301	0.1906
GEV	0.2279	0.4724	0.1227	1.3263
Gumble	0.4714	0.2353	0.0328	0.3616

From Tables 3.7 and 3.8, it is seen that for discharge data of river Mundeswari, only PT III distribution fits well on the basis of χ^2 test at 5% level of significance, and it is also seen that D-Index value is minimum for PT III distribution. It is also seen that for gauge data of river

Mundeswari both PT III and LPT III distributions fit well on the basis of χ^2 test at 1% level of significance and it is seen that D-Index value is minimum for PT III distribution. All the distributions except GEV distribution fit the gauge data of river Ajoy on the basis of χ^2 test at 5% level of significance and D-Index value is minimum for lognormal distribution.

On the basis of χ^2 test, it is seen that all the distributions fit well on the data set of river Churni. It is also seen that D-Index value is lowest for LPT III distribution. Standard error of estimates of T-year floods for different distributions of each river is given in Table 3.9

Table 3.9 Standard error of estimates of T-year floods for different distributions

River site	Distribution	Return periods				
		20	50	100	200	300
Mundesewari(Q)	LN	1087.9	1632.1	2127.1	2703.2	3077.6
	PT III	219.7	312.82	390.96	470.93	517.39
Harinkhola	LPT III	950.4	1427.18	1720.90	1950.76	2054.15
	Gumbel	475.5	607.30	707.39	807.82	866.70
Mundesewari(h)	LN	0.61	0.74	0.84	0.94	1.04
	PT III	0.41	0.62	0.78	0.92	1.004
Harinkhola	LPT III	0.58	0.88	1.09	1.28	1.38
	Gumbel	0.80	1.02	1.18	1.35	1.45
Ajoy (h)	LN	1.42	1.64	1.79	1.95	2.03
	PT III	2.00	2.76	3.40	4.07	2.48
Darbardanga	LPT III	2.01	2.78	3.43	4.12	4.54
	Gumbel	2.33	2.98	3.47	3.96	4.25
Churni (Q)	LN	40.72	53.43	63.69	74.55	81.13
	PT III	33.70	46.32	56.89	68.14	74.92
Hanskhali	LPT III	37.53	63.47	89.52	121.58	130.85
	Gumbel	40.65	51.91	60.47	69.05	74.08

Conclusion

It is seen from the Turning point test and Anderson's Correlogram test that all the observed data series may be considered random at 5% confidence level. From Kendal's rank correlation test, it is concluded that there is no trend. Hence, probability distribution can be fitted to these data series.

For discharge data of river Mundeswari only PT III distribution fits well on the basis of χ^2 test at 5% level of significance. The D-Index value is minimum for PT III distribution. PT III distribution also gives the lowest standard error of estimate for the river Mundeswari. Hence, it can be inferred that PT III distribution is the best-fitted distribution to estimate the return period flood for discharge data of river Mundeswari.

For gauge data of river Mundeswari, both PT III and LPT III distributions fit well on the basis of χ^2 test at 1% level of significance. The D-Index value is minimum for PT III distribution. PT III distribution also gives lowest standard error of estimate for the river Mundeswari. Hence, it can be inferred that PT III distribution is the best-fitted distribution to estimate return period flood for gauge data of river Mundeswari.

Hence, it can be concluded that PT III distribution is the best-fitted distribution to estimate return period flood for both gauge and discharge data of river Mundeswari. It may not always be possible to measure discharge data at the remote place during floods. However, it is possible to measure gauge data. Using PT III distribution, we can estimate gauge for different return periods. And from the gauge-discharge curve, it is possible to estimate flood data of river Mundeswari at different return periods.

Similarly, on the basis of all three test criteria, it may be inferred that lognormal distribution is the best-fitted distribution to estimate different return period water level data of the river Ajoy.

There is reason to believe that for the discharge data of the river Churni all the five distributions fit equally well on the basis of χ^2 test at 5% level of significance. Though LPT III distribution gives the best fit for upper tail region on the basis of D-Index value, the standard error of estimate of T-year flood is lowest for PT III distribution. Since our aim is to estimate the design flood, it is better to use PT III distribution to estimate the return period flood for river Churni.

References

1. Chow,V.T., (1964): Handbook of Applied Hydrology, Mc GRAW- HILL BOOK COMPANY (ISBN 07-010774-2)
2. Kite, G.W., (1978): Frequency and risk analysis in Hydrology, Water Resources Publications.
3. Bulletin 17B Hydrological Sub Committee (1981): Guidelines for Determining Flood Flow Frequency.
4. Lieblein, J., (1954): A New Method of Analysing Extreme Value Data, Natl. Advisory Comm. Aeronaut. Tech. Note 3053.

CHAPTER 4

Evaluation of Paint Wastewater Treated by Anaerobic Batch Reactor

Manju. E. S[1]

Research Scholar, Department of Civil Engineering, National Institute of Technology, Karnataka Surathkal, India

Basavaraju Manu[2]

Associate Professor, Department of Civil Engineering, National Institute of Technology, Karnataka Surathkal, India

Abstract

Paint plays an important role in the overall aesthetics of a building. Wastewater from paint industries has high chemical oxygen demand (COD) and contains harmful chemicals. Many cases of allergies have been reported with exposure of the human body to paint. Hence, this wastewater has to be treated before disposal. The manufacturing process of paint does not produce much wastewater as such. It is mainly produced by cleaning the equipment and vessels used in the production processes. The method discussed and used here is sequential anaerobic treatment and the paint used is Acronal 295 D ap, which is collected from a factory in Mangalore. The reactor was run with an HRT of 15 days and was fed with a starch solution every 24 hours. COD removal efficiency, ORP, gas production and pH were monitored throughout the process. The reactor was fed with varying ratios of paint every 15 days. pH was found to be almost constant, around 5. The COD removal efficiency was around 80%–98%, which shows there was no toxicity effect of paint wastewater on the anaerobic bacteria in the reactor. The gas production in the anaerobic reactor showed similar results as control reactor, which implies that there is no inhibitory effect in the methanogenic activity. ORP throughout the process was highly negative, which showed the reductive nature of the reactor. It also indicates sulphide formation, biological phosphorus release and methane formation. The high COD removal suggests that a sequential anaerobic reactor can be considered an effective method for the treatment of paint wastewater.

Keywords: anaerobic reactor, COD removal efficiency, paint wastewater

Introduction

Water is one of the most essential resource on earth. Water is available in abundance but the freshwater source is only around 3%. Water is required in almost all sectors, including agriculture and industries. The increase in water usage has made it important to take care of the usage of water.

[1]manjuues@gmail.com; [2]bmanu8888@gmail.com

The dumping of industrial wastewater into the waterbodies makes it toxic. Hence, proper treatment is required before releasing these effluents into water bodies. It also pollutes the soil and thus making it difficult for the microorganisms in the soil to survive (Kaur et al., 2010).

The paint manufacturing industries and textile industries are the largest consumers of water and hence the reusability of treated wastewater in these units also has much importance (da Silva et al., 2016). Paint consists of resins, solvents, pigments, fillers and additives. Paint can be classified as water-based paint and solvent-based paint. Paint wastewater is not the wastewater generated during the making process of paint. It is mainly from the cleaning operations of the equipment used in the preparation process. It accounts for about 80% of the paint industry wastewater (Shazly et al., 2010). An average the paint industry in Istanbul produced around 20 tons per day and wastewater produced was 4.6 m^3/day (Germirli Babuna and Orhon n.d.). There are many physical and chemical treatments for paint wastewater mainly coagulation/flocculation processes (Aboulhassan et al., n.d.; Dovletoglou et al., 2002; Eremektar et al., 2006; Verma et al., 2012). The chemical treatment uses chemicals that may be expensive. Also, the by-product formed at the final stage may be a new and more harmful component which itself may require treatment before disposal (Show et al., 2020).

The paint contains various heavy metals. Some of which are copper, zinc, cadmium and nickel. These heavy metals can be toxic to microorganisms even in minute concentrations (Güven et al., 2017). Copper is found to have more inhibitory effects than zinc. Even in small concentrations, these can be harmful to the flora and fauna (Juliastuti et al., 2003).

Biological methods have been found effective for the treatment of industrial effluents in recent years. It is found that it is also effective for paint wastewater (Show et al., 2020). Anoxic aerobic sequential batch reactor gave good results and was proved to be an energy-efficient system for the treatment of paint wastewater (Jena et al., 2020). SBR reactors are economically feasible and highly efficient (Mahesh and Manu, 2019).

Materials and Methods

Wastewater for the Study

Wastewater was synthetically prepared by mixing paint and water in a laboratory scale. The paint was collected from a paint factory in Mangalore. The reactor was fed with this prepared paint solution every 15 days. COD, pH and ORP were checked every 3 days.

Characteristics of Paint Wastewater

Paint wastewater contains high COD due to the presence of various organic and inorganic matter present. The inorganic matter may be from the pigments and dye used in paint.

Table 4.1 Wastewater characteristics for the present study

Parameters	Acronal 295 paint
COD	19,200 mg/L
Nitrates	96 mg/L
pH	7–8

Experimental Design

A 2.5-L capacity glass bottle was used for making the reactors. There are two reactors, one that will be fed with paint wastewater and the other is a control reactor. Both the reactor volume was maintained at 2 litres. The bottle was 12 cm in diameter and 20 cm in height. For gas collection, the reactor was attached to a similar bottle by a small tubular pipe attached to its cork. The reactors were designed to operate manually to avoid operational issues. The procedure involves manual feeding and decanting. The reactor was completely air-tightened to collect and quantify the gas produced.

Seed sludge for the anaerobic reactor was collected from the wastewater treatment plant in Mangalore. The wastewater was sieved through a 250-mm sieve to remove all the large particles. MLVSS is calculated for the sludge to get the organic fraction. It had an MLVSS concentration of 35.86 g in 1 litre. The COD of the sludge was found. The anaerobic reactor is fed with sludge. Starch solution with a concentration of 2 g/L was added to the reactor as the carbon source for the microbes. NaHCO3 of concentration 500 mg/L was also added to this solution to make the pH stable. The experiment occurred in four stages, filling, reaction, settling and decanting. That is, the wastewater is filled in the reactor, reactions take place, then settling of the sludge occurs after that the clear supernatant will be removed from the reactor. It is a 24-hour process. After every 24 hours, the reactor is fed with the starch solution and every 15 days, the reactor is fed with the paint wastewater together with the starch. The clear supernatant is taken and tested for COD, ORP and pH.

Analysis Methods

Nitrate concentration in paint was determined by a UV-spectrophotometer. The reagent used was 1 N hydrochloric acid. A standard nitrogen curve was drawn by using the standard solutions. The paint wastewater as well as the treated wastewater was taken for evaluation. Both samples were mixed with 1 N hydrochloric acid. Absorbance at a wavelength of 220 nm was measured in comparison with the blank solution. Nitrates nitrogen in the paint wastewater was hence determined from the graph.

The presence of some of the heavy metals copper and zinc was determined by using atomic absorption spectroscopy. Samples were kept for the atomic absorption spectrophotometer and the values were determined from the formerly calibrated curves of copper and zinc.

COD was determined by the closed reflux titrimetric method where the wastewater reacts with a strong oxidizing agent. pH was determined by the digital pH metre and ORP was determined by using a digital ORP meter. ORP indicates the oxidation-reduction potential of the sample. That is whether it is oxidative or reductive in nature.

Two samples were taken for HPLC, one the paint wastewater and the other treated wastewater. The graph gives the details of the disappearance or lowering of any compound present in the wastewater (da Silva et al., 2016). The mobile phase used in this technique is 0.01 mol per litre of potassium monohydrogen phosphate of pH 5.

Results and Discussions

Chemical Oxygen Demand

The anaerobic reactor showed high COD removal efficiency, mostly around 80%–98%. After adding the paint wastewater, the COD removal efficiency was not much affected, which may be because the anaerobic bacteria have taken the paint wastewater as a feeding source together with the starch in the solution.

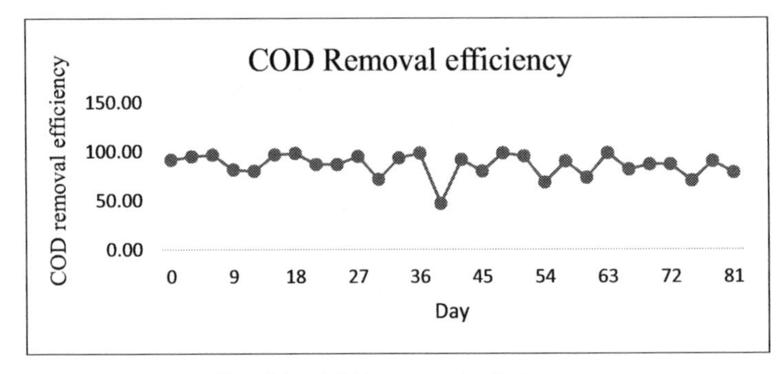

Fig. 4.1 COD removal efficiency

Gas Production

Gas production in an anaerobic reactor was found to be increasing at the initial stage of acclimatisation. It is almost constant throughout the control reactor. The gas production in the anaerobic reactor was found to be varying in between but still, it was giving gas production all day. This implies that the anaerobic bacteria do not show much inhibition to the paint wastewater. Figure 4.2 shows the gas production in the anaerobic reactor (series 1) and gas production in the control reactor (series 2).

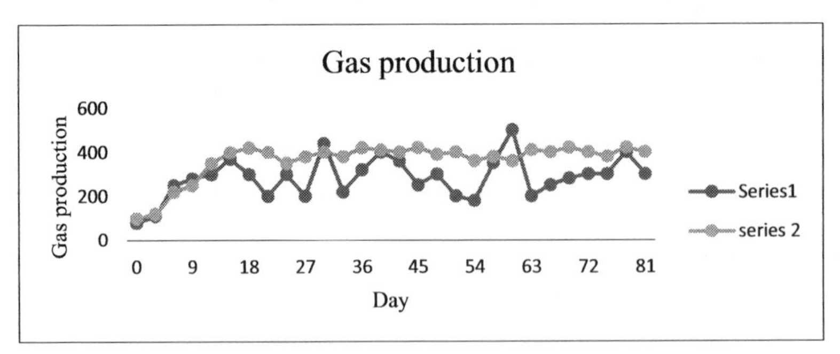

Fig. 4.2 Gas production in the anaerobic reactor and the control reactor

pH

pH of the wastewater in the reactor was around 4.5–5 throughout the working days of the reactor.

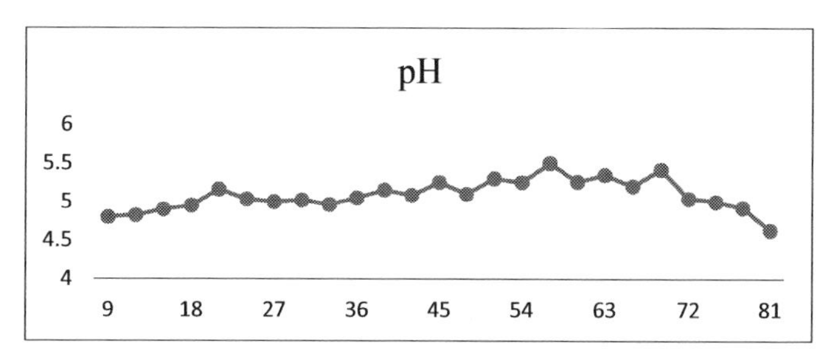

Fig. 4.3 pH distribution in 81 days

Oxidation Reduction Potential

The ORP metre showed high negative values around -100 to -200. It shows the reductive nature of the reactor. Also, it shows that methane production is happening in the reactor.

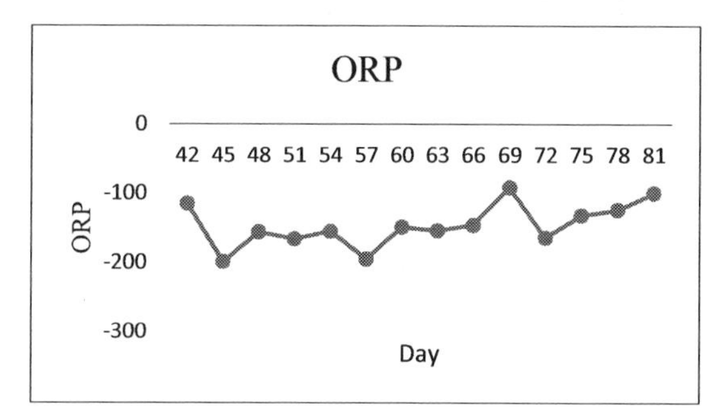

Fig. 4.4 ORP variation

Nitrates

Nitrogen is a nutrient source for algae. But too much nitrogen in wastewater may lead to the overgrowth of algae in rivers and ponds, thus reducing the oxygen content in water. The nitrogen removal efficiency was found to be 76% in the first 3 days itself, which indicates that this method can be effectively used for nitrogen removal in paint wastewater.

Heavy Metals

The presence of copper and zinc was determined in the paint wastewater by using AAS. There was no presence of copper and zinc in this paint and that may be the reason it did not show any inhibition in the reactor.

High-Performance Liquid Chromatography

The HPLC results showed that there is a reduction in certain components concentration after the treatment. One component at retention time of 5.22 got reduced after the treatment and the formation of two components occurred at 0.11 and 0.41. This showed that the component which was found in paint wastewater might have converted into other compounds after treatment. Many small peaks were completely removed after the treatment.

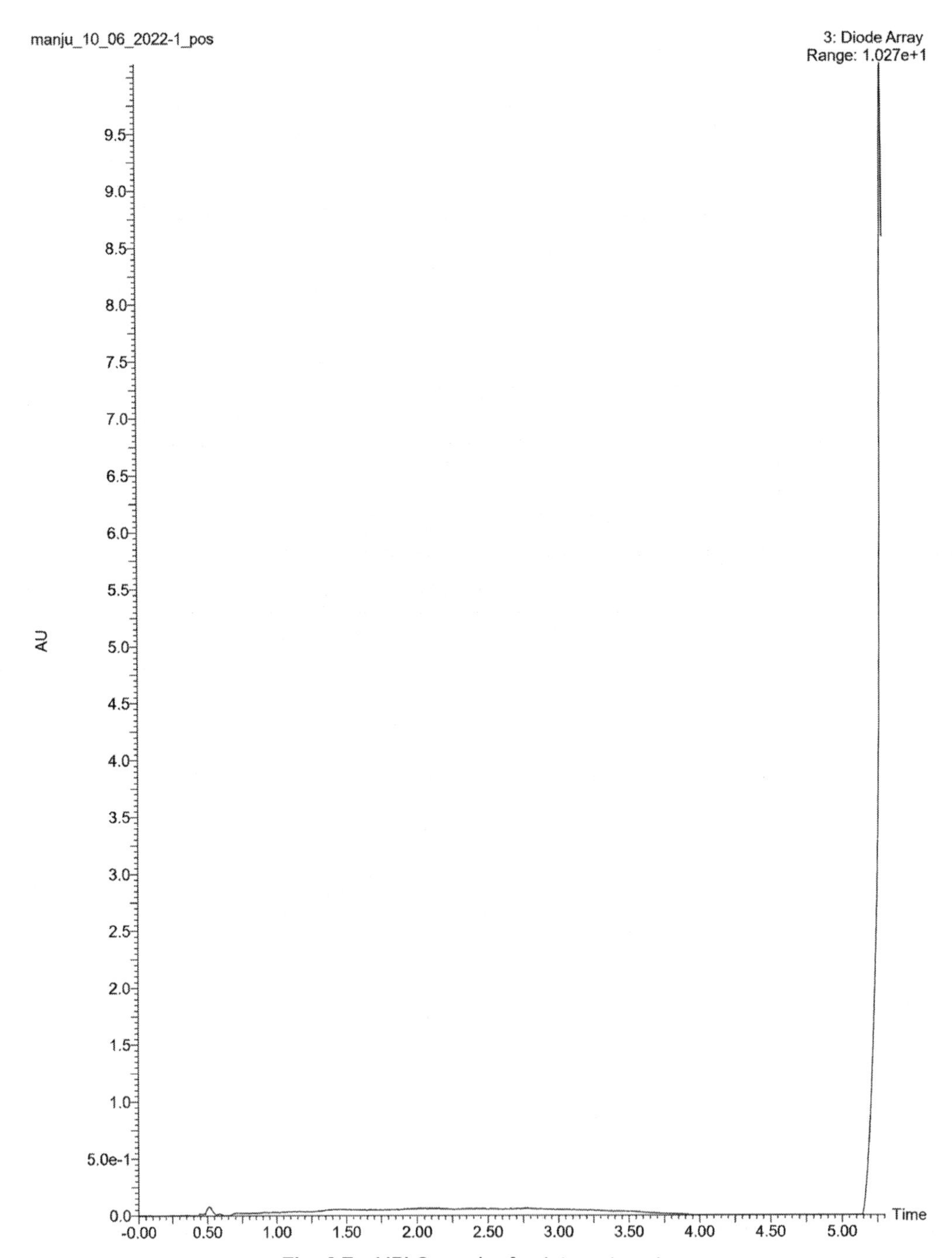

Fig. 4.5 HPLC graph of paint wastewater

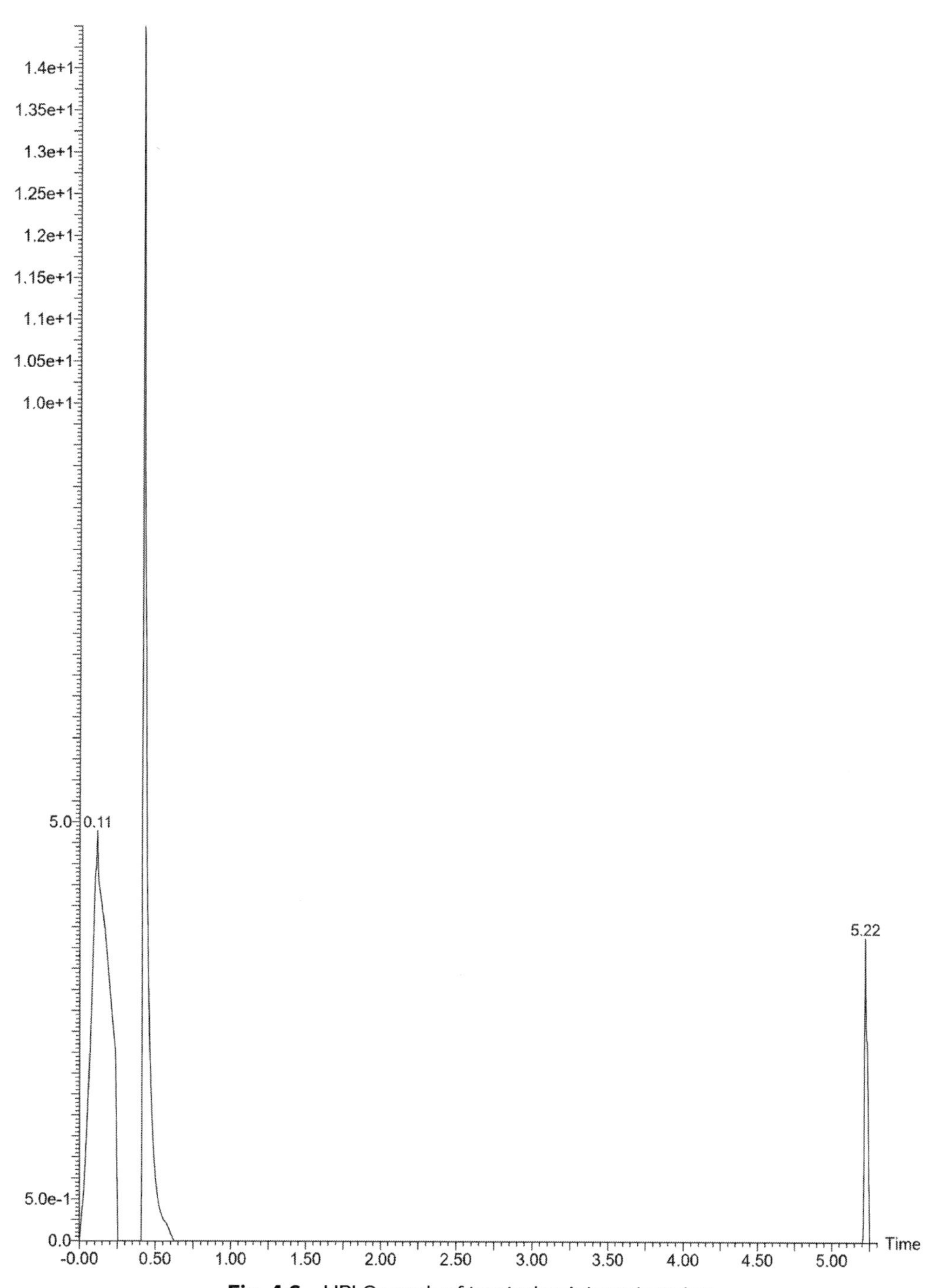

Fig. 4.6 HPLC graph of treated paint wastewater

Conclusions

The paint wastewater was treated by an anaerobic reactor and from the results following conclusions were obtained:

- The COD removal efficiency in the anaerobic reactor was high, which makes the sequential anaerobic method an effective method for the treatment of this paint wastewater.
- The gas production was found to be similar to that of the control reactor. That is, the paint wastewater showed no inhibition to the anaerobic reactor There was no inhibition to the methanogenic activity.
- ORP was highly negative, which shows the reductive nature of the reactor.
- The absence of copper and zinc shows that there is less toxicity of heavy metals compared to other paints.
- The high nitrogen removal efficiency shows this method can be used for the removal of nitrates from the wastewater.

Acknowledgment

The author sincerely thanks the Ministry of Human Resources and Development, Government of India for providing an institute fellowship to carry out this research.

References

1. Aboulhassan, M. A., Souabi, S., Yaacoubi, A., and Baudu, M. (n.d.). *TREATMENT OF PAINT MANUFACTURING WASTEWATER BY THE COMBINATION OF CHEMICAL AND BIOLOGICAL PROCESSES.*
2. Dovletoglou, O., Philippopoulos, C., and Grigoropoulou, H. (2002). "Coagulation for treatment of paint industry wastewater." *J. Environ. Sci. Heal. - Part A Toxic/Hazardous Subst. Environ. Eng.*, 37(7), 1361–1377.
3. Eremektar, G., Goksen, S., Babuna, F., and Dogruel, S. (2006). "Coagulation-flocculation of wastewaters from a water-based paint and allied products industry and its effect on inert COD." *J. Environ. Sci. Heal. - Part A Toxic/Hazardous Subst. Environ. Eng.*, 1843–1852.
4. Germirli Babuna, F., and Orhon, D. (n.d.). *Treatability of Water-Based Paint Effluents APPROPRIATE TECHNOLOGIES FOR THE MINIMIZATION OF ENVIRONMENTAL IMPACT FROM INDUSTRIAL WASTEWATERS-TEXTILE INDUSTRY, A CASE STUDY View project GIS for Sustainable Environment View project.*
5. Güven, D., Hanhan, O., Aksoy, E. C., Insel, G., and Çokgör, E. (2017). "Impact of paint shop decanter effluents on biological treatability of automotive industry wastewater." *J. Hazard. Mater.*, 330, 61–67.
6. Jena, J., Narwade, N., Das, T., Dhotre, D., Sarkar, U., and Souche, Y. (2020). "Treatment of industrial effluents and assessment of their impact on the structure and function of microbial diversity in a unique Anoxic-Aerobic sequential batch reactor (AnASBR)." *J. Environ. Manage.*, 261.
7. Juliastuti, S. R., Baeyens, J., Creemers, C., Bixio, D., and Lodewyckx, E. (2003). "The inhibitory effects of heavy metals and organic compounds on the net maximum specific growth rate of the autotrophic biomass in activated sludge." *J. Hazard. Mater.*, 100(1–3), 271–283.

8. Kaur, A., Vats, S., Rekhi, S., Bhardwaj, A., Goel, J., Tanwar, R. S., and Gaur, K. K. (2010). "Physico-chemical analysis of the industrial effluents and their impact on the soil microflora." *Procedia Environ. Sci.*, 2(5), 595–599.
9. Mahesh, G. B., and Manu, B. (2019). "Biodegradation of ametryn and dicamba in a sequential anaerobic-aerobic batch reactor: A case study." *Water Pract. Technol.*, 14(2), 423–434.
10. Shazly, M. A. El, Hasanin, E. A., and Kamel, M. M. (2010). "Appropriate Technology for Industrial Wastewater Treatment of Paint Industry." *J. Agric. Environ. Sci*, 8(5), 597–601.
11. Show, K. Y., Ling, M., Guo, H., and Lee, D. J. (2020). "Laboratory and full-scale performances of integrated anaerobic granule-aerobic biofilm-activated sludge processes for high strength recalcitrant paint wastewater." *Bioresour. Technol.*, 310.
12. Silva, L. F. da, Barbosa, A. D., Paula, H. M. de, Romualdo, L. L., and Andrade, L. S. (2016). "Treatment of paint manufacturing wastewater by coagulation/electrochemical methods: Proposals for disposal and/or reuse of treated water." *Water Res.*, 101, 467–475.
13. Verma, A. K., Dash, R. R., and Bhunia, P. (2012). "A review on chemical coagulation/flocculation technologies for removal of colour from textile wastewaters." *J. Environ. Manage.*

CHAPTER 5

Assessment and Mapping of Urban Road Traffic Noise at Hospital Buildings: The Case of Surat City, India

Ramesh B Ranpise[1]
Ph.D Scholar, Department of Civil Engineering,
Sardar Vallabhbhai National Institute of Technology, Surat, India

Dhaval Sindhav[2]
M. Tech Scholar, Department of Civil Engineering,
Sardar Vallabhbhai National Institute of Technology, Surat, India

B. N. Tandel[3]
Assistant Professor, Department of Civil Engineering,
Sardar Vallabhbhai National Institute of Technology, Surat, India

Abstract

The exponential rise in noise levels in all types of zones, such as residential, commercial, industrial and silent, is a major cause of concern. Exposure to noise *may affect the quality of life. It causes various health impacts. The prime focus* of this study is to observe the noise levels and their exposure (propagation) at the hospital buildings in an urban area. This study represents the current scenario of noise levels in an urban area through noise assessment and mapping. Noise mapping is a classic tool for conducting noise impact assessment (NIA) in an urban region. Four hospital buildings were selected for data collection and classified as silent area as per the Central Pollution Control Board (CPCB) India. Daytime noise measurements were conducted by using KIMO DB 300 sound level metre. The Leq levels vary between 68.4 dBA to 79 dBA, which is higher than the limits given by the government of India for the silence zone. The assessed data were mapped by using ArcGIS 10.8 software, which allowed the clear visualisation of the distribution of sound levels across the study location. The study discloses that current noise levels at all locations exceed the norms. Based on the result, it can be brought up that the people in this area are exposed to high traffic noise. Hence, control measures are needed to reduce the problem.

Keywords: noise pollution, urban road traffic, noise monitoring, noise maps, GIS

Introduction

The ownership noise map is an effective tool that helps in the identification and understanding of the noise levels of a particular region. It is helpful for one who does not understand the values

[1]ranpiseramesh6588@gmail.com; [2]p20en002@ced.svnit.ac.in; [3]bnt@ced.svnit.ac.in

and their significance. Noise maps are nothing but the representation of an image that provides the information in the pictorial form so one can easily understand by seeing it (Zambon et al., 2021). The noise maps are developed using Arc Gis 10.8 software, which gives—a clear picture of the intensity of noise and its propagation in the selected study locations (Nasim Akhtar, Kafeel Ahmad, 2012).

The idea is extremely basic—a noise map is only a visual portrayal of noise levels in a working environment as a contour map. Contours are generally shaded to demonstrate the intensity of noise, presence of high or low frequencies, and so forth, and the map is typically overlaid on a plan of the area or working environment.

In metropolitan planning, a noise map is utilised to dissect the sound characteristics of the soundscapes in a particular metropolitan region to create proposals for the metropolitan plan of the soundscapes (Zannin & Sant'Ana, 2011).

Literature Review

Early with the growing number of vehicles and the continuous improvement in technology, street traffic noise has turned into a difficult issue regarding ecological protection (Pan & Zhu, 2011). It is important to conduct regular measurements to identify urban areas where there are maximum noise levels that exceed permissible noise limits (Manojkumar et al., 2019). On-site measurements or through predictions, using specially designed software can be used for assessment (Mioduszewski et al., 2011). Noise mapping softwares are executive tool, broadly utilised by numerous specialists from various backgrounds and with extraordinary involvement with different applications, information and programming. The way of utilising the product is vital on the grounds that it can influence the quality of the outcomes in the noise mapping process (Petrovici et al., 2015).

F. Farcas et al. developed a noise map of skåne region located in the south region of Sweden. For mapping, they used ArcGIS desktop software for the preparation of maps. They explained the importance of data accuracy under a data quality heading. The maximum noise level achieved was 78 dB. They concluded that 9 m high building barrier could minimise the sound level by 15 dB (F. Farcas, n.d.).

Naji et al. studied the effect of urban land use on the level of noise pollution by utilising GIS and satellite technologies in Tehran city of Iran. Total of 170 stations were selected for measurement of noise levels. They observed that the morning peak of noise was 76.29 dB and the evening peak was 76.46 dB (Naji et al., 2020).

Sonaviya and Tandel (2019) prepared 2D noise maps for Surat city in India. They selected urban roads to carry out research. They used SoundPLAN software to generate noise maps. They recorded noise levels at five different locations. All recording stations noted higher noise levels than CPCB limits (Sonaviya & Tandel, 2019).

D. Banerjee et al. (2009) presented a study on the evaluation and mapping of the propagation of urban road traffic noise, Asansol city of West Bengal India. They calculated the temporal and spatial traffic noise distribution. A total of 35 locations, including residential, commercial, industrial and sensitive, were selected for noise monitoring. Daytime noise levels varied between 51.2 dB to 89 dB, whereas during night time it varied between 43.5 dB to 81.9 dB (Banerjee et al., 2009).

In this study, noise mapping was adopted to analyse noise levels and their propagation at hospital buildings situated along the traffic stream of the city region.

Methodology

Instrument Used

KIMO DB 300 sound level metre was used to monitor the noise levels around the hospital building. To know the speed of the vehicle, the Falco HR radar gun was used. Along with the noise monitoring survey, the traffic volume count as per the class of vehicle was carried out (Ranpise & Tandel, 2022a). To determine the location of exposed noise, the mobile GPS tracker was used and the latitude and longitude of the monitoring points were noted down. Finally, to prepare the noise map, ArcGIS 10.8 software was utilised.

Study Area

Surat city is known for textile as well as a diamond industrial city situated in Gujarat, India. Surat is a commercial and economic centre in south Gujarat, and one of the fastest-growing urban areas of western India (Ranpise, Tandel, & Darjee, 2021). Nowadays, the city has experienced heavy growth in urbanisation and industrialisation, which leads to an increase in the population and ultimately an increase in the use of vehicles. The study area includes four hospital buildings (Table 5.1) which are located near the urban road traffic streams chosen for data collection. Table 5.1 shows the list of hospital buildings along with coordinates and noise levels.

Table 5.1 List of study locations along with geographic coordinates and Leq levels

Sr. no.	Name of location	Latitude	Longitude	Leq levels in dBA
1	Maitreya Multispecialty Hospital, Udhana-Madgulla, Road, Surat, Gujarat.	21.153823	72.778274	79
2	Sushrut Hospital, SVNIT Campus, Surat, Gujarat 395007	21.169488	72.78759	68.4
3	Tristar Hospital, Athwa, Surat, Gujarat 395007	21.186546	72.811897	68.8
4	Reliance Hospital, Dumas Road, Surat	21.161941	72.776642	73.8

Sampling and Data Collection

A sound level metre, Kimo DB 300 class-II with frequency weighting network having measurement scale 0-140 dB was utilised for this study. Noise monitoring was done at all four hospital buildings for a continuous 12-hour duration. The measured noise level was extracted from the instrument and recorded in an MS excel worksheet. During the noise monitoring, the longitude and latitude of each location were marked with the help mobile GPS tracker.

Development of Noise Map

In the development of the noise contour map, ArcGIS 10.8 tool was used. The shape file consisting road network of the study location was downloaded from the geo informatics world database and then it is imported into GIS tool. To plot the noise levels concerning contours, the interpolation (Inverse Distance Weighted) method has been adopted. This interpolation technique displays noise levels at 11 monitoring points by different colour codes. The maps indicate more sensitive areas to the threat of noise pollution and provide more useful graphic details of the areas with greater noise levels and traffic jams (Alam, 2011; Wu et al., 2014). The 2.5 dBA bandwidths have been used to construct the noise maps. Each band displays a distinct colour, which on the map represents a different noise level. The sound scale ranges from 50 dBA to 77 dBA. The coordinate system projection for Surat city is UTM-WGS 1984-Zone 45N. The prepared maps are further explained in the next section.

Results and Discussion

Noise maps were prepared using ArcGIS 10.8 software to determine noise propagation due to urban traffic surrounding the hospital buildings in Surat city. Noise maps of 12-hour duration were developed in order to demonstrate the variation in the noise environment at the hospital premises. For each map eleven monitoring points were chosen as shown in Table 5.2

Table 5.2 Monitoring points, latitude and latitude along with noise level at Maitrey hospital, Surat

Sr. no.	MP	Noise level (Leq)	Latitude	Longitude
1	O1	66.7	21.153560	72.778426
2	O2	79	21.153529	72.778345
3	O3	72.2	21.153500	72.778260
4	I1	66.7	21.153618	72.778399
5	I2	67.1	21.153646	72.778339
6	I3	65	21.153650	72.778200
7	L1	64.7	21.153789	72.778419
8	L2	63.1	21.153917	72.778414
9	R1	63.9	21.153791	72.778208
10	R2	62.2	21.153927	72.778220
11	C1	50	21.153864	72.778446

Source: Author's compilation

(Note: I1, I2 = Inside noise level, O1, O2 = outside noise level, R1, R2 = right side of the building, L1, L2 = left side of the building and C = centre of the building)

Figure 5.1 shows the noise map of Maitrey hospital Surat. The noise level at Maitrey Hospital was higher in the morning due to increased traffic flow, patient vehicles and other surrounding activities. The outside noise levels are higher than the inside premises of the hospital. Near points, O2 and O3, the highest noise level on the roadside were observed and it was measured at 72 and 77 dBA, respectively. In the afternoon, traffic flow was reduced so that the colour

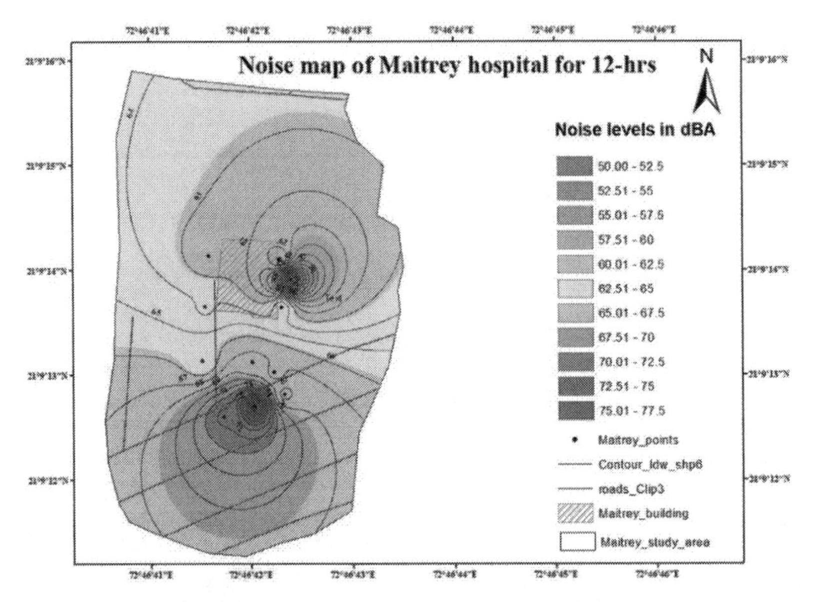

Fig. 5.1 Noise map for Maitrey hospital, Surat

approached yellow to green at all points. The observed noise levels were more than 50 dBA at each location, exceeding the CPCB's silence zone standard.

At Sushrut hospital, the noise level from morning to evening was observed higher, there is no big change in colour codes on the map as show in Fig. 5.2. The noise level during morning peak

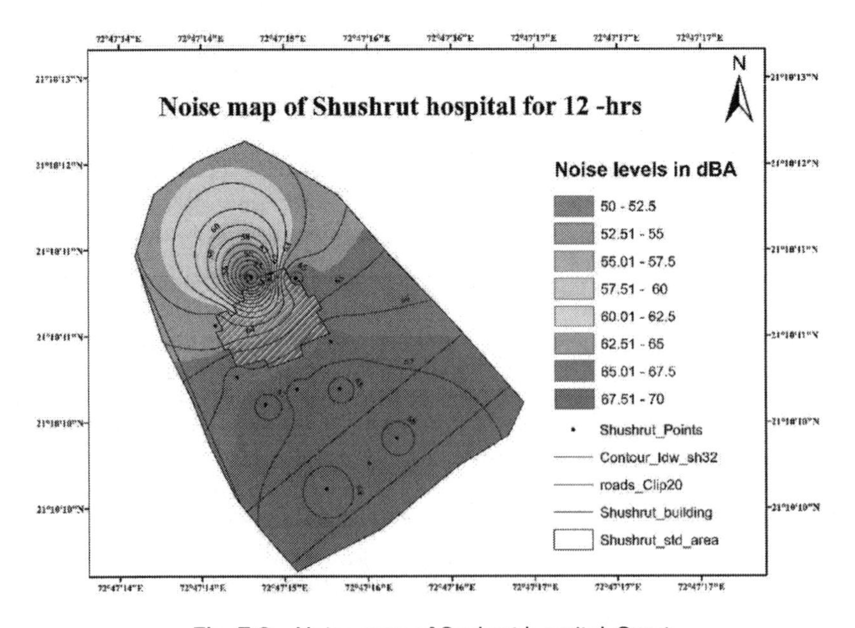

Fig. 5.2 Noise map of Sushrut hospital, Surat

period was above 70 dBA. It is observed that Leq for the entire daytime is almost constant. The daytime noise levels surpassed the acceptable noise level of 50 dBA. The reason could be the continuous traffic flow, major arterial road and the commercial complexes near the Sushrut hospital.

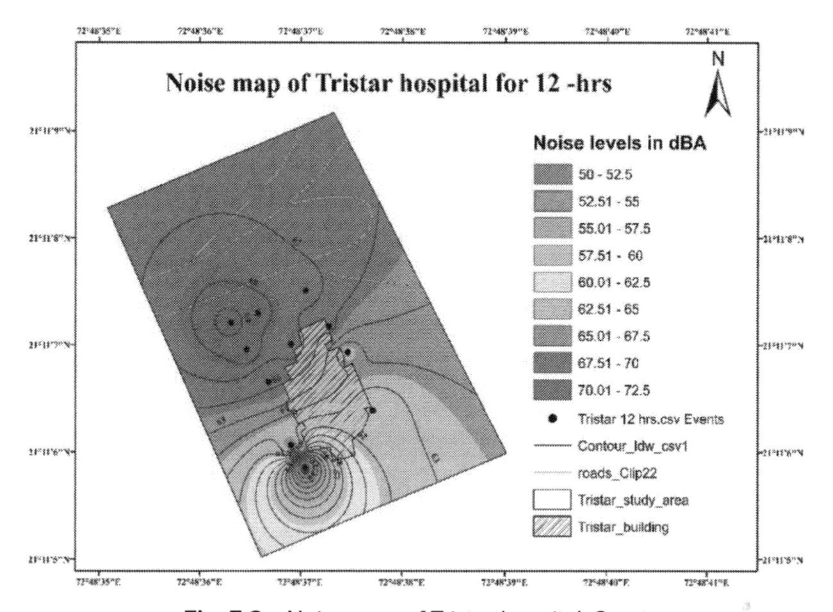

Fig. 5.3 Noise map of Tristar hospital, Surat

Out of all four hospital buildings, the maximum intensity of traffic was seen at Tristar hospital building located at the Athwa gate circle. Figure 5.3 represents the noise propagation from outside to inside the hospital building. From the propagation, it is clearly understood that the traffic on the road is significantly contributing to the noise level in the Tristar hospital environment. It is observed that the area occupied by red and orange colour is maximum than green and yellow colour. Yellow colour itself indicates the noise level above 60 dBA which is higher than the limit given by CPCB.

Figure 5.4 shows the noise map of Reliance hospital Surat. At, Reliance hospital, the movement of vehicles on the road was observed very smooth. There is no big difference in speed of the vehicle. But still, there was a maximum noise level was observed near outside the premises. Inside, noise levels also range from 55 to 65 dBA. The noise level along the roadside ranges from nearly 67.5 dBA to 77.5 dBA.

Conclusion

This research work presents the study of the development of noise maps and the evaluation of the noise level of hospital buildings based on the noise monitoring surrounding the building. At each site, eleven monitoring locations are fixed to measure the noise levels. The prepared noise map indicates that locations near the roads, near the entrance of the building and the passage of parking have the highest noise levels.

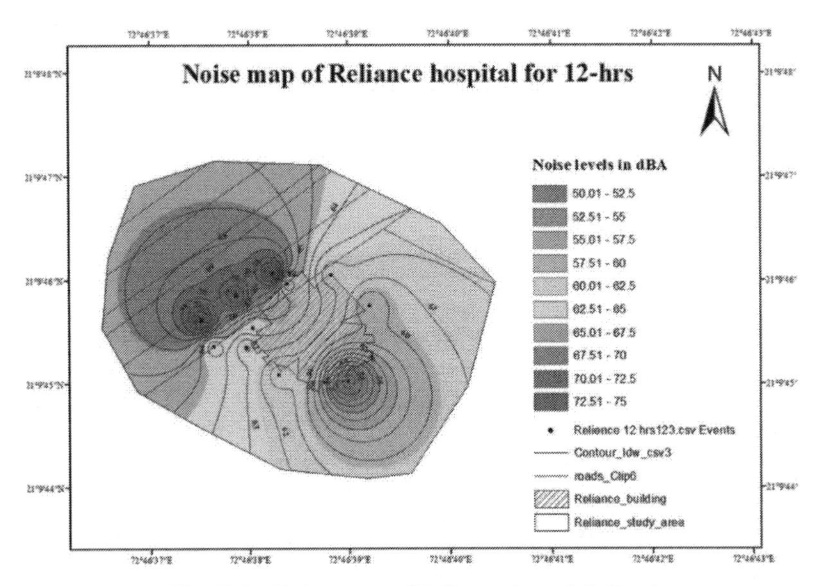

Fig. 5.4 Noise map of Reliance hospital, Surat

The study reveals that noise levels at all four monitoring points surpass the recommended limit of 50 dBA. Hence, the noise exposure in these premises cause discomfort and annoyance to the patients and doctors. Moreover, exposure to traffic noise is a hazard for people's health and considerably impacts their quality of life (Ranpise, Tandel, & Singh, 2021; Tandel et al., 2011).

This hospital needs immediate attention for noise mitigation in the form of the noise barrier, as higher noise levels inside the hospital premises can impact patients' ability to rest, heal and recover.

Due to the exponential growth of population, less availability of public transportation facilities and continuous addition of private vehicles in the metropolitan region heavy traffic noise around the schools, colleges and emergency clinics, which falls under the silence zone as endorsed by the Central Pollution Control Board India (Ranpise & Tandel, 2022b). The effective step to control noise pollution in a tier II city, such as Surat, is the development of noise maps. The noise maps themselves give the description of noise levels in a simple way. It provides the standard data for city planners, engineers and other town planning authorities. Noise maps help in the planning and execution of project works. Most of the cities in India have not developed noise pollution maps. Hence, it is strongly suggested that noise maps should be prepared for every metropolitan city in India to serve as a noise control measure.

References

1. Alam, W. (2011). GIS based Assessment of Noise Pollution in Guwahati City of Assam, India. International Journal of Environmental Sciences, 2(2), 731–740.
2. Banerjee, D., Chakraborty, S. K., Bhattacharyya, S., & Gangopadhyay, A. (2009). Appraisal and mapping the spatial-temporal distribution of urban road traffic noise. International Journal of Environmental Science and Technology, 6(2), 325–335. https://doi.org/10.1007/BF03327636

3. F. Farcas. (n.d.). Gis Tools for Noise Mapping and a Case Study for Skåne Region.
4. Manojkumar, N., Basha, K., & Srimuruganandam, B. (2019). Assessment, Prediction and Mapping of Noise Levels in Vellore City, India. Noise Mapping, 6(1), 38–51. https://doi.org/10.1515/noise-2019-0004
5. Mioduszewski, P., Ejsmont, J. A., Grabowski, J., & Karpiński, D. (2011). Noise map validation by continuous noise monitoring. Applied Acoustics, 72(8), 582–589. https://doi.org/10.1016/j.apacoust.2011.01.012
6. Naji, F. K., Djahed, B., Mahvi, A. H., Nowrouz, P., Nazmara, S., & Kardooni, H. (2020). Urban land use effect analysis on the level of noise pollution using satellite and GIS technologies: A case study in Tehran city. Global NEST Journal, 22(3), 361–368. https://doi.org/10.30955/gnj.003337
7. Nasim Akhtar, Kafeel Ahmad, S. G. (2012). Road Traffic Noise Mapping and a Case Study for Delhi. International Journal of Applied Engineering and Technology, 2(4), 39–45. http://www.cibtech.org/J ENGINEERING TECHNOLOGY/PUBLICATIONS/2012/Vol_2_No_4/06-015...Nasim...Road...Region...39-45.pdf
8. Pan, Z., & Zhu, J. (2011). Study on traffic noise map of old city zone in increasing modern city in China. Geotechnical Special Publication, 223 GSP, 214–219. https://doi.org/10.1061/47634(413)27
9. Petrovici, A., Claudia, T., Gey, F., Nedeff, F., & Oana, I. (2015). Noise Prediction, Calculation and Mapping Using Specialized Software. Journal of Engineering Studies and Research, 21(3), 59–64. https://doi.org/10.29081/jesr.v21i3.19
10. Ranpise, R. B., & Tandel, B. N. (2022a). Assessment and Appraisal of Morning Peak Time Urban Road Traffic Noise at Selected Locations of Major Arterial Roads of Surat City, India. Asian Journal of Water, Environment and Pollution, 19(1), 81–86. https://doi.org/10.3233/AJW220012
11. Ranpise, R. B., & Tandel, B. N. (2022b). Noise Monitoring and Perception Survey of Urban Road Traffic Noise in Silence Zones of a Tier II City—Surat, India. Journal of The Institution of Engineers (India): Series A. https://doi.org/10.1007/s40030-021-00598-x
12. Ranpise, R. B., Tandel, B. N., & Darjee, C. (2021). Assessment and mlr modeling of traffic noise at major urban roads of residential and commercial areas of surat city. Lecture Notes in Civil Engineering, 93, 181–191. https://doi.org/10.1007/978-981-15-6887-9_21
13. Ranpise, R. B., Tandel, B. N., & Singh, V. A. (2021). Development of traffic noise prediction model for major arterial roads of tier-II city of India (Surat) using artificial neural network. Noise Mapping, 8(1), 172–184. https://doi.org/10.1515/noise-2021-0013
14. Sonaviya, D. R., & Tandel, B. N. (2019). 2-D noise maps for tier-2 city urban Indian roads. Noise Mapping, 6(1), 1–7. https://doi.org/10.1515/noise-2019-0001
15. Tandel, B., Macwan, J., & Ruparel, P. N. (2011). Urban Corridor Noise Pollution: A case study of Surat city, India. Ipcbee , 12, 144–148. http://ipcbee.com/vol12/28-C10015.pdf
16. Wu, E. M. Y., Tsai, C. C., Cheng, J. F., Kuo, S. L., & Lu, W. T. (2014). The Application of Water Quality Monitoring Data in a Reservoir Watershed Using AMOS Confirmatory Factor Analyses. Environmental Modeling and Assessment, 19(4), 325–333. https://doi.org/10.1007/s10666-014-9407-5
17. Zambon, G., Confalonieri, C., Angelini, F., & Benocci, R. (2021). Effects of COVID-19 outbreak on the sound environment of the city of Milan, Italy. Noise Mapping, 8(1), 116–128. https://doi.org/10.1515/noise-2021-0009
18. Zannin, P. H. T., & Sant'Ana, D. Q. De. (2011). Noise mapping at different stages of a freeway redevelopment project - A case study in Brazil. Applied Acoustics, 72(8), 479–486. https://doi.org/10.1016/j.apacoust.2010.09.014

CHAPTER 6

A Positive Pressure Approach Against Air Pollution using Sensorless Brushless DC Motor Powered Respirator

Arijit Majumder[1]
Applied Electronics and Instrumentation Engineering,
Heritage Institute of Technology, Kolkata, India;

Santanu Mondal[2]
Electronics and Instrumentation Engineering, Techno Main, Kolkata, India

Debjyoti Chowdhury[3]
Applied Electronics and Instrumentation Engineering,
Heritage Institute of Technology, Kolkata, India

Madhurima Chattopadhyay[4]
Applied Electronics and Instrumentation Engineering,
Heritage Institute of Technology, Kolkata, India

Abstract

The recent increase in air pollution has made air quality in many cities extremely hazardous. This has necessitated the use of respirators as a safeguard against the impact of air pollution. This study is concerned with modelling a forced air respirator using a matrix-converter-fed sensorless brushless DC motor (BLDCM) drive. Here, the electrical model of human lungs along with the mathematical model of the said converter is considered for the simulation. The forced air respiration system is modelled to study the characteristics of both normal and adverse condition of volumetric flow parameters with different respiratory rates. The model is implemented in MATLAB/Simulink for stable airflow and volumetric pressure. The simulation model borrows the BLDCM output motor parameters (i.e. RPM, max output torque), which are fed to the lungs model as input to find volumetric pressure and airflow. Finally, a detailed study on implemented matrix converter reveals significant improvement over a non-converter-fed one. The output results show that the scheme described in this scope of work handles DC power using a matrix converter and delivers a stable rotation per min (RPM) and output torque to the respirator for stable airflow and volumetric air pressure in the lungs parameter.

Keywords: sensorless brushless DC motor, flyback converter, positive pressure respirator, human lungs modelling

[1]arijit.majumder28@gmail.com; [2]santanu_aec1984@yahoo.co.in; [3]debjyoti.chowdhury@heritageit.edu; [4]madhurima.chattopadhyay@heritageit.edu

Introduction

The present scenario has shown the increased threats represented by respiratory illnesses, such as chronic obstructive pulmonary disease (COPD), and asthma. That risk has increased due to an increase in air pollutants, such as PM2.5 and PM10. A respirator can be utilised as an immediate countermeasure on an individual level safety measure as bringing down pollution levels requires a much longer time than the severity of the problem is allowing. Normally used N95 respirators are of negative pressure variant, that is, they require the wearer's lungs to inhale air through the resistive membranes of the filter layers. This is strenuous and uncomfortable to wear for a long duration. This is non-existent in positive air pressure respirators as they use external filters and has a motorised air supply system. The pandemic in recent scenarios also necessitates respiration apparatus as a part of its treatment. Respirators that are commonly used are negative pressure system, which require the power of the lungs to draw-in purified air which is not suitable and sometimes not possible if the person lacks sufficient lungs strength, or if they suffer from respiratory illness. This work proposes a forced air (positive air pressure) solution to the problem.

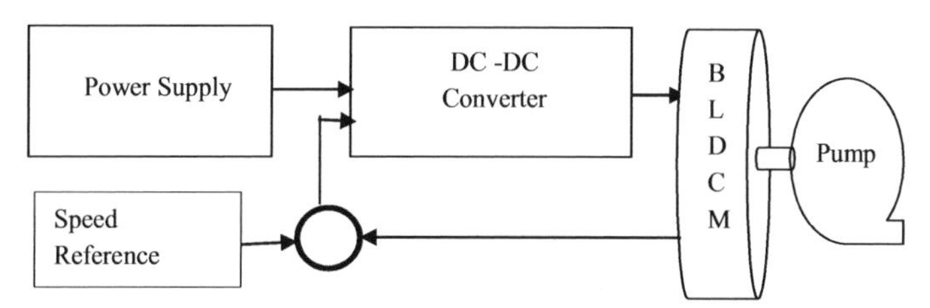

Fig. 6.1 Block diagram of the proposed system

In this work, brushless DC motor (BLDCM) is chosen to drive the system due to being compact in size and having the ability to produce a high range of RPM with a decent amount of torque. The ability to run on DC power directly or modified AC power source facilitates the flexibility of either making the device mobile or compact enough to field in limited space constraints, aiding its portability. The inherently non-commutating nature of BLDCM necessitates the use of external commutating circuitry along with position sensing and six switches three-phase inverter to perform phase commutation. This work uses the back-EMF zero crossing topology to determine rotor position [1-5]. A flyback converter [6-8] is utilised to smoothen the RPM and output torque of the motor as it smoothens the input of ripples.

The RPM of BLDCM is fed to the impeller of the pump model [9-17]. The resultant air flow is fed to the human lungs model [18-20] to perform continuous positive airway pressure (CPAP) respirator application [21-26].

The objective of this work is to model a flyback-converter-fed BLDCM driven respirator both mathematically and using MATLAB/SIMULINK software for simulation and also build a real-time hardware prototype of the system (as shown in Fig. 6.1) using the aforesaid modelling data.

The paper arrangement is as follows. The introduction is given in Section 1. In Section 2, we have described the mathematical model of the system. Section 3 puts forwards the experimental results of the developed system. Finally, Section 4 concludes the work with the essence of the final outcome of this study by means of experimental results in MATLAB/Simulink platform.

Mathematical Model

Mathematical Model of BLDCM

The mathematical model of BLDCM, a sensorless, can be modelled and implemented by the means of mathematical modelling (transfer functions) shown in the equations shown below. The winding of a three-phase BLDC motor can be modelled as a series circuit consisting of a resistance R, an inductance L and a speed-dependent voltage source which is known as the back EMF voltages due to the rotor magnet. While designing a BLDC motor, a few parameters, such as induced current in the rotor due to stator harmonics fields and iron and stray losses are neglected. Self and mutual inductances are considered constant [1-5]. The BLDC motor is supplied with three-phase voltage represented in Eq. 1–3.

$$V_A = R_S I_A + \frac{d}{dt} F_A + E_A \tag{1}$$

$$V_B = R_S I_B + \frac{d}{dt} F_B + E_B \tag{2}$$

$$V_C = R_S I_C + \frac{d}{dt} F_C + E_C \tag{3}$$

Here, V_A, V_B and V_C represent the three stator currents, R_S represents resistance of stator winding, I_A, I_B and I_C represent the three-phase stator current, E_A, E_B and E_C represent three-phase back EMF, F_A, F_B and F_C represent the stator flux linkage.

$$F_A = L_{AA} I_A + L_{AB} I_B + L_{AC} I_C \tag{4}$$

$$F_B = L_{BA} I_A + L_{BB} I_B + L_{BC} I_C \tag{5}$$

$$F_A = L_{CA} I_A + L_{CB} I_B + L_{CC} I_C \tag{6}$$

L_{AA}, L_{BB} and L_{CC} are stator self-inductance and L_{AB}, L_{AC}, L_{BC}, L_{CB} and L_{BA} are mutual inductance between the stator winding.

The above equation can be written as:

$$\begin{bmatrix} V_A \\ V_B \\ V_C \end{bmatrix} = R_S \begin{bmatrix} 1 & 0 & 0 \\ 0 & 1 & 0 \\ 0 & 0 & 1 \end{bmatrix} \begin{bmatrix} I_A \\ I_B \\ I_C \end{bmatrix} + \begin{bmatrix} L-M & 0 & 0 \\ 0 & L-M & 0 \\ 0 & 0 & L-M \end{bmatrix} \frac{d}{dt} \begin{bmatrix} I_A \\ I_B \\ I_C \end{bmatrix} + \begin{bmatrix} E_A \\ E_B \\ E_C \end{bmatrix} \tag{7}$$

Here, L represents the self-inductance of the stator winding and M represents the mutual inductance of the stator winding. The direct relationship between applied source voltages to phase terminal (V) and induced back EMF (E) can be derived for constant self-inductance and mutual inductance around air gap, given by Eq. 10.

$$E \propto V \tag{8}$$

Mathematical Model of Flyback Converter

The flyback converter (as shown in Fig. 6.2) is an isolated converter ideal for low to medium power level dependent on flyback topology. Isolation between load and control sections is provided by an isolated transformer. The control switching is done by a fast-switching MOSFET. The popularity of flyback converter is due to its relatively low-cost and simple construction.

Though flyback converter has both continuous conduction mode (CCM) and discontinuous conduction mode (DCM), for this we have considered DCM as it has a faster transient response and better efficiency [6,7].

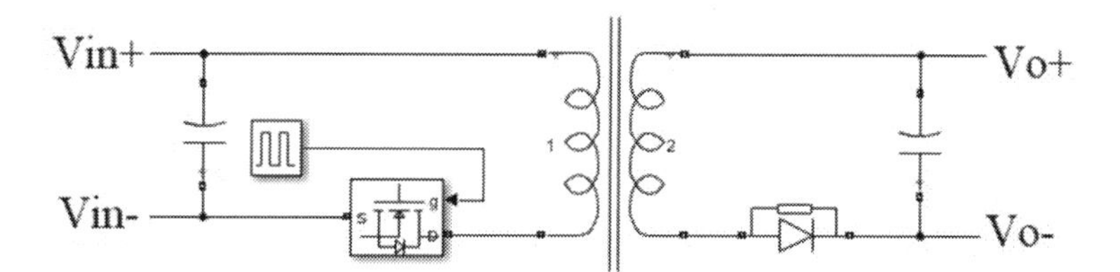

Fig. 6.2 MATLAB/Simulink Model of flyback converter

The output voltage can be calculated as Eq. 9.

$$V_o = \frac{D}{(1-D)} \times \left(\frac{N_2}{N_1}\right) \times V_{in} \tag{9}$$

The output capacitor can be calculated as Eq. 10.

$$C_o = \frac{D}{Rf_S} \times \left(\frac{V_o}{\Delta V_{co}}\right) \tag{10}$$

The secondary inductance can be calculated as Eq. 11.

$$L_s = \frac{(V_o + V_f) \times (1-D)^2}{2 \times I_{o_{max}} \times f_{s_{max}}} \tag{11}$$

The primary inductance can be calculated as Eq. 12.

$$L_P = L_S \times \left(\frac{N_S}{N_P}\right)^2 \tag{12}$$

Mathematical Model of Pump System

The airflow towards the lungs is generated by a pump, mathematical model shown in Fig. 6.3 that is driven by the BLDCM. The transfer function of the centrifugal pump is designed with the help of controlled and manipulated variables to match the biomedical parameters of human lungs. The flow rate of the pump Q_n is determined by Eq. 13.

$$Q_n = K_p \cdot N \cdot (D_f)^3 \tag{13}$$

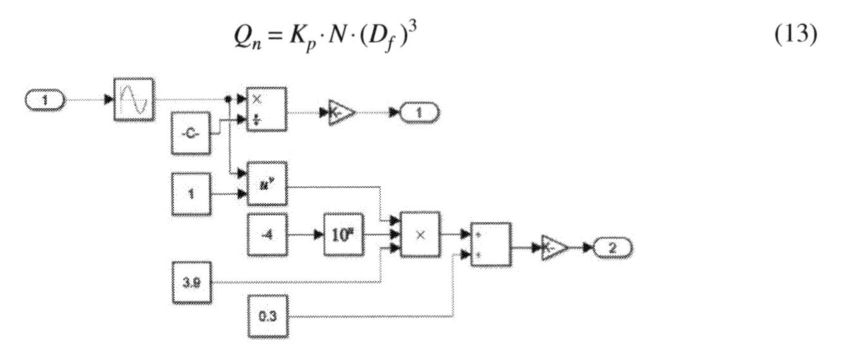

Fig. 6.3 MATLAB/Simulink model of pump

The specific diameter of the pump is determined by Eq. 14.

$$D_s = \frac{D(gH_n)^{\frac{1}{4}}}{\sqrt{Q_n}} \tag{14}$$

The specific speed of the pump is determined by Eq. 15.

Here,

$$N_s = \frac{\omega\sqrt{Q_s}}{(gH_n)^{\frac{3}{4}}} \tag{15}$$

The pressure can be determined by Eq. 16.

$$C_r = 0.3 + \omega^{1.8} \times 10^{-4} \times \sqrt[4]{(3.9)} \tag{16}$$

The control variables are as follows: g is acceleration due to gravity in m/s^2, ω is rotor speed in rad, K_p is the pump constant, N is RPM and H_n is the specific density of the fluid.

Mathematical Model of Human Lungs

The human lungs can be modelled mathematically (shown in Fig. 6.4 and 6.5) using electrical equivalent parameters. Here, we have assumed that the frictional components are termed as resistance and storage elements as capacitance [17, 18], over all model is shown in Fig. 6.6.

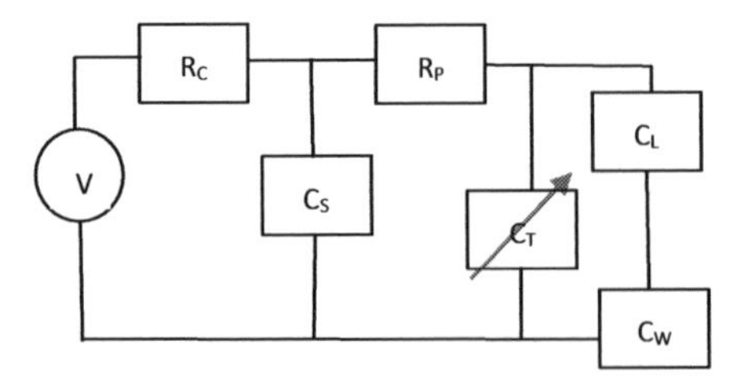

Fig. 6.4 Electrical impedance model of human lungs

Source: Ref. [17, 18, 19, 21]

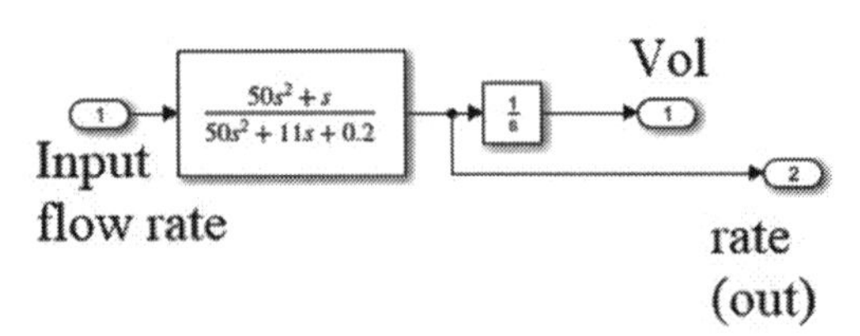

Fig. 6.5 MATLAB/Simulink model of human Lungs

Eq. 17 denotes the transfer function model of the human lungs.

$$\frac{I(s)}{P(s)} = \frac{R_C \times S^2 + \dfrac{S}{R_P \times C_T}}{R_C \times S^2 + \left(\dfrac{1}{C_S} + \dfrac{R_C}{R_P \times C_T}\right) \times S + \left(\dfrac{1}{R_P \times C_S}\right)\left(\dfrac{1}{C_L} + \dfrac{1}{C_W}\right)} \tag{17}$$

The control variables are, I(S) is the air flow rate and P(S) represents periodic function, values of the parameters are shown by Table 6.1.

Table 6.1 Performance parameters of the electrical model of human lungs [20]

Parameter	Remarks	Values
R_p	represents airflow resistance	0.5 cm/s
R_C	represents peripheral airways resistance	1 cm/s
C_w	represents chest wall capacity	200 mL
C_L	represents alveoli capacity	200 mL
C_S	represents dead space capacity	5 mL
C_T	represents total compliance of airways	100 mL

Overall Model of BLDCM Drive Respirator

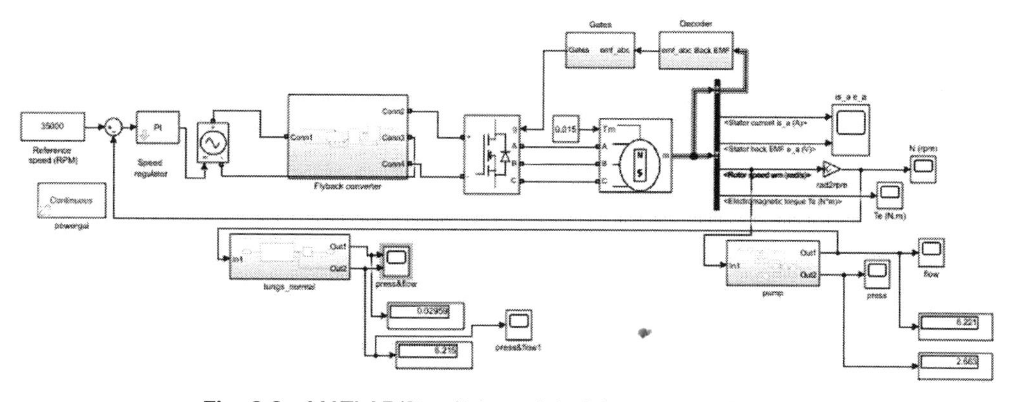

Fig. 6.6 MATLAB/Simulink model of the proposed system

Results and Discussions

In this work, we have studied the characteristics of a flyback-converter-fed sensorless BLDCM driven respirators effect on the transfer function model of human lungs. Figure 6.7 and Fig. 6.8 show the flow rate characteristics of pump and lungs, respectively. In which we can verify that the simulated model is able to deliver around 3 lit/breath flow rate in the lungs, which is the required amount [19].

In Figs. 6.9 and 6.10, the pressure waveform of both pump and lungs are observed, respectively. In this, it is found that in the lungs model during inhalation the required alveoli pressure of 250–350 mPa is exerted [19]. In Fig. 6.11 the RPM characteristics is observed to be stable.

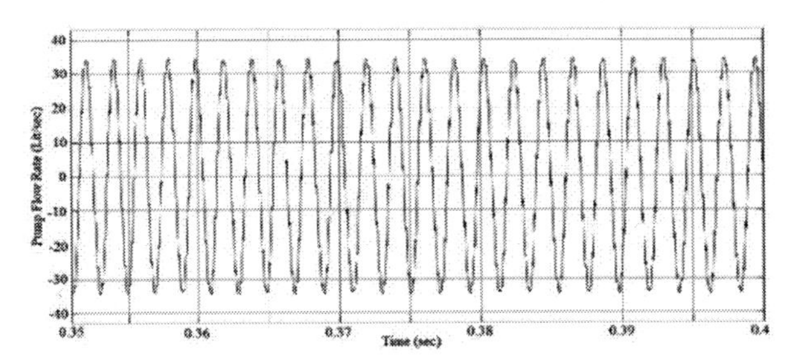

Fig. 6.7 Pump outlet flowrate

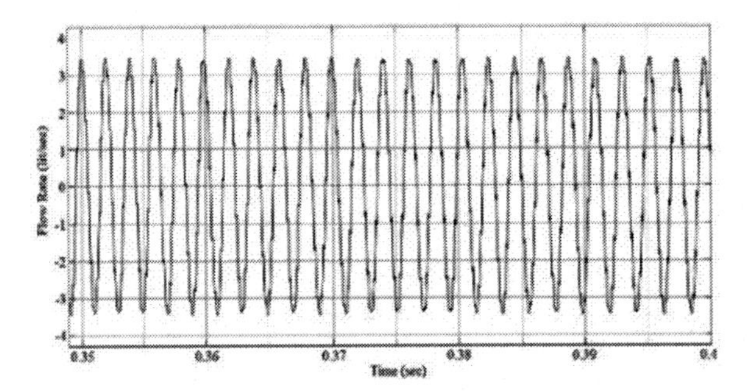

Fig. 6.8 Lungs flowrate waveform

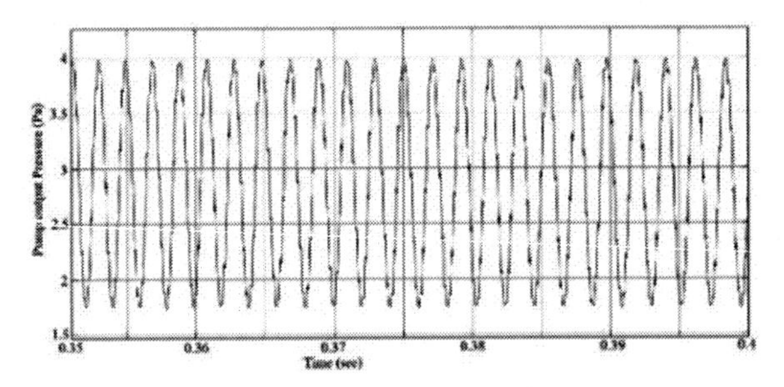

Fig. 6.9 Pump outlet pressure

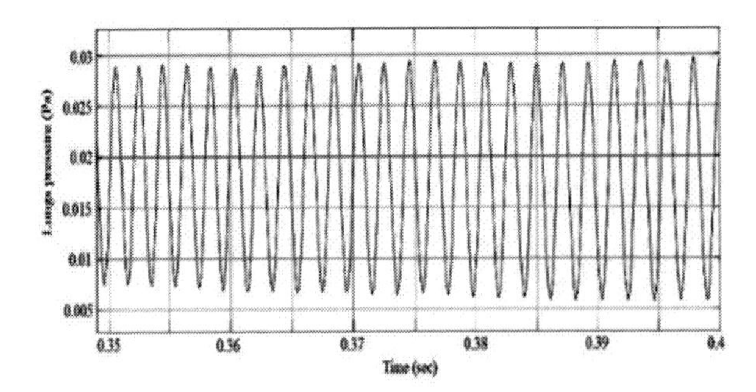

Fig. 6.10 Lungs pressure waveform

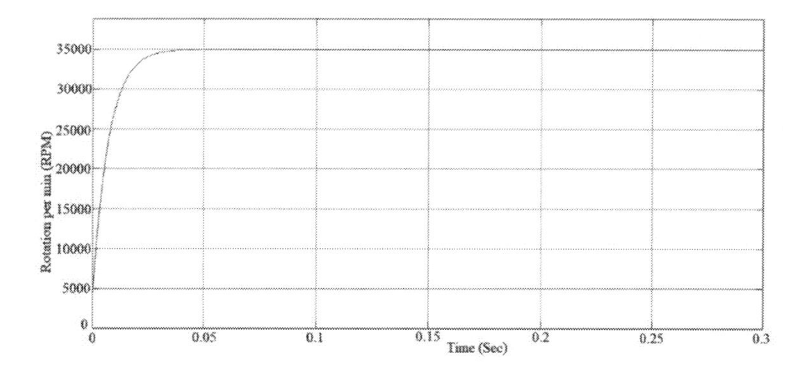

Fig. 6.11 RPM waveform of BLDC motor

Conclusion

In this study, a flyback-converter-fed BLDCM powered respirator in MATLAB/Simulink platform is presented. It is found that the addition of a flyback converter reduces RP M slip and torque ripple of the BLDCM significantly enough, so that the pump can supply a stable airflow at a reasonable pressure to the lungs. Minor oscillation is still present due to the drag coefficient of a different component. The pressure and flow rate parameters of the anatomical necessities are maintained in the required ranges are maintained in this study.

As shown in this study, with proper filter systems at the input, this system can be implemented to combat the hazardous impact of air pollution.

References

1. Uniyal S., Sikander A., "A Novel Design Technique for Brushless DC Motor in Wireless Medical Applications" Wireless Personal Communications volume 102, 369 – 381 (2018). doi: 10.1007/s11277-018-5845-8
2. Alex S S, Daniel A E, "An Efficient Position Tracking Smoothing Algorithm for Sensorless Operation of Brushless DC Motor Drives", Hindawi Modelling and Simulation in Engineering, Volume 2018, 1-9. doi: 10.1155/2018/4523416.
3. Karnavas Y L, Topalidis A, Drakaki M, "Development and Implementation of a Low Cost µC-Based Brushless DC Motor Sensorless Controller: A Practical Analysis of Hardware and Software Aspects", Electronics 2019, 8, 1456. doi: 10.3390/electronics8121456
4. Santhosh, P. Vijayakumar P., "Performance Study of BLDC Motor Used in Wireless Medical Applications", Wireless Personal Communications: An International Journal June 2017. doi: 10.1007/s11277-016-3561-9
5. Mondal S, Majumder A, Chowdhury D, Chattopadhyay M, "An efficient power delivering scheme for sensorless drive of Brushless DC motor", Microsystem Technologies 26, no. 10 (2020). 3113-3120. doi: 10.1007/s00542-018-3715-7
6. Mondal Santanu, Chattopadhyay Madhurima. "Comparative study of three different bridge-less converters for reduction of harmonic distortion in brushless DC motor." Indonesian Journal of Electrical Engineering and Computer Science 20, no. 3 (2020): 1185-1193.

7. Sasikala R.,.Seyezhai R, "Review of AC-DC power electronic converter topologies for power factor correction" International Journal of Power Electronics and Drive System (IJPEDS) Vol. 10, No. 3, Sep 2019. 1510-1519. ISSN: 2088-8694. doi: 10.11591/ijpeds.v10.i3.1510-1519
8. Ramkmar R., Vikram A. A., "Wind energy based asymmetrical half bridge flyback converter for BLDC motor," 2017 Third International Conference on Science Technology Engineering & Management (ICONSTEM), 2017. 605-610. doi: 10.1109/ICONSTEM.2017.8261394.
9. Hamdi H, Regaya C B, Zaafouri A, "A sliding-neural network control of induction-motor-pump supplied by photovoltaic generator", Protection and Control of Modern Power Systems 5, 1 (2020). doi: 10.1186/s41601-019-0145-1
10. More K C., Dongre S, Deshmukh G P., "Experimental and numerical analysis of vibrations in impeller of centrifugal blower", SN Appl. Sci. 2, 82 (2020). doi: 10.1007/s42452-019-1853-x
11. Harja, G., Nascu, I., Muresan, C," Improvements in Dissolved Oxygen Control of an Activated Sludge Wastewater Treatment Process", Circuits Syst Signal Process 35, 2259–2281 (2016). doi: 10.1007/s00034-016-0282-y
12. Breunung, T., Dohnal, F., Pfau, B., "An approach to account for interfering parametric resonances and anti-resonances applied to examples from rotor dynamics", Nonlinear Dyn 97, 1837–1851 (2019). doi: 10.1007/s11071-019-04761-9
13. Vasylius M, Janutėnienė J, Grigonienė J, "Mathematical modeling of dynamics of air blower rotor", 20th EURO Mini Conference "Continuous Optimization and Knowledge-Based Technologies" (EurOPT-2008). ISBN 978-9955-28-283-9
14. Jiang C., Habetler T. G., Cao W., "Improved condition monitoring of the faulty blower wheel driven by brushless DC motor in air handler unit (AHU)," 2016 IEEE Energy Conversion Congress and Exposition (ECCE), Milwaukee, WI, 2016, 1-5, doi: 10.1109/ECCE.2016.7855188.
15. Tu J, Yeoh G H, Liu C, "Computational Fluid Dynamics A Practical Approach Third Edition", Butterworth-Heinemann publication Elsevier
16. Aref H, Balachandar S., "A First Course in Computational Fluid Dynamics", Cambridge University Press
17. Sharma S, Danckers M, Sanghavi D, Chakraborty R K, "High Flow Nasal Cannula", NCBI Bookshelf. A service of the National Library of Medicine, National Institutes of Health. StatPearls [Internet]. Treasure Island (FL): Stat Pearls Publishing; 2020 Jan
18. E. Elmaati, "Modeling and simulation of a new artificial breathing system," 2018 IEEE 4th Middle East Conference on Biomedical Engineering (MECBME), 2018. 45–50. doi: 10.1109/MECBME.2018.8402404.
19. Goldman M. D., "Electrical circuit models of the human respiratory system reflect small airway impairment measured by impulse oscillation (IOS)," 2010 Annual International Conference of the IEEE Engineering in Medicine and Biology, 2010. 2467–2472. doi: 10.1109/IEMBS.2010.5626611.
20. Barrett K E, Barman S M, Boitano S, Brooks H L, "Ganong's review of medical physiology" McGraw Hill publication.
21. Bagchi S., Chattopadhyay M., "Electrical modeling of respiratory system and identification of two common COPD diseases through stability analysis technique," 2012 IEEE International Conference on Advanced Communication Control and Computing Technologies (ICACCCT), 2012. 403–408. doi: 10.1109/ICACCCT.2012.6320811.
22. Meenu M M, Hariharan S., "Position sensorless control of BLDC motor in continuous positive airway pressure device," 2015 International Conference on Control Communication & Computing India (ICCC), Trivandrum, 2015. 230–235. doi: 10.1109/ICCC.2015.7432897.
23. Al-Naggar, N. "Modelling and Simulation of Pressure Controlled Mechanical Ventilation System". Journal of Biomedical Science and Engineering, 8, 707–716. doi: 10.4236/jbise.2015.810068.

24. https://www.who.int/news-room/questions-and-answers/item/coronavirus-disease-covid-19-masks
25. Jayan V, Ajan A, Mohan H, Manikutty G, Sasi D, Kappanayil M, Vijayaraghavan S, Rao R B, "Design and development of a low-cost powered air-purifying respirator for frontline medical workers for COVID-19 response," 2020 IEEE 8th R10 Humanitarian Technology Conference (R10-HTC), 2020. 1-6. doi: 10.1109/R10-HTC49770.2020.9356954.
26. Zhao X, Li X, Chai Z, Song H, Kang J, "Overview of the Application of Powered Intelligent Air-Purifying Respirator in the Health Field," 2021 IEEE 3rd International Conference on Civil Aviation Safety and Information Technology (ICCASIT), 2021. 555-562. doi: 10.1109/ICCASIT53235.2021.9633740.

CHAPTER 7

Evaluation of Construction-Demolition Noise and its Impact on Annoyance in the Academic Zone: Case Study of Surat, India

Avnish Shukla* and Anirudh Mishra[1]
Research Scholars, Department of Civil Engineering,
Sardar Vallabhbhai National Institute of Technology, Surat, India

B. N. Tandel[2]
Assistant Professor, Department of Civil Engineering,
Sardar Vallabhbhai National Institute of Technology, Surat, India

Abstract

Sound pollution has been considered a threat to the environment as it emerged as a severe problem in the past few years. Continuous increase in traffic, urbanisation and industrialisation were considered as a major source of noise pollution. It has been observed that construction and its related activities cause an increase in air pollution. Very few studies have been carried out to evaluate the impact of construction activities on noise pollution. So, this study was done to know the impact of construction activities on the noise level. In order to carry out this study, the S.V. National Institute of Technology, Surat, was selected as the study area, which comes under the silence zone. The noise level was observed in the physics laboratory. A sound level metre Kimo dB300 (class-2) was used to measure noise pollution indices. Data on noise level (LAeq) were collected for four days time period (8 hr per day) and observed at four different distances (5, 8, 11 and 14 m) from the construction activity. SPPS, ArcGIS and MS Excel software were used for the statistical analysis and data visualisation. The work showed that the noise level was higher than the permissible limit most of the time. Averaged equivalent values of noise pollution indices at different distances showed that noise level decreases as the distance increases from the construction activities. Maximum reduction in L_{eq} was observed for the 14 m distance, and minimum reduction was observed for the 8 m distance. A minimum reduction of 3.9 dB (A) and a maximum reduction of 28.07 dB (A) was observed. Results revealed that maximum noise level was observed during the morning and evening time periods. It was found that the noise level was minimum for the afternoon time period. The percentage of higher annoyance was also calculated to know the impact of noise pollution on surrounding people. Results showed that a percentage of higher annoyance was found at 51.46% for the 5 m distance and 49.17% for the 14 m distance. The current study concluded that the noise level is in alarming situation in the physics laboratory due to ongoing construction-demolition

**Corresponding Author:* avnishshukla1706@gmail.com
[1]anirudhmishra1996@gmail.com; [2]bnt@ced.svnit.ac.in

activities. The study revealed that noise propagation varies with distance, and optimum distance from construction activities reduces the noise level. So, a sufficient distance can be considered a natural barrier to the noise level.

Keywords: annoyance level, construction-demolition noise, noise pollution, sound propagation

Introduction

Constant exposure to noise might have negative impact on human health and well-being (Singh & Davar, 2017; Valente et al., 2012). Excessive noise exposure may cause physiological symptoms, such as hearing loss (Basner et al., 2014), impatience, insomnia and ischemic heart disease, which could lead to an increase in blood pressure (Foraster et al., 2016). The human central nervous system's physiological reaction to excitement may be affected significantly over long-term low-frequency noise exposure (Rahmat et al., 2015). It has a detrimental effect on persons' decision-making (Shea et al., 2017) and quality of life (Baliatsas et al., 2016); it causes hearing loss, raises the risk of cardiovascular disease (Gupta et al., 2015) and induces stress responses in the neurological system (Kujala & Brattico, 2009). Its impact on human health has lately been recognised as a significant and crucial topic, which inspires scientists and interested organisations worldwide. The World Health Organization (WHO) has recently performed more studies on noise level monitoring and its detrimental consequences. It also identifies community noise or environmental noise as one of the most prevalent societal disturbances (Berglund et al., 2000). Aside from industrial noise, additional sources include road (Tandel & Macwan, 2011), rail, aircraft (Clark et al., 2021), construction-demolition (Lokhande et al., 2021; Rahmat et al., 2015), public works and leisure activities. According to research, it is also a significant environmental component that negatively impacts the behavior (Forouzanfar, 2016; Gupta et al., 2015) and retention ability of mind in urban and peri-urban areas (Golmohammadi et al., 2020; Ranpise & Tandel, 2022). Due to its negative impact on human health, the majority of the urban population is compelled to relocate to secondary roads located distant from the sources of noise (Kamp et al., 2013; Tiesler et al., 2013). Several studies (A Martini, M Mazzoli, 1997; Francis et al., 2009) indicate that the levels of environmental noise pollution released by human-made sources, mainly construction-demolition and vehicular activity (Fiedler & Zannin, 2015), have increased significantly.

The majority of construction-demolition noise is created by machine noise (Suter, 2010) and/or impulsive noise resulting from the hammer effect (Ng, 2000). It was determined that excessive daytime exposure to this noise is connected with irritation (Massonnié et al., 2020), headache, nausea, fatigue, direct hearing loss and hearing impairment (Forns et al., 2016; Klatte et al., 2010; Lim et al., 2018). In addition, noise irritation induces a range of unpleasant emotional states, such as anger, disappointment, sadness, anxiety and even depression (Nassiri et al., 2011). According to the Noise Observation and Information Service for Europe (NOISE), following road traffic noise, construction-demolition noise gives the highest amount of exposure in Europe, both inside and outside urban areas, when compared to the other noise-generating sources (Crombie et al., 2011). The current research attempts to further increase our

Table 7.1 Ambient noise standards (CPCB)

Area Code	Category of Zone	Limits in dBA (L_{eq})	
		Daytime	Night-time
A	Industrial Area	75	70
B	Commercial Area	65	55
C	Residential Area	55	45
D	Silence Area	50	40

Source: Central Pollution Control Board (CPCB)

understanding of the physio-psychological effects of noise by continuing the path of inquiry established by the aforementioned writers. According to the Central Pollution Control Board (CPCB), the permitted noise level in quiet zones is 50 decibels during the day and 40 decibels at night, however, it has been observed that in almost all situations of the current study, the noise level exceeds these limitations. Table 7.1 shows ambient noise standards.

Study Area and Methodology

Surat, popularly known as the 'city of bridges', is one of Gujarat's most developed cities, with a total area of 474.2 km^2. According to the 2011 census, the city has a population density of 10052 people per square kilometre. It is also considered as the business hub for diamonds and textiles. It is situated at 210°12'00"N and 72°52'00"E on the western bank of Gujarat at the Tapi river. The study was carried out in the department of physics at SVNIT campus. The department consists of three floored building with nearly 20 room on each floor with occupancy of 200 individuals. The detailed sketch of the study area is shown in Fig. 7.1.

The study location was investigated for the possibility of data collection and approval from departmental authority to conduct the preliminary survey. It is found that the third floor was best suitable for data collection in the department. The class 2 precision sound level metre (Kimo dB300) was placed in the corridor for collection of data for 4 days with 8 hr per daytime period. The sound level metre was placed at a distance of 1.3 m from the walls with 1.2 m ground height. The face of the sound level metre was kept on the side of noise generation. The 8 hr data were collected at 5 m from the noise source on the first day followed by 8 m, 11 m and 14 m distance from source on the second, third and fourth day. The L_{eq}, L_{95} and L_{10} were calculated from computer-based software of Kimo group. For the analysis of data and data visualisation MS Excel, Statistical Package for Social Science (SPSS), ArcGIS 10.8 and Google Earth Professional software were used. To analyse annoyance (%), as per taking reference of (Hunyadi, 2014) the proposed model was adopted for predicting annoyance based on noise level measurements:

$$\%HA = 50 + \frac{\pi}{100} arctan[0.0985(x - 77.8)] \quad (1)$$

where %HA is the high annoyance rate, and 'x' is the total noise level in dB(A).

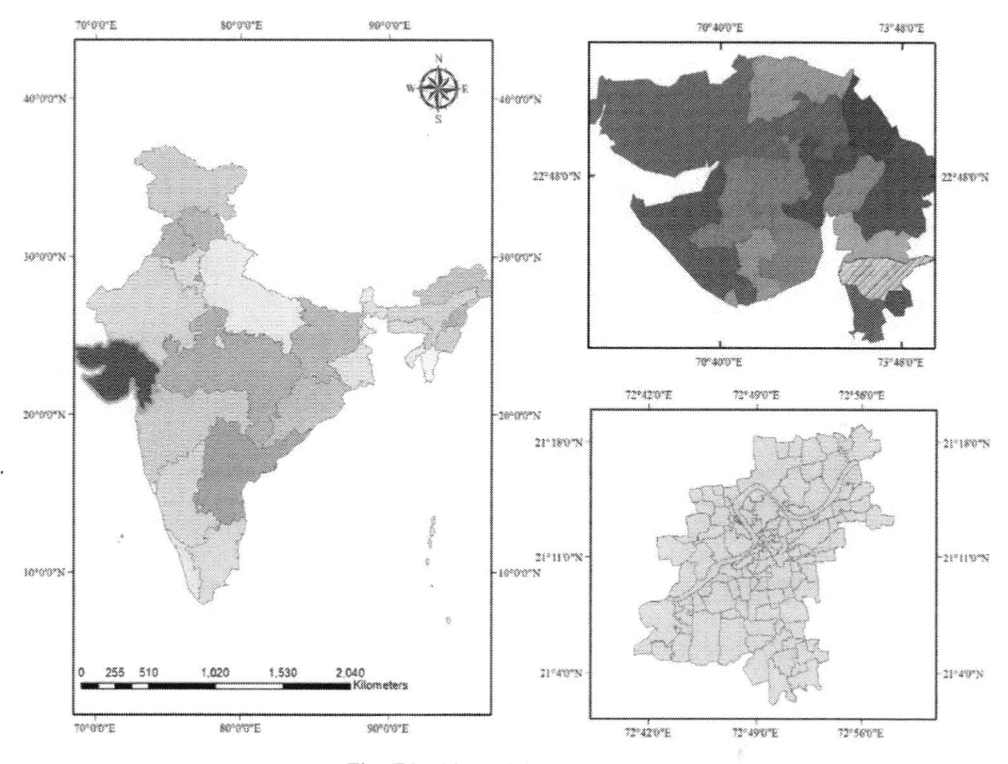

Fig. 7.1 Map of the study area

Source: ArcGIS and Google Earth Pro

Total noise levels dB(A) was calculated as:

$$(x)\ dB(A) = 10 * \log_{10}\left(\frac{Si}{So}\right)^2 \tag{2}$$

$$\left(\frac{Si}{So}\right)^2 = 10^{\frac{pi}{10}} \tag{3}$$

where p_i is the equivalent noise level (L_{eq}) (from Table 7.2), 'S' is the sound pressure level.

Results and Discussion

Data on emitted noise from the demolition and construction activities going in the physics laboratory were collected and analysed. Data of noise pollution index L_{eq} were observed at different locations and on different day. Noise pollution data were not observed for the night-time as there was no construction activity in night-time. Data sets were extracted and subjected to statistical analysis. The results of statistical analysis show the alarming condition of noise pollution and are depicted in Table 7.2. Results of Table 7.2 shows that as a distance of sound

Table 7.2 Measurement of noise levels on site dBA

Time	L_{eq}(dB)	L_{eq}(dB)	L_{eq}(dB)	L_{eq}(dB)
	5 m	8 m	11 m	14 m
09 am–10 am	79.2	76.8	68.7	65.5
10 am–11 am	78.6	75.2	67.1	64.2
11 am–12 am	77.8	74.6	66.1	63.8
12 pm–01 pm	79.5	72.3	65.0	61.5
01 pm–02 pm	78.5	71.8	64.2	60.6
02 pm–03 pm	78.1	72.4	64.8	62.5
03 pm–04 pm	73.8	73.8	69.6	64.1
04 pm–05 pm	83.9	75.2	68.7	65.6

measuring increases, the reduction in noise level was observed. From the result shown in Table 7.2 and Fig. 7.2, it was seen that peak of noise pollution was observed for the morning and evening time period. Furthermore, there is a drop in the noise pollution level was observed in the afternoon hours as most of the workers were having lunch at that time. A reduction of 2 dBA and 4 dBA was observed in the afternoon period from the morning and evening period, respectively. If a person is considered to be sitting in the physics laboratory from 9:00 am to 5:00 pm (usually lab assistant, faculty and research scholars stays at lab from 9:00 am to 5:00 pm), he/she is going to receive equivalent noise of 78.7 dBA, 74 dBA, 66.8 dBA and 63.5 dBA at a distance of 5 m, 8 m, 11 m and 14 m, respectively, in one day. This amount of noise is highly distracting and irritating for the individual. Maximum noise pollution level of 83.9 dBA, 76.8 dBA, 69.6 dBA and 65.6 dBA were found at distance of 5 m, 8 m, 11 m and 14 m, respectively. Minimum noise level of 73.2 dBA, 71.8 dBA, 64.2 dBA and 60.6 dBA were observed at a distance of 5 m, 8 m, 11 m and 14 m, respectively.

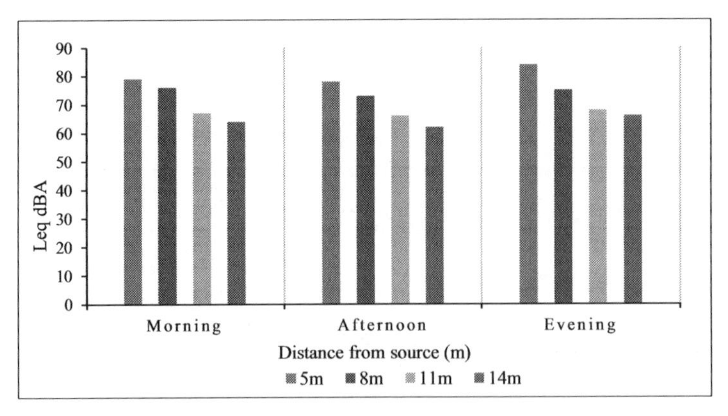

Fig. 7.2 Average data of noise pollution level at different locations

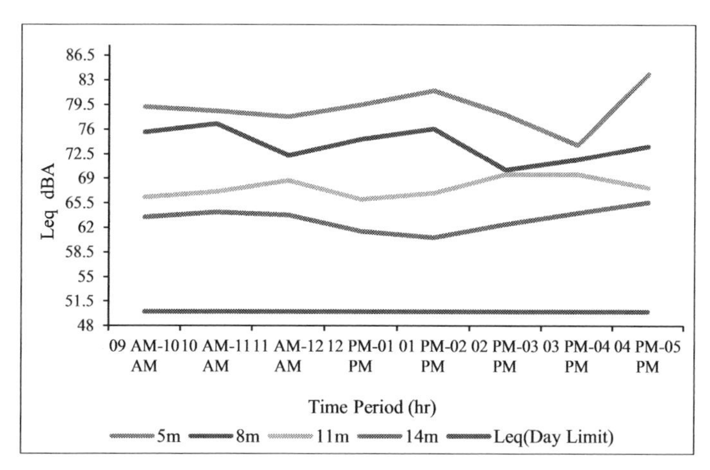

Fig. 7.3 Time-wise variation of noise pollution level

It can be observed from Fig. 7.2 and Table 7.2 that the minimum noise pollution level exceeded the permissible limit of noise level given by the CPCB. CPCB has set the permissible limit of daytime noise level is 50 dBA for areas coming in the silence zone. A comparison of observed noise level was done with the permissible noise limit of CPCB and the result is depicted in Fig. 7.3. The highest variation of noise level from the standard limit of CPCB was found at 5 m distance followed by 8 m, 11 m and 14 m, respectively. Maximum and minimum variation of 23 dBA and 33 dBA was observed for the noise pollution level at 5 m. A slight reduction in noise level was observed at 8 m and maximum and minimum difference of 21 dBA and 26 dBA was observed from the permissible limit of daytime noise level. It was observed that a significant reduction was found when level of noise pollution was recorded at 11 m and 14 m. At distance of 11 m, 14 dBA and 20 dBA were the minimum and maximum variation in noise pollution level from the standard limit of CPCB. Minimum and maximum variation of 10 dBA and 15 dBA in noise pollution index from the permissible limit was observed at a distance of 14 m.

From the analysis, it can be observed that level of noise pollution is reduced on increasing the distance from the sources (demolition activities). In order to know the significant difference in the noise pollution at different distance, a null hypothesis was assumed that there is no significant difference in the measured noise level at different distance. For the study of significant difference, a statistical t-test was performed. The t-test was performed at a significant level of 5%. Also, the t-test was conducted on noise pollution level recorded at a distance of 5 m and 8 m, 8 m and 11 m and 11 m and 14 m, respectively. The findings of the t-testare depicted in Table 7.3. It was observed from Table 7.3 that in all locations p-value was found less than 0.05. The t-value was also found more than critical values in all the locations. The results of analysis of the statistical t-test showed a significant difference in the level of noise pollution level from one distance to other distance.

As it was discussed earlier the CPCB has set the daytime standard limit of the silence zone. As the sampling site come under the silence zone, the recorded noise level is compared

Table 7.3 Results of statistical analysis at different distance

Measuring distance	*t*-value	*p*-value	Remarks
5 m and 8 m	4.56 > **2.36**	0.0012 (<.05)	Significant difference
8 m and 11 m	15.00 > **2.36**	0.0007 (<.05)	Significant difference
11 m and 14 m	9.21 > **2.36**	0.0003 (<.05)	Significant difference

*Bold features indicating the critical values of t-test.

with daytime standard limit, that is, 50 dBA. Percentage exceedance analysis of noise level recorded data with 1-hour frequency interval from the daytime permissible limit was done and shown in Fig. 7.4. Maximum percentage exceedance from the permissible limit was found for the evening time period at all different locations. Maximum and minimum % exceedance of 68% and 31% was observed in noise levels at 5 m and 14 m, respectively, for the evening time period. In the morning time period maximum and minimum average % exceedance was found 57% and 28% at a distance of 5 m and 14 m, respectively, from the daytime permissible limit. It is observed that less reduction (2 and 4 dBA) for the afternoon period from the morning and evening period.

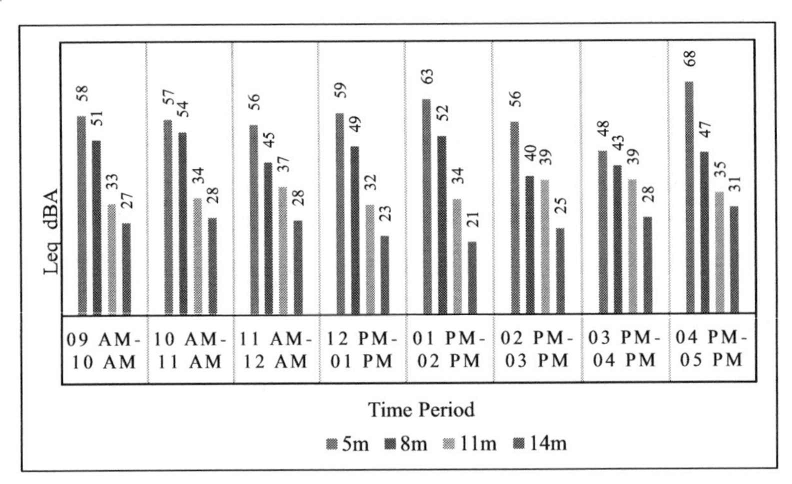

Fig. 7.4 Percentage exceedance of noise level from the daytime standard limit

Analysis of exceedance (%) shows that maximum and minimum average % exceedance at a distance of 5 m and 14 m was observed 59% and 24%, respectively. Less reduction in noise pollution level for the afternoon was observed as demolition work was not stopped. The annoyance (%) was investigated for noise exposure at 5 m, 8 m, 11 m and 14 m distance with the help of Eq. (i), (ii) and (iii). It is found that a working individual in physics lab is facing a 51.46% annoyance exposure at 5 m distance from the source of noise generation (construction-demolition site). The annoyance percent found to decrease as the measuring distance increases, that is, maximum annoyance percent was found at 5 m and minimum annoyance percent was found to be 49.17% at 14 m distance. The variation in annoyance (%) with measured noise levels at different distance of noise level measurements was shown in Fig. 7.5.

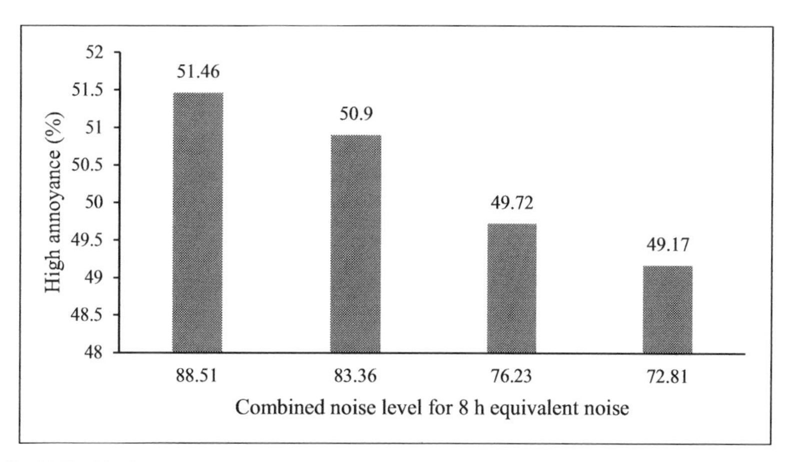

Fig. 7.5 High annoyance (%) at 5 m, 8 m, 11 m and 14 m distance from source

Conclusion

Analysis and assessment of noise pollution due to construction-demolition activities were carried out and it is observed that level of noise pollution is in the alarming condition in the physics laboratory. The highest level of noise pollution was observed in the evening time period followed by the morning and afternoon period, respectively. Occupants (faculties and students) are exposed to noise levels of 78.7 dBA and 63.5 dBA at a distance of 5 m and 14 m, respectively. Studies of noise pollution in the laboratory of physics revealed that L_{eq} exceeds by a great amount from the daytime permissible limit set by CPCB. Analysis of the hypothesis (t-test) revealed that a significant difference was not observed in the measured index of noise pollution at different. Furthermore, an analysis of percentage exceedance was performed, and it was observed that noise pollution level exceeded from the permissible limit for all the distance and time period. Maximum % exceedance was observed for the evening time period followed by the afternoon period and morning time period, respectively. The annoyance (%) level also indicates that as distance increase from point of noise generation the noise level decreases, due to which percent annoyance also decreases and make the working environment more suitable for students and staff of that place. Distance from the source can be considered as a barrier for the reduction in noise pollution level. So, this study will help the policymakers to take some solid measures to mitigate noise pollution level before starting the demolition activities. To mitigate the noise generated due to this demolition activity, it is suggested to use vertical gardens on the walls, double-glazed doors and windows, and sound-absorbing curtains to prevent penetration of noise inside the room. The usage of earmuffs and earplugs by the occupants would help in reducing the severity of noise to a significant level.

Acknowledgements

The authors gratefully acknowledge the students, staff and authority of the physics department for their cooperation in the research.

Funding Details

No financial grant from any authority or organisation has been received for this work.

References

1. Baliatsas, C., van Kamp, I., Swart, W., Hooiveld, M., & Yzermans, J. (2016). Noise sensitivity: Symptoms, health status, illness behavior and co-occurring environmental sensitivities. *Environmental Research, 150*, 8–13. https://doi.org/10.1016/J.ENVRES.2016.05.029
2. Basner, M., Babisch, W., Davis, A., Brink, M., Clark, C., Janssen, S., & Stansfeld, S. (2014). Auditory and non-auditory effects of noise on health. *The Lancet, 383*(9925), 1325–1332. https://doi.org/10.1016/S0140-6736(13)61613-X
3. Berglund, B., Lindvall, T., & Schwela, D. H. (2000). New Who Guidelines for Community Noise: *Noise & Vibration Worldwide, 31*(4), 24–29. https://doi.org/10.1260/0957456001497535
4. Clark, C., Head, J., Haines, M., van Kamp, I., van Kempen, E., & Stansfeld, S. A. (2021). A meta-analysis of the association of aircraft noise at school on children's reading comprehension and psychological health for use in health impact assessment. *Journal of Environmental Psychology, 76*, 101646. https://doi.org/10.1016/J.JENVP.2021.101646
5. Crombie, R., Clark, C., & Stansfeld, S. A. (2011). Environmental noise exposure, early biological risk and mental health in nine to ten year old children: A cross-sectional field study. *Environmental Health: A Global Access Science Source, 10*(1). https://doi.org/10.1186/1476-069X-10-39
6. Fiedler, P. E. K., & Zannin, P. H. T. (2015). Evaluation of noise pollution in urban traffic hubs—Noise maps and measurements. *Environmental Impact Assessment Review, 51*, 1–9. https://doi.org/10.1016/J.EIAR.2014.09.014
7. Foraster, M., Eze, I. C., Vienneau, D., Brink, M., Cajochen, C., Caviezel, S., Héritier, H., Schaffner, E., Schindler, C., Wanner, M., Wunderli, J. M., Röösli, M., & Probst-Hensch, N. (2016). Long-term transportation noise annoyance is associated with subsequent lower levels of physical activity. *Environment International, 91*, 341–349. https://doi.org/10.1016/J.ENVINT.2016.03.011
8. Forns, J., Dadvand, P., Foraster, M., Alvarez-Pedrerol, M., Rivas, I., López-Vicente, M., Suades-Gonzalez, E., Garcia-Esteban, R., Esnaola, M., Cirach, M., Grellier, J., Basagaña, X., Querol, X., Guxens, M., Nieuwenhuijsen, M. J., & Sunyer, J. (2016). Traffic-Related air pollution, noise at school, and behavioral problems in barcelona schoolchildren: A cross-sectional study. *Environmental Health Perspectives, 124*(4), 529–535. https://doi.org/10.1289/EHP.1409449
9. Forouzanfar, M. H. (2016). *Global, regional, and national comparative risk assessment of 79 behavioural, environmental and occupational, and metabolic risks.*
10. Francis, C. D., Ortega, C. P., & Cruz, A. (2009). Noise Pollution Changes Avian Communities and Species Interactions. *Current Biology, 19*(16), 1415–1419. https://doi.org/10.1016/J.CUB.2009.06.052
11. Golmohammadi, R., Darvishi, E., Faradmal, J., Poorolajal, J., & Aliabadi, M. (2020). Attention and short-term memory during occupational noise exposure considering task difficulty. *Applied Acoustics, 158*, 107065. https://doi.org/10.1016/J.APACOUST.2019.107065
12. Gupta, D., Gulati, A., & Gupta, U. (2015). Impact of socio-economic status on ear health and behaviour in children: A cross-sectional study in the capital of India. *International Journal of Pediatric Otorhinolaryngology, 79*(11), 1842–1850. https://doi.org/10.1016/J.IJPORL.2015.08.022
13. Hunyadi, D. (2014). Calculation model of road traffic noise annoyance rate in urban areas. *Pollack Periodica, 9*(1), 41–48. https://doi.org/10.1556/POLLACK.9.2014.1.5

14. Kamp, I., Gidlöf-Gunnarsson, A., & Persson Waye, K. (2013). The effects of noise disturbed sleep on children's health and cognitive development. *The Journal of the Acoustical Society of America, 133*(5), 3506–3506. https://doi.org/10.1121/1.4806243
15. Klatte, M., Lachmann, T., & Meis, M. (2010). Noise and Health. *Noise and Health, 12*(49), 270. https://doi.org/10.4103/1463-1741.70506
16. Kujala, T., & Brattico, E. (2009). Detrimental noise effects on brain's speech functions. *Biological Psychology, 81*(3), 135–143. https://doi.org/10.1016/J.BIOPSYCHO.2009.03.010
17. Lim, J., Kweon, K., Kim, H. W., Cho, S. W., Park, J., & Sim, C. S. (2018). Negative Impact of Noise and Noise Sensitivity on Mental Health in Childhood. *Noise & Health, 20*(96), 199. https://doi.org/10.4103/NAH.NAH_9_18
18. Lokhande, S. K., Motwani, D. M., Dange, S. S., & Jain, M. C. (2021). Abatement of Traffic Noise Pollution on Educational Institute and Visualization by Noise Maps Using Computational Software: A Case Study. *Lecture Notes on Data Engineering and Communications Technologies, 55*, 93–103. https://doi.org/10.1007/978-981-15-8677-4_8/COVER
19. Martini,A., Mazzoli, M., W. K. (1997). An introduction to the genetics of normal anddefective hearing. *Annals of the New York Academy of Sciences, 830*, 361–374. https://doi.org/10.1111/j.1749-6632.1997.tb51908.x
20. Massonnié, J., Frasseto, P., Mareschal, D., & Kirkham, N. Z. (2020). Learning in Noisy Classrooms: Children's Reports of Annoyance and Distraction from Noise are Associated with Individual Differences in Mind-Wandering and Switching skills: *https://Doi.Org/10.1177/0013916520950277, 54*(1), 58–88. https://doi.org/10.1177/0013916520950277
21. Nassiri, P., Azkhosh, M., Mahmoodi, A., Alimohammadi, I., Zeraati, H., Shalkouhi, P. J., & Bahrami, P. (2011). Assessment of noise induced psychological stresses on printery workers. *International Journal of Environmental Science and Technology, 8*(1), 169–176. https://doi.org/10.1007/BF03326206
22. Ng, C. F. (2000). EFFECTS OF BUILDING CONSTRUCTION NOISE ON RESIDENTS: A QUASI-EXPERIMENT. *Journal of Environmental Psychology, 20*(4), 375–385. https://doi.org/10.1006/JEVP.2000.0177
23. Rahmat, A., Hilmy, A., & Hamid, A. (2015). STUDY OF NOISE PRODUCED BY CONSTRUCTION ACTIVITIES IN ACADEMIC AREA. *Jurnal Teknologi, 75*(5), 107–111. https://doi.org/10.11113/JT.V75.5079
24. Ranpise, R. B., & Tandel, B. N. (2022). Noise Monitoring and Perception Survey of Urban Road Traffic Noise in Silence Zones of a Tier II City—Surat, India. *Journal of The Institution of Engineers (India): Series A, 103*(1), 155–167. https://doi.org/10.1007/S40030-021-00598-X
25. Shea, B. J., Reeves, B. C., Wells, G., Thuku, M., Hamel, C., Moran, J., Moher, D., Tugwell, P., Welch, V., Kristjansson, E., & Henry, D. A. (2017). AMSTAR 2: a critical appraisal tool for systematic reviews that include randomised or non-randomised studies of healthcare interventions, or both. *BMJ, 358*, 4008. https://doi.org/10.1136/BMJ.J4008
26. Singh, N., & Davar, S. C. (2017). Noise Pollution-Sources, Effects and Control. *Kamla Raj Enterprises, 16*(3), 181–187. https://doi.org/10.1080/09709274.2004.11905735
27. Suter, A. H. (2010). Construction Noise: Exposure, Effects, and the Potential for Remediation; A Review and Analysis. *Https://Doi.Org/10.1080/15428110208984768, 63*(6), 768–789. https://doi.org/10.1080/15428110208984768
28. Tandel, B. N., & Macwan, J. (2011). Urban Corridor Noise Pollution: A case study of Surat city, India. *International Conference on Environment and Industrial Innovation (ICEII 2011) (Vol. 12, Pp. 144-148).*

29. Tiesler, C. M. T., Birk, M., Thiering, E., Kohlböck, G., Koletzko, S., Bauer, C. P., Berdel, D., Von Berg, A., Babisch, W., Heinrich, J., Wichmann, H. E., Schoet-zau, A., Mosetter, M., Schindler, J., Höhnke, A., FrankeK., Laubereau, B., Gehring, U., Sausenthaler, S., … Martin, F. (2013). Exposure to road traffic noise and children's behavioural problems and sleep disturbance: Results from the GINIplus and LISAplus studies. *Environmental Research, 123*, 1–8. https://doi.org/10.1016/J.ENVRES.2013.01.009
30. Valente, D. L., Plevinsky, H. M., Franco, J. M., Heinrichs-Graham, E. C., & Lewis, D. E. (2012). Experimental investigation of the effects of the acoustical conditions in a simulated classroom on speech recognition and learning in children. *The Journal of the Acoustical Society of America, 131*(1), 232–246. https://doi.org/10.1121/1.3662059

CHAPTER 8

Degradation of Phenalkamine Condensate and its Constituents by Fenton's Oxidation

Aswathy K. R.[1]
Research Scholar, Department of Civil Engineering,
National Institute of Technology, Karnataka Surathkal, India

Nithin Varghese John[2] **and Febeena C. K.**[3]
P. G. Students, Department of Civil Engineering,
National Institute of Technology, Karnataka Surathkal, India

Basavaraju Manu
Associate Professor, Department of Civil Engineering,
National Institute of Technology, Karnataka Surathkal, India

Abstract

Phenalkamine is an emerging new class of curing agent produced by the reaction of cardanol, formaldehyde (FA) and diethylenetriamine (DETA). These are generally used in coatings for ships and offshore constructions, marine and industrial maintenance coatings, etc. Cardanol is a distillation product of cashew nut shell liquid (CNSL). Even though phenalkamine has many applications, the compound used for the manufacturing and the effluent generated during the production process are highly toxic and harmful to the environment. The effluent from cashew nut industry consists of phenolic compounds having high stability, which are carcinogenic. FA can produce an irritating effect on the human body, damage the immune system and cause cancer. The phenalkamine condensate (PAC) has high pH, high chemical oxygen demand (COD) nearly 1,50,000 mg/L, and low biodegradability. The aim of the study is to evaluate the degradation efficiency of PAC and its constituent compounds FA and DETA using Fenton's oxidation process. The progress of the degradation was monitored by the reduction in COD content in the treated solution. The influences of reaction parameters, such as initial pH, H_2O_2/COD ratio and H_2O_2/Fe^{2+} ratio, on COD removal were also analysed. For PAC, maximum COD removal efficiency of 80% was achieved at operating conditions of pH-3.5, H_2O_2/COD-1, H_2O_2/ Fe^{2+}- 20 and reaction time of 24 hours. It is shown that Fenton's oxidation was efficient for the degradation of FA and DETA with concentration of 400 g/: and 930 g/L, respectively. After 24 hours of reaction time, it was found that 92.6% of the initial COD was reduced for FA and 73.84% was reduced for DETA. All the results indicate that Fenton's oxidation is an effective treatment for PAC and its constituents.

Keywords: degradation, DETA, Fenton's oxidation, FA, PAC, phenalkamine

---

[1]aswathykr12@gmail.com; [2]nithinvj97@gmail.com; [3]fabeenack03@gmail.com; [4]bmanu8888@gmail.com

Introduction

Phenalkamine is an industrial product synthesised from cardanol, formaldehyde (FA) and some amine, such as diethylenetriamine (DETA). The structure of phenalkamine is shown in Fig. 8.1. Cardanol is a distillation product of cashew nut shell liquid (CNSL).

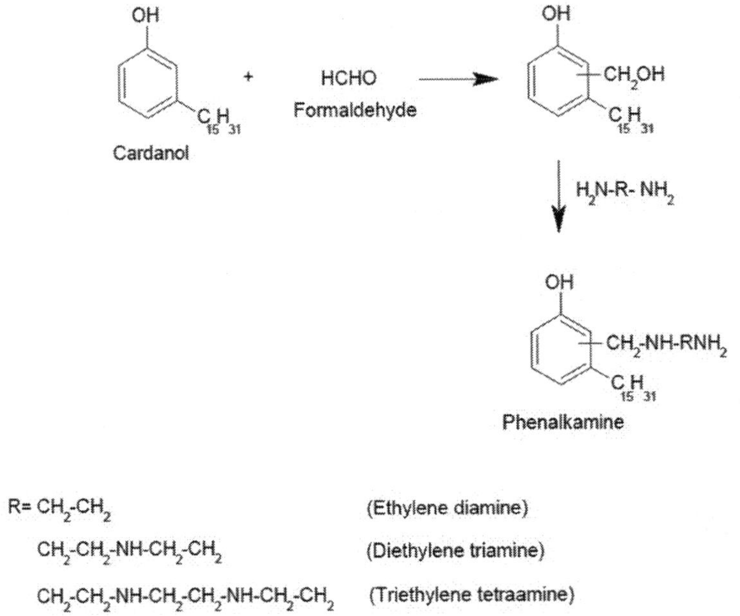

R= CH_2-CH_2 (Ethylene diamine)

CH_2-CH_2-NH-CH_2-CH_2 (Diethylene triamine)

CH_2-CH_2-NH-CH_2-CH_2-NH-CH_2-CH_2 (Triethylene tetraamine)

Fig. 8.1 Structure of phenalkamine

Phenalkamine has gained a lot of interest in recent years due to its exceptional qualities, such as extremely fast and low-temperature cure (even below 0°C), excellent water and saltwater resistance and anticorrosive nature. It was developed to satisfy all requirements for usage as a curing agent in high-performance coating and marine sectors (Rao and Pathak, 2006), (Pathak and Rao, 2006), (Gonçalves et al., 2021).

Even though phenalkamine is a new class of product having wide industrial applications, the phenalkamine condensate (PAC) produced during the production process of phenalkamine consists of high organic compounds, such as phenols, aldehydes and amines, which are very toxic to the environment and also to aquatic life. These effluents are purely organic in nature, with very high chemical oxygen demand (COD), high pH and low biodegradability (K R et al. 2019), (Sahoo et al. 2019). The compounds used for the production of phenalkamine are also toxic and highly reactive. Cardanol is reported as a highly toxic organic compound (Martins et al. 2009). Short-term exposure to FA can induce dermatitis and mucous membrane irritation, while long-term exposure can lead to a variety of cancers (Mei et al. 2019), (Lodyga et al. 2013). DETA at high dose has corrosive properties and can induce acute poisoning. The skin, mouth and lungs are the major potential exposure pathways of DETA. It can also cause dermatitis and asthma-like reactions in humans (Furnell et al. 2021).

Different treatment methods have been employed for the degradation of cardanol, CNSL and cashew industrial effluent by electrochemical oxidation (da Costa et al. 2019), (Oliveira et al. 2018) and bioremediation (Cheriyan and Abraham 2010). Heterogeneous Fenton's oxidation using iron nanoparticles and aluminium dross could reduce COD by 88.91% and 75.83%, respectively (Sahoo et al. 2019). It is reported that FA can be degraded by various advanced oxidation processes (AOPs) like photo Fenton with high efficiency (Guimarães et al. 2012), (Mohammadifard et al. 2019). However, in cases of high concentration, the treatment needs large energy for efficient COD removal and degradation of DETA has not been reported anywhere. These compounds constitute a concern to environment if effluents containing them are not treated properly. It is not feasible to employ biological treatments due to their high toxicity and low biodegradability.

AOPs have proven to be effective methods for degradation of organic compounds. They are based on the generation of highly oxidative radicals, particularly hydroxyl radicals, which are capable of degrading variety of compounds. There are several AOP's, such as Fenton's oxidation, photocatalysis, electrochemical oxidation and ozonation. Among these, Fenton's oxidation is an interesting alternative for wastewaters with high organic contents, which are simple, effective and economical. The oxidation mechanism in the Fenton process involves the generation of reactive hydroxyl radicals under acidic conditions by the catalytic decomposition of hydrogen peroxide (Eq. 1). Fenton's oxidation is a very effective and promising treatment technique for the removal of many hazardous organic pollutants from water and wastewaters (Neyens and Baeyens, 2003). The efficiency of the process largely depends on pH, the concentration of hydrogen peroxide and the catalyst. The reaction mechanism is shown in the following steps.

$$Fe^{2+} + H_2O_2 \rightarrow Fe^{3+} + OH- + OH\bullet \tag{1}$$

The generated $OH\bullet$ oxidises organics (RH) and produces organic radicals ($R\bullet$) as shown in Eq. 2, which are highly reactive and can be oxidised further (Eq. 3)

$$OH\bullet + RH \rightarrow R\bullet + H_2O \tag{2}$$

$$R\bullet + H_2O \rightarrow ROH + OH\bullet \tag{3}$$

The produced Fe^{3+} can react with H_2O_2 and form a hydroperoxyl radical as in Eq. 4. This reaction is known as a Fenton-like reaction, which leads to regeneration of Fe^{2+} in an effective cyclic mechanism (Eqs 4 and 5). The hydroperoxyl radicals may also attack organic contaminants, but they are less sensitive than hydroxyl radicals. Fe^{2+} regeneration is also possible by reacting with organic radical intermediates (Eq. 6).

$$Fe^{3+} + H_2O_2 \rightarrow Fe^{2+} + HO_2\bullet + H^+ \tag{4}$$

$$Fe^{3+} + HO_2\bullet \rightarrow Fe^{2+} + O_2 + H^+ \tag{5}$$

$$Fe^{3+} + R\bullet \rightarrow Fe^{2+} + R^+ \tag{6}$$

The generated hydroxyl radical may be scavenged by excess hydrogen peroxide or iron as shown in Eqs. 7 and 8.

$$OH\bullet + H_2O_2 \rightarrow HO_2\bullet + H_2O \tag{7}$$

$$OH\bullet + Fe^{2+} \rightarrow Fe^{3+} + OH^- \tag{8}$$

However, Fenton oxidation also has some drawbacks, which are the production of sludge and strict pH range. The conventional Fenton process can also improve by reducing the sludge production by using heterogeneous catalysts, such as iron minerals, zero-valent iron andiron nanoparticles. Iron nanoparticles are now widely used in wastewater treatment due to their high surface area and high reactivity.

In this work, the COD removal of PAC and its constituents FA and DETA by Fenton's oxidation was investigated. The effect of initial pH and reagent dosage on COD removal was studied.

Materials and Methods

Materials

The chemical reagents FA, DETA, 50% w/w H_2O_2, hydrochloric acid (38%), sulphuric acid (98%), ferrous (II) sulphate heptahydrate ($FeSO_4.7H_2O$) and sodium hydroxide (98%), are analytical grade and purchased from Merck manufactured in India. Iron nanoparticles purchased from the commercial market.

Wastewater Characteristics

PAC used in this study was collected from the nearby cashew nut industry and was stored at a low temperature of 4°C. It was used without any pretreatment in this experiment. The effluent was brownish-yellow in colour and have high COD. The raw wastewater was characterized by measuring ph. BOD, COD and total suspended solids (TSS), and the measured values are shown in Table 1. The FA used in the study was having initial concentration 400 g/L with COD of 565 g/L and DETA of 930 g/L initial concentration with COD of 940 g/L.

Experiment Methodology

Fenton's oxidation was conducted in PAC and its components FA and DETA. Fenton's oxidation generally depends on different parameters, such as pH, H_2O_2 concentration and catalyst concentration. Initially, preliminary experiments were conducted to know the approximate range of values for all parameters (H_2O_2/COD, H_2O_2/Fe and pH). Then, the actual experiments were started by considering the amount of reagents that were determined according to the initial COD value of the effluent. All the experiments were conducted in batch reactors at ambient temperature (27±30°C) and pressure. First, 10 mL of sample was transferred into a 500 mL of a conical flask, and pH of the sample was adjusted to the desired value using hydrochloric acid sodium hydroxide. Then required amount of catalyst was added. Finally, H_2O_2 was added drop by drop carefully to initiate Fenton reaction.

The experimental setup for Fenton's oxidation is shown in Fig. 8.2. After the reaction time of 24 hrs, the supernatant was decanted and filtered using Whatman no: 42 filter paper for further analysis. The optimum Fenton dosage for maximum degradation was found by varying the H_2O_2/COD and H_2O_2/Fe^{2+} ratios. In FA and DETA, heterogeneous Fenton oxidation was also conducted using iron nanoparticles.

Fig. 8.2 Experimental setup for Fenton's oxidation

Analytical Methods

The pH of the sample was measured using a digital pH electrode with an integrated temperature sensor (HI763100). COD was estimated using the closed reflux titrimetric method, 5-day BOD and total suspended solids were measured according to standard methods (American Public Health Association. Andrew D., American Water Works Association, Water Environment Federation., 2005).

Results and Discussions

Characterization of PAC

PAC was analysed for different parameters, including pH, BOD, COD and total suspended solids, and the results are given in Table 8.1.

Table 8.1 Characterization of PAC

Parameter	Measured Value
pH	11.5
BOD	3000–3500 mg/L
COD	1,20,000–1,50,000 mg/L
Total Suspended Solids (TSS)	7800–8200 mg/L

Fenton's Oxidation

Fenton's oxidation was conducted in PAC and its components DETA and FA. The effect of reaction parameters, such as H_2O_2/COD, H_2O_2/Fe and pH, on COD removal efficiency, was studied for all compounds. Since the effect of temperature on COD degradation was reported as negligible, the temperature was not taken into account as a reaction parameter in Fenton's oxidation.

Effect of pH on COD Removal

Fenton's oxidation process is highly dependent on the pH of the sample. The production rate of hydroxyl radicals is directly affected by the initial pH value. Fenton's oxidation was carried out in all the compounds at different pH values and all other reaction parameters were kept constant for each compound.

Since DETA was strongly alkaline in nature, altering the pH was a difficult procedure. 1 N H_2SO_4 and 1 N HCl were added to DETA but rather than dilution, the change in pH was not noticeable. And there was formation of precipitate during the addition of both concentrated H_2SO_4 and HCl acids. So, the experiment was decided to be conducted at the prevailing pH of the sample itself. The reaction parameters values of other two compound in this part of the experiment is shown in Table 8.2.

Table 8.2 Reaction parameters for the optimisation of Ph

Compound	PAC	FA
pH	2.5–4.5	2–4
H_2O_2/COD ratio	2.13	1
H_2O_2/ Fe^{2+}	20	150

Research reported that the optimum pH for Fenton's process was 3 and 3.5 irrespective of water characteristics (Basavaraju et al. 2011). Similarly, optimum pH of 3.5 was obtained for both PAC and FA. Maximum COD removal efficiency at optimum pH was 70% and 92.7% for PAC and FA, respectively as shown in Figs 8.3 and 8.4.

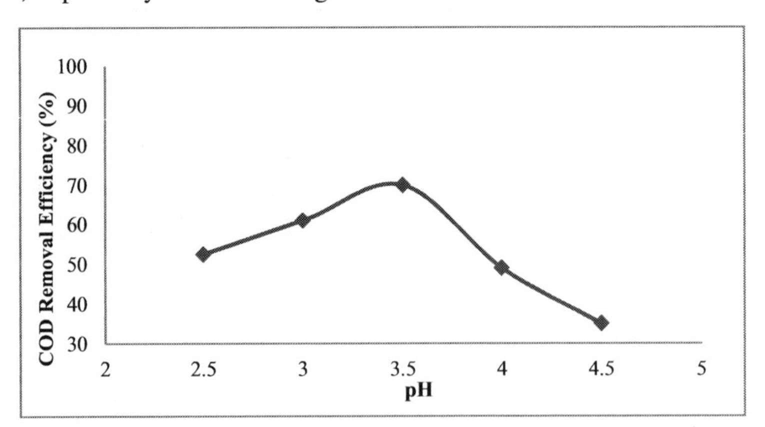

Fig. 8.3 Effect of pH on COD removal efficiency of PAC

From both graphs, it can be seen that the COD removal rate was reduced as pH increased. This might be due to the self-decomposition of hydrogen peroxide and formation of ferric hydroxide precipitate at higher pH which will deactivate the iron available to react with peroxide to produce hydroxyl radicals (Gar and Ahmed, 2017), (Tyagi et al. 2020). The efficiency also decreases at lower pH, which might be due to the formation of stable oxonium ions. It reduces the reactivity of peroxide with ferrous iron (Kwon et al. 1999).

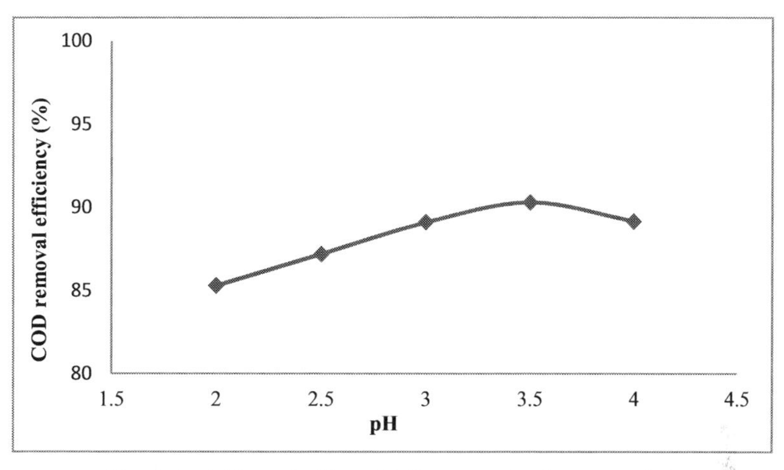

Fig. 8.4 Effect of pH on COD removal efficiency of FA

Effect of H_2O_2/COD Ratio on COD Removal

Since hydrogen peroxide is the source of hydroxyl radicals that lead to breakdown of organic compounds, its dosage is one of the major factor influencing the efficiency of Fenton's process. The dosage mainly depends on the organic content of the sample. The amount of hydrogen peroxide required to treat an effluent increases as its organic content increases. However, excess dosage of hydrogen peroxide increases operational costs as well as the scavenging action of hydroxyl radical by H_2O_2 (Eq. 7), which inhibits the breakdown of organic pollutants (Domingues et al. 2021). The effect of H_2O_2/COD ratio on COD removal was studied by conducting a set of experiments with varying H_2O_2/COD ratio. The range of H_2O_2/COD ratio for each compound was selected according to the organic load present in the compound. The various reaction parameters for three compounds are shown in Table 8.3.

Table 8.3 Reaction parameters for the optimisation of H_2O_2/COD ratio

Compound	PAC	FA	DETA
pH	3.5	3.5	12
H_2O_2/COD ratio	0.6–1.6	0.75–1.75	0.5–3
H_2O_2/Fe^{2+}	20	150	10

The effect of H_2O_2/COD ratio on COD removal for PAC, FA and DETA is shown in Figs 8.5, 8.6 and 8.7, respectively. The result shows that the COD removal rate increase with an increase in H_2O_2/COD ratio to a certain limit above that ratio efficiency reduces. Maximum COD removal efficiency of 78% and 63% was obtained at H_2O_2/COD ratio of 1 for PAC and FA, respectively. For DETA, the optimum ratio was 2.5 and the corresponding COD removal rate was 92%. The COD removal efficiency was reduced at higher ratio for all the compounds. This could be because hydrogen peroxide does not specifically transform into hydroxyl radicals but instead auto decomposes into water and oxygen after radicals have been generated. And they are exhausted through scavenging action by hydrogen peroxide itself to form hydroperoxyl radicals with lower oxidising power (Ali et al. 2013).

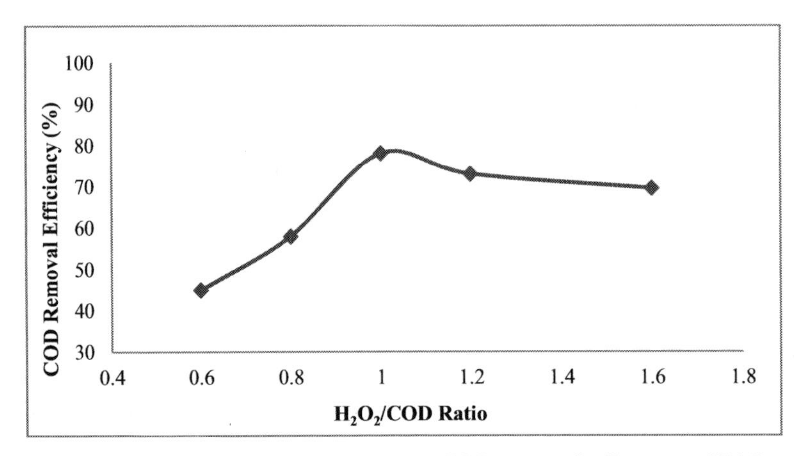

Fig. 8.5 Effect of H_2O_2/COD ratio on COD removal efficiency of PAC

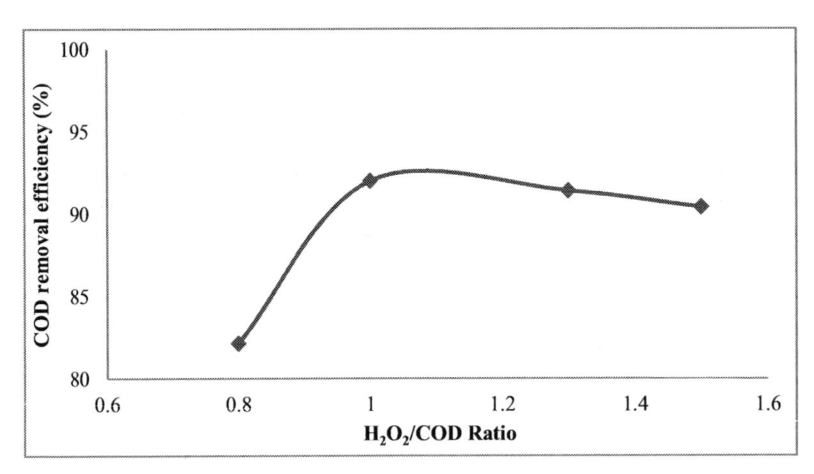

Fig. 8.6 Effect of H_2O_2/COD ratio on COD removal efficiency of FA

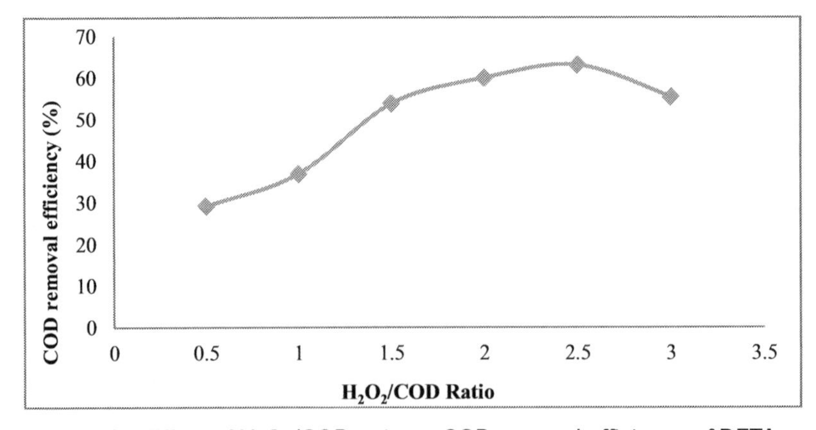

Fig. 8.7 Effect of H_2O_2/COD ratio on COD removal efficiency of DETA

Effect of H_2O_2/ Fe^{2+} Ratio on COD Removal

The generation of hydroxyl radical also depends on the existence of a catalyst. Generally, the COD removal rate increases with an increase in iron load. However, hydrogen peroxide, iron dosage also has a limit. Addition beyond the limit reduces the degradation efficiency. Therefore, it is crucial to optimise the iron dosage for maximum efficiency. A set of experiments was conducted at different H_2O_2/Fe^{2+} molar ratio with constant pH and H_2O_2/COD ratio, and the values of different parameters for each compound are shown in Table 8.4. The effect of H_2O_2/ Fe^{2+} ratio on COD removal of PAC is shown in Fig. 8.8.

Table 8.4 Reaction parameters for the optimisation of H_2O_2/ Fe^{2+} ratio

Compound	PAC	FA	DETA
pH	3.5	3.5	12
H_2O_2/COD ratio	1	1	2.5
H_2O_2/ Fe^{2+}	20–100	20–200	10–100

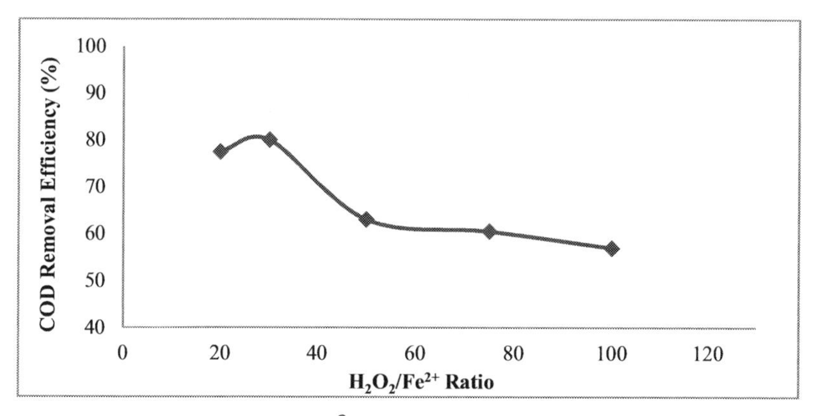

Fig. 8.8 Effect of H_2O_2/ Fe^{2+} ratio on COD removal efficiency of PAC

For PAC, maximum COD removal of 80% was obtained at H_2O_2/Fe^{2+} ratio of 30 and above a slight reduction in COD removal was observed.

In FA and DETA, the experiments were conducted using iron as well as FeNP as a catalyst at constant reaction parameters. From Fig. 8.9, it can be observed that the heterogeneous using FeNP Fenton's oxidation has shown a similar trend of COD removal in function of catalyst concentration compared to classical Fenton's oxidation in terms of efficiency but the optimum dosage has been found to be a bit low compared to that of classical Fenton's oxidation. For an H_2O_2/ FeNP ratio of 100, the COD removal efficiency was only 64.8%. When the ratio was enhanced from 100 to 20, COD removal increased from 64.8 to 89.8%. This is explained by the increase of iron concentration resulting in an increase of catalytic centres, which improves oxidation and eventually better COD removal occurs. An optimum H_2O_2/catalyst ratio of 20 was obtained with a COD removal efficiency of 89.8% for FeNP. Whereas the maximum efficiency was 92% achieved for H_2O_2/catalyst ratio of 150 for classical Fenton's process.

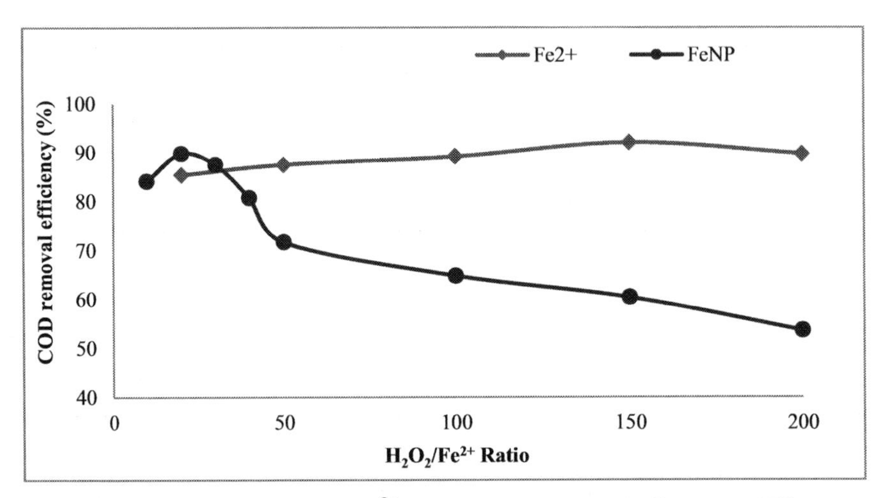

Fig. 8.9 Effect of H_2O_2/ Fe^{2+} ratio on COD removal efficiency of FA

In the case of DETA, a similar trend of COD removal was observed for both homogenous and heterogeneous Fenton's oxidation (Fig. 8.10). An optimum H_2O_2/catalyst ratio of 30 was obtained for both processes. And the corresponding COD removal efficiency was 71.12% and 73.84% for iron sulphate and FeNP, respectively.

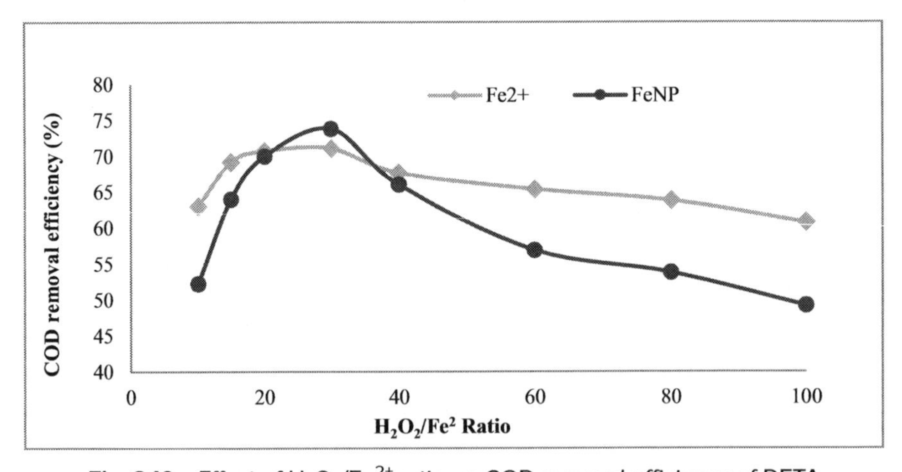

Fig. 8.10 Effect of H_2O_2/Fe^{2+} ratio on COD removal efficiency of DETA

For all the compounds, COD removal efficiency increases with the increase in dosage till optimum value. When the amount was exceeded the optimum, the COD removal efficiency decreases. This might be due to the scavenging action of hydroxyl radical by excess iron (Eq. 8). Compared to conventional Fenton's oxidation, sludge production was less when iron nanoparticles were used as the catalyst instead of iron sulphate.

Conclusion

PAC and its constituent's cardanol, DETA and FA are toxic compounds with high amount of organic matter. This type of pollutant poses a high risk to environment. The present work deals with the degradation of PAC and its constituents FA, DETA by Fenton's oxidation. The following conclusions were made on the basis of the results and discussions arrived from the experimental results obtained from the studies.

- A maximum of 80% COD removal rate was obtained for PAC at optimum reaction conditions (pH-3.5, H_2O_2/COD-1, H_2O_2/ Fe^{2+}- 20).
- A maximum COD removal efficiency of 92.6% was obtained for FA at the optimised conditions (pH-3.5, H_2O_2/COD -1, H_2O_2/Fe^{2+}- 150) for classical homogeneous Fenton's oxidation using ferrous sulphate heptahydrate ($FeSO_4.7H_2O$) as the catalyst.
- Maximum COD removal efficiency of 73.84% was obtained for DETA at the optimised conditions (pH-12, H_2O_2/COD-2, H_2O_2/FeNP-30) for heterogeneous Fenton's oxidation using iron nanoparticle.
- It can be found that the efficiency of treatment is dependent upon the amounts of hydrogen peroxide and catalyst. It is because when either of them has overdosed, both react with hydroxyl radical and therefore inhibit the oxidation reaction.

In this context, Fenton's oxidation seems to be an effective treatment method for PAC and its constituents with high concentration.

Acknowledgement

The authors sincerely thank the MHRD Government of India for the institute fellowship to carry out this research.

References

1. Ali, M. E. M., Gad-Allah, T. A., and Badawy, M. I. (2013). "Heterogeneous Fenton process using steel industry wastes for methyl orange degradation." *Appl. Water Sci.*, 3(1), 263–270.
2. American Public Health Association. Andrew D., American Water Works Association., Water Environment Federation., E. (2005). *Standard methods for the examination of water and wastewater.* Washington, D.C.: APHA-AWWA-WEF.
3. Cheriyan, S., and Abraham, E. T. (2010). "Enzymatic bioremediation of cashew nut shell liquid contamination." *J. Hazard. Mater.*, 176(1–3), 1097–1100.
4. Costa, P. R. F. da, Emily, E. C. T., Castro, S. S. L., Fajardo, A. S., and Martínez-Huitle, C. A. (2019). "A sequential process to treat a cashew-nut effluent: Electrocoagulation plus electrochemical oxidation." *J. Electroanal. Chem.*, 834(October 2018), 79–85.
5. Domingues, E., Fernandes, E., Gomes, J., Castro-Silva, S., and Martins, R. C. (2021). "Olive oil extraction industry wastewater treatment by coagulation and Fenton's process." *J. Water Process Eng.*, 39(December 2020).
6. Furnell, E., Tian, X., and Bobicki, E. R. (2021). "Diethylenetriamine as a selective pyrrhotite depressant: Properties, application, and mitigation strategies." *Can. J. Chem. Eng.*, 99(6), 1316–1333.

7. Gar, M., and Ahmed, A. (2017). "Investigation of optimum conditions and costs estimation for degradation of phenol by solar photo-Fenton process." *Appl. Water Sci.*, 375–382.
8. Gonçalves, F. A. M. M., Santos, M., Cernadas, T., Ferreira, P., and Alves, P. (2021). "Advances in the development of biobased epoxy resins: insight into more sustainable materials and future applications." *Int. Mater. Rev.*, 0(0), 1–31.
9. Guimarães, J. R., Turato Farah, C. R., Maniero, M. G., and Fadini, P. S. (2012). "Degradation of formaldehyde by advanced oxidation processes." *J. Environ. Manage.*, 107, 96–101.
10. K R, A., K V, C., Damodar, V. C., and Manu, D. B. (2019). "Coagulation/Flocculation and Aerobic Treatment of Fenton Pre-treated Phenalkamine Condensate." *SSRN Electron. J.*, (May), 336–344.
11. Kwon, B. U. M. G. U. N., Lee, D. S. O. O., Kang, N., and Yoon, J. (1999). "CHARACTERISTICS OF P -CHLOROPHENOL OXIDATION BY FENTON ' S REAGENT." 33(9), 2110–2118.
12. Lodyga, A., Minda-Data, D., KozioL, M., and Tyński, P. (2013). "Using the deep oxidation process with fenton's reagent to remove formaldehyde from industrial wastewater." *Chemik*, 67(7), 648–653.
13. Manu, B., Mahamood, S., Vittal, H., and Shrihari, S. (2011). "International Journal of Research in Chemistry and Environment A novel catalytic route to degrade paracetamol by Fenton process." 1(1).
14. MARTINS, R., BEATRIZ, A., SANTAELLA, S., and LOTUFO, L. (2009). "Ecotoxicological analysis of cashew nut industry effluents, specifically two of its major phenolic components, cardol and cardanol." *Panam. J. Aquat. Sci.*, 4(3), 363–368.
15. Mei, X., Guo, Z., Liu, J., Bi, S., Li, P., Wang, Y., Shen, W., Yang, Y., Wang, Y., Xiao, Y., Yang, X., Liu, Y., Zhao, L., and Wang, Y. (2019). "Treatment of formaldehyde wastewater by a membrane-aerated biofilm reactor (MABR): The degradation of formaldehyde in the presence of the cosubstrate methanol." *Chem. Eng. J.*, 372(April), 673–683.
16. Mohammadifard, Z., Saboori, R., Mirbagheri, N. S., and Sabbaghi, S. (2019). "Heterogeneous photo-Fenton degradation of formaldehyde using MIL-100(Fe) under visible light irradiation." *Environ. Pollut.*, 251, 783–791.
17. Neyens, E., and Baeyens, J. (2003). "A review of classic Fenton's peroxidation as an advanced oxidation technique." 98, 33–50.
18. Oliveira, E. M. S., Silva, F. R., Morais, C. C. O., Oliveira, T. M. B. F., Martínez-Huitle, C. A., Motheo, A. J., Albuquerque, C. C., and Castro, S. S. L. (2018). "Performance of (in)active anodic materials for the electrooxidation of phenolic wastewaters from cashew-nut processing industry." *Chemosphere*, 201, 740–748.
19. Pathak, S. K., and Rao, B. S. (2006). "Structural effect of phenalkamines on adhesive viscoelastic and thermal properties of epoxy networks." *J. Appl. Polym. Sci.*, 102(5), 4741–4748.
20. Rao, B. S., and Pathak, S. K. (2006). "Thermal and viscoelastic properties of sequentially polymerized networks composed of benzoxazine, epoxy, and phenalkamine curing agents." *J. Appl. Polym. Sci.*, 100(5), 3956–3965.
21. Sahoo, D. K., M, V., and Manu, D. B. (2019). "Fenton's Oxidation of Phenalkamine Condensate Using Aluminium Dross and Laterite Iron Nanoparticle as a Catalyst." *SSRN Electron. J.*, (May), 352–362.
22. Tyagi, M., Kumari, N., and Jagadevan, S. (2020). "A holistic Fenton oxidation-biodegradation system for treatment of phenol from coke oven wastewater: Optimization, toxicity analysis and phylogenetic analysis." *J. Water Process Eng.*, 37(June), 101475.

CHAPTER 9

Solar Tracking System to Harvest Photovoltaic Energy with Higher Efficiency

Indrajit Bose[1], Arkendu Mitra[2], Subhra Mukherjee[3]
Department of Electrical Engineering,
Narula Institute of Technology, Kolkata, India

Sujoy Bhowmik[4]
Department of Electrical Engineering,
Kalyani Government Engineering College, Nadia, India

Abstract

Nowadays, fossil fuels are getting depleted and cannot be replenished, so an alternative source of energy can be opted for due to enhancement of energy demand day by day. Renewable energy is derived from different natural sources, which are inexhaustible, pollution-free and can be restored easily. Among various types of renewable energies, the photovoltaic system is quite simple and widely used. Due to less efficiency of standard photovoltaic cells, the locus of the sun is required to be tracked by the panel(s) to obtain most of the energy from the solar cells.

This proposed work investigates a simpler design and construction of a solar tracker system that follows the direction of the sun to increase the output of the solar-powered applications. To implement the same, light-depended resistors (LDRs) are used to detect the position of the sun at different time. The entire system will be driven by a servo-mechanism which is controlled through Arduino-NANO. The panel will be reset after sunset and the output is stored in a battery during daytime. The overall prototype is working satisfactorily, and the energy is harvested with higher efficiency.

Keywords: LDR, microcontroller, photovoltaic cell, solar tracker, servo-mechanism

Introduction

The population growth over the world is increasing gradually and the energy scarcity problems faced by the world, more especially the third world countries, are propelling researchers to find an alternative source of energy that would complement the conventional fossil fuel which is getting depleted and cannot be replenished to mitigate the energy crisis. Thereby, coal-based power plants (thermal), which supply bulk amount of energy to the nation, are now getting

[1]indrajit24y@gmail.com; [2]arkendu83@gmail.com; [3]subhra.reek@gmail.com; [4]sujay.bhowmik654@gmail.com

exhausted and so that renewable energy is derived from different natural sources, such as solar, wind, hydel, and biomass, which are inexhaustible, pollution-free and can be restored easily. The installation of a wind system is generally area specific and the frequency of the generation is dependent on the velocity of the wind. Either a battery can be used to store the energy or a doubly-fed induction generator (DFIG) can be installed to deliver the energy to the grid. The generation of hydel energy requires a large area and the installation cost is relatively high. Also, costly salient pole generators are involved, which requires low-speed operation. The hydel energy can be mostly utilised during the rainy season and remains in less operational condition rest of the time. Biomass energy can be easily produced and operational cost is lower compared to others, but it is not hygienic, environment free and may lead to deforestation. Among all of these, the photovoltaic system is quite simple, widely used and adequate in nature. Due to less efficiency of standard photovoltaic cells, the locus of the sun is required to be tracked by the panel(s) to obtain most of the energy from the solar cells.

Literature Review

A microcontroller-based prototype with a multi-functional solar tracking system is developed, which keeps the solar panels aligned with the sun's direction in order to maximise efficiency. Maximum power point tracking (MPPT) can be applied to solar systems to meet up energy demand. Along with this algorithm, microcontroller-based PV tracker is to be implemented to absorb more solar energy and feed to the load [1]. A two-axes heat collecting sun tracking system is designed using light-depended resistors (LDR) sensor mounted on a Fresnel lens to control and track the sun's radiation. Depending upon the position of heating head and sunlight, the maximum temperature from the Stirling engine can be obtained. If the heat loss area is decreased, the temperature of the heating head will increase faster and higher and the Stirling engine will get more power [2]. Sun tracking systems can be broadly classified as single-axis and dual-axis trackers, they can also be classified as active and passive trackers depending on the actuator and their mode of rotation. The azimuth and altitude dual-axis tracking system are more efficient in comparison to other tracking systems. Looking from the cost and flexibility angle single-axis tracking system is much more feasible than dual-axis [3]. A modified autonomous solar tracking system has been designed with a tracking mechanism, a battery and an inverter that is resistant to weather, temperature and minor mechanical stresses, thereby maximising the output and increasing the efficiency [4]. A programmable logic controller (PLC) unit has been employed to control and monitor the mechanical movement of the PV module and to collect and store data related to the sun's movement and direction of shift of the radiation which improves the efficiency of solar panels [5]. Single axis tracking system has been designed to detect the movement of the sun by three LDRs, one of which detects whether the solar panel is focused, the second resistor determines whether there is cloud cover and the third sensor senses whether it is daytime or night. This mechanism can also be implemented for tracking solar parabolic trough collectors of medium to high concentration ratios under any weather conditions [6]. A low-cost solar tracking system has been integrated using a calibration algorithm to perform motion stage camera by accepting minimal data from the earth-sun geometry, decoder motor values and solar image locations. This performs the accurate translation from pixel displacements to spherical motion parameters [7].

This proposed work investigates a simpler design and construction of a solar tracker system that tracks the direction of the sun to increase the output of solar-powered applications. The solar energy converted into electrical energy by solar panels or photovoltaic cells requires less maintenance and does not have any environmental pollution. To implement the same, photo resistors also known as LDRs are used to detect the position of the sun at different time whose resistance decreases as the intensity of light they are exposed to increases. The entire system will be driven by servo-mechanism, which is controlled through Arduino-NANO. The servo motor is preferably used which can rotate 360 degrees of operation due to its gear operation. Generally, Tower-pro SG90 motor with 2.5 kg/cm torque may be chosen as per requirement. The panel will be reset after sunset and the output is stored in a battery during day time.

Photo-Voltaic Technology

Different Types of Solar Panel

The most efficient (17%–22%) and expensive solar panels are made with mono-crystalline cells. These solar cells are fabricated with very pure silicon and involve a complicated crystal growth process to reduce the reflection of solar rays. Long silicon rods are produced, which are cut into slices of 0.2 to 0.4 mm thick discs or wafers which are then processed into individual cells that are wired together in the solar panel. When sunlight falls on the top of the solar panel, an electric field is created which can absorb the power and it may utilise as DC source in power devices.

Multi-crystalline solar panels are generally made with polycrystalline cells, which are less expensive and slightly less efficient (14%–16%) than mono-crystalline cells consist of a large block of many crystals. This is what gives them that striking shattered glass appearance. Like mono-crystalline cells, they are also sliced into wafers to produce the individual cells that make the solar panel. Amorphous solar panels consist of a thin layer of silicon deposited on a base material, such as metal or glass, to create the solar panel. These solar panels are much cheaper, and their energy efficiency is also much less (6%–7%), so more area is required to produce the same amount of power compared to mono-crystalline or polycrystalline type of solar panels. It can even be made into long sheets of roofing material to cover large areas of a south-facing roof surface.

Single Axis Tracker

In numerous applications that require tracking, one-axis tracking is utilised and this becomes justifiable when cost cutting is considered, which is shown in Fig. 9.1. Although, for higher concentrations and high-precision applications, two orthogonal tracking may be required. The rotation of the earth is the major action in solar radiation collectors, while the impact of other movements has very gradual effects which are mostly neglected in many applications without introducing much error. This makes single-axis tracking more acceptable, as it represents a relatively efficient investment, with application in areas, such as research work and high-precision tracking applications. Ideally, the stand should be made from aluminium angle as it is strong, durable and suitable for outdoor use, but it can also be made from wood, plywood or PVC piping.

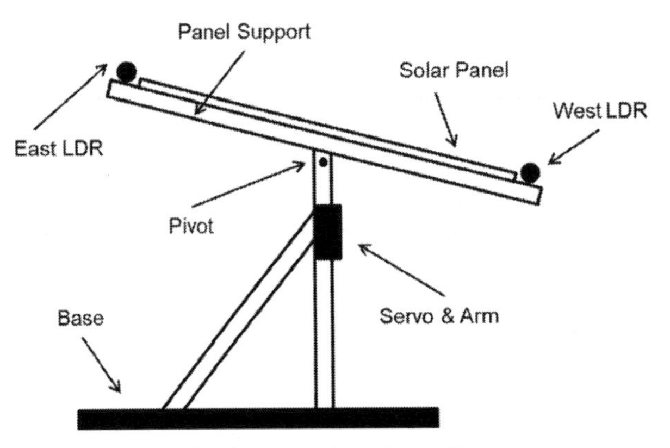

Fig. 9.1 Single axis tracker

The stand is essentially made in two parts, the base and the panel support. They are joined around a pivot point on which the panel support rotates. The servo is mounted onto the base and the arm actuates the panel support. The panel should protrude from the panel support as little as possible to keep the out-of-balance load on the servos to a minimum. Ideally, the pivot point should be placed at the centre of gravity of the panel and panel support together so that the servo has an equal load placed on it no matter which direction the panel is facing, although this is not always practically possible.

Control and Processing Unit

Choosing a control unit in this case, the microcontroller is one that must meet the task at hand efficiently and cost-effectively. In analysing the needs of a microcontroller-based project, an Arduino-NANO was used as it satisfied the following requirements.

- It comes with exactly the same functionality as in Arduino UNO but is quite in small size.
- It comes with an operating voltage of 5 V, however, the input voltage can vary from 7 to 12 V.
- Arduino-NANO pinout contains 14 digital pins, 8 analog pins, 2 reset pins and 6 power pins.
- Each of these digital and analog pins is assigned multiple functions but their main function is to be configured as input or output.
- The SRAM can vary from 1 KB or 2 KB and EEPROM is 512 bytes or 1 KB for Atmega168 and Atmega 328, respectively.
- It is programmed using Arduino IDE, which is an integrated development environment that runs both offline and online, which is quite user friendly and programming and debugging code in such a space is very much easy and efficient.
- The cost per unit is very less compared to other Arduino boards in the market.
- Tiny size and breadboard-friendly nature make this device an ideal choice for most of the applications where sizes of the electronic components are of great concern.

Communication Features

The Arduino-NANO device comes with the ability to set up communication with other controllers and computers. The serial communication is carried out by the digital pins Rx and Tx, which stands for receiving and transmitting data, respectively. The serial monitor is added on the Arduino Software, which is used to transmit textual data to or from the board. Future Technology Devices International Limited (FTDI) drivers are also included in the software, which behaves as a virtual communication port to the software. The Tx and Rx pins come with a light emitting diode (LED), which blinks as the data is transmitted between FTDI and universal serial bus (USB) connection to the computer.

Servo-mechanism

The pulse width modulation (PWM) signal is required to govern the motor, which can be generated using a multivibrator circuit or other microcontroller platforms. To make this motor operational, +5 V power supply is required. From Fig. 9.2, it is shown that the PWM signal produced should have a frequency of 50 Hz, that is, the PWM period should be 20 ms. Out of which the on-time can vary from 1 ms to 2 ms. So, when the on-time is 1 ms the motor will be 0° 10 and when 1.5 ms the motor will be 90°, similarly when it is 2 ms it will be 180°. So, by varying the on-time from 1 ms to 2 ms the motor can be controlled from 0° to 180°.

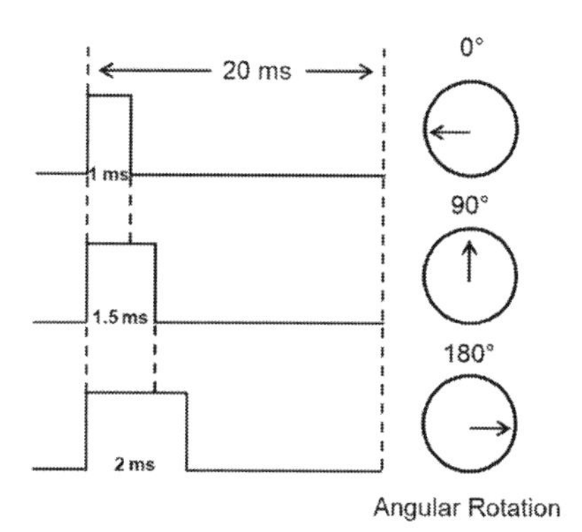

Fig. 9.2 Operation of servomotor

Battery Charging Module

A battery charging module TP4056 is used, which is a completely constant current-voltage linear charging module for single-cell 3.7 V lithium batteries. It will continuously monitor the voltage level of the battery during charging and discharging. The module operates with 5 V, 1 A DC voltage, can be provided by the USB mini cable, and is commonly used in smartphone chargers. Due to the low number of external component count, make the TP4056 module ideally suited for our portable electronics application. The TP4056 module has many key features that make it the desirable choice, it includes current monitor, under voltage lockout, automatic recharge charger and protection circuit in a single module. It has two status pins to indicate charge termination. The module further indicates the presence of an input voltage

preset 4.2 V charge voltage with 1.5% accuracy. The schematic of a TP4056, 3.7 V lithium battery charging and discharging module circuit is shown in Fig. 9.3 and its specifications are shown in Table 9.1.

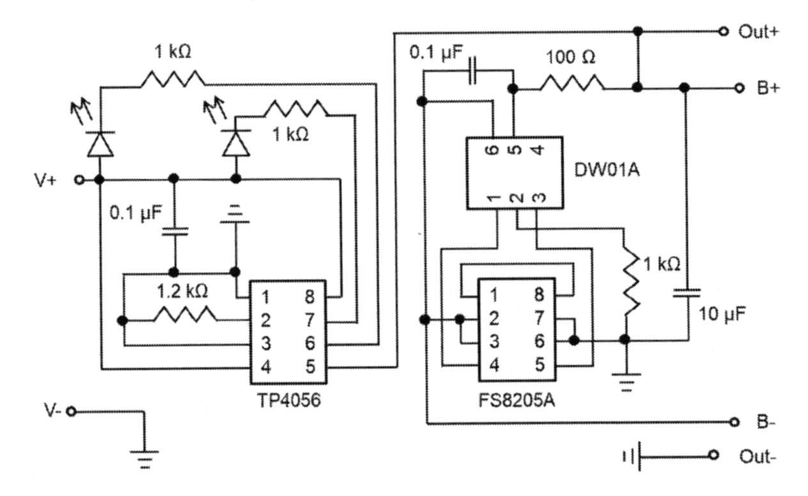

Fig. 9.3 TP4056 connection diagram

Table 9.1 TP4056 module specifications

Type	Charger, Protection Board
Module	TP4056
Battery Type	Li-Ion
Battery Voltage	3.7 V to 4.2 V
Maximum Input Current	1 Amp
Input Voltage	4–8 Volt
Output Voltage	4.2 Volt
Connector Type	USB Mini
Charging Method	Line Charging
Charging Precision	1.5%
Package	SMD

Methodology

Solar Tracking using LDR Sensor

The prime task, solar tracking that is tracking the sun's direction, is done by LDRs, which are set up at two ends of the solar panel, east and west end points. In the dark, an LDR can have a high resistance like a few mega ohms (MΩ). In the light, it can have very low resistance, like a few hundred ohms. When a constant voltage is applied across the LDR and the intensity of light is increased, the current starts increasing. The resistivity of the LDRs will vary according to the intensity of the light falling on them, when light is incident on the LDR of one side (east) the LDR of other side (west) is in the shade region or low light region. During the afternoon, the sun goes towards the west gradually. The intensity of the incident light on the LDR in the west will be high and the east LDR will be in the shade portion or low light portion. The servo will rotate the solar panel towards the LDR whose resistance will be low, which means towards the LDR on which light is falling, that way it will keep following the light. And if there is same amount of light falling on both the LDR, then servo will not rotate. The servo will try to move the solar panel in the position where both LDRs will have the same resistance which means at the position where the same amount of light will fall on both the resistors and if the resistance of one of the LDR will change, then it rotates towards the lower resistance LDR.

Charging of Lithium-Ion Battery

Inside the module, IC TP4056A, DW01A and P-type MOSFET FS8205A are used. The charging process is controlled by the TP4056A liner voltage IC, charge current is set by connecting a 1.2 KΩ resistor from RPROG (Pin: 2) to GND. The DW01A battery protection IC is designed to protect lithium-ion/polymer batteries from damage or degrading lifetime due to overcharge and over-discharge. No blocking diode is required due to the FS8205A internal PMOSFET architecture and has prevented negative charge current circuit. Lithium-ion and Lithium-polymer cells might explode if shorted, overcharged, charged or discharged with too high currents, thus TP4056 module is used which is a combination of charger and protection for single cell 3.7 V lithium batteries. Thereby, this module will monitor the voltage level of the lithium battery during charging and discharging. The recommended operating voltage for the TP4056 module is 4-8 V, 1 A DC supply. When the charger is turned ON, the red LED will go high, indicating that the battery is being charged. Once the module charges the battery completely, it will automatically stop charging and the red LED will turn OFF and the blue LED will turn ON to indicate the completion. The circuit diagram of the proposed system is shown in Fig. 9.4.

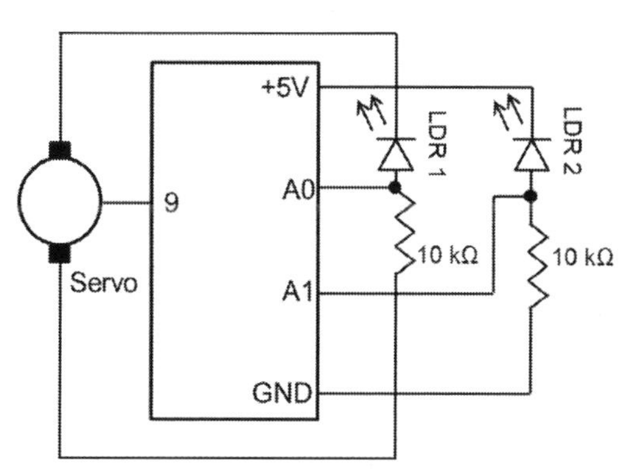

Fig. 9.4 Circuit diagram

Developed Prototype

The hardware of the proposed system consists of an Arduino-NANO board, a solar panel, one servo motor, two LDRs, a charging module and a lithium-ion battery. The Arduino solar tracker with servo motor is employed by means of Arduino ATmega328p microcontroller. The essential software is developed via Arduino-NANO which is shown in Fig. 9.5.

Fig. 9.5 Developed prototype

Conclusion

The overall prototype is working satisfactorily, and the energy is harvested with higher efficiency. An automated solar tracking system has been designed which will increase the efficiency of the solar panel system available.

The tracking system successfully sketched the light source even if it is a small torchlight, in a dark room, or it is the sunlight's rays. The Arduino solar tracker with a servo motor is employed by means of Arduino ATmega328p microcontroller. The essential software is developed via Arduino-NANO. The cost and reliability of this solar tracker create it suitable for rural usage. The purpose of renewable energy this paper offered a new and advanced idea to help people.

References

1. Ali Al-Mohamad, Efficiency improvements of photo-voltaic panels using a Sun-tracking system,Applied Energy, Volume 79, Issue 3,2004, Pages 345–354, ISSN 0306-2619, https://doi.org/10.1016/j.apenergy.2003.12.004.
2. Juang, J. and Radharamanan, R. (2014) "Design of a solar tracking system for renewable energy," Proceedings of the 2014 Zone 1 Conference of the American Society for Engineering Education, pp. 1–8, doi: 10.1109/ASEEZone1.2014.6820643.
3. Kassem, A and Hamad, M. (2011)." Microcontroller-based multi-function solar tracking system," in Proc. IEEE International on Systems Conference, pp. 13–16, 2011.
4. Racharla, S and Rajan, K. (2017): Solar tracking system—a review, International Journal of Sustainable Engineering.

5. Saldaña-GonzÂ´lez, I. Ayaquica-Martínez, R. Lemuz and M. Romero, "A Calibration Algorithm for Solar Tracking System," in 2013 12th Mexican International Conference on Artificial Intelligence, Puebla, Mexico, 2011 pp. 139–143.doi: 10.1109/MICAI.2011.22.
6. Soteris A. Kalogirou,Design and construction of a one-axis sun-tracking system, Solar Energy,Volume 57, Issue 6, 1996, Pages 465–469, ISSN 0038-092X, https://doi.org/10.1016/S0038-092X(96)00135-1.
7. Yang, C and Cheng, T. (2016) "A low consuming power solar heat collection tracker for solar thermal applications," in 2016 3rd International Conference on Green Technology and Sustainable Development (GTSD), Kaohsiung, Taiwan, pp. 49–50.doi: 10.1109/GTSD.2016.21.

CHAPTER 10

Implementation of Post-COVID Automated UV Sanitisation System in a Classroom—A Case Study

Shubhankar Sardar, Disha Mukerjee, Arnab Ganguly*, Srijan Banerjee and Amartya Roy

Electrical Engineering Department, Gargi Memorial Institute of Technology, Baruipur, Kolkata, West Bengal, India

Abstract

In this paper, an automated UV sanitisation system is developed for a classroom, which also can be applicable to any office room. The sanitisation process has gained its importance significantly after the worldwide effect of coronavirus. In the present scenario of technological advancement, it is very much obvious to develop a smart system, which will work in an automated way to overcome the needs of human supervision and also to achieve good efficiency in working a dynamic system. Classrooms of any institute generally are equipped with a lot of students and faculty members throughout the day. So, periodic sanitisation is very much needed for safety and security from the adverse effect of the virus. In this design, a UV-C LED lamp is used for the sanitization process, which is fitted inside the classroom. UV light is widely used as sterilising and disinfectant agent for rooms and surfaces, as well as it is used for air filtration purposes. Here, infrared (IR) sensors are used to detect the human incoming and outgoing inside the room, and a counting system is developed to display the total count inside the room in a display fitted outside the classroom. As UV ray causes an adverse effect on any living organisms, this is must to initiate the sanitisation process only after the class becomes empty. So, after the end of a class, the entire students leave the class to make it vacant and the sanitisation process starts. The entire system is controlled with Arduino UNO microcontroller board, based on ATmega328P microcontroller to ensure that the UV sanitisation will not start until the class becomes fully vacant.

Keywords: UV sanitisation, UV-C LED lamp, infrared sensor, microcontroller

Introduction

UV Sanitisation system is a process that is helpful to implement a chemical-free disinfection system for air, water and other surfaces. Sanitisation means to decrease or eliminate bacteria and other pathogenic agents from the surface. The use of UV light for sterilising areas and

**Correspondence Author:* arnab.ee_gmit@jisgroup.org

reducing the transmission of pathogens was first proposed in 1878 by Arthur Downes and Thomas P. Blunt (Parise, F. Lamonaca and D.L. Carnì., 2015). The first recorded use of UV light as a disinfection agent was reported in Marseilles, France, in 1910 (Ranjit Singh and others, 2009), where this method was used to sterilise drinking water in a prototype plant. Today, UV light is widely used as a sterilisation agent for rooms and surfaces (Dheeraj Bansal and others, 2017).

To make the entire process of UV sanitisation efficient and free from human supervision, the proposed design is a development of a smart and automated system. UV-C lamp is used for the sanitisation purpose as the emitted UV ray acts as disinfectant to make the surface free from bacteria and other gems. This concept of sanitisation is very much useful for the classroom of educational institute as well as in any office buildings, where a good amount of people are present throughout the day. In post-COVID-19 time, it is essential to keep the building premises clean and free from bacteria and other germs. Here, the scheme is developed for a classroom, where a good number of students are assembled for classes. So, to make the classroom bacteria-free, periodic sanitisation is required. As UV ray is harmful to human health and any living organisms, the foremost condition of this design scheme is to make the room mandatorily vacant before starting the sanitisation process. To make the entire process efficient, a sensor-based automated design scheme has been proposed. Infrared (IR) sensors are used to detect the incoming and outgoing of human beings (Amrita Saha and others, 2018). Here, the IR sensors have been used for detecting the incoming and outgoing human beings in a particular classroom. A LCD display is fitted outside the classroom to display the real-time count of people inside the classroom. After the conclusion of a particular class, students and teacher leave the classroom to make it vacant and the sanitisation process starts. The entire process is controlled by Arduino UNO microcontroller board, based on ATmega328P microcontroller.

Proposed Design Scheme

In this paper, we have proposed a design scheme for automated UV Sanitisation system for a classroom. UV ray is electromagnetic radiation that has a wavelength longer than X-rays but shorter than visible light (Amartya Roy and others, 2020). UV light is categorised into different wavelengths, including UV-C, which is short-wavelength UV light that is often referred to as "germicidal" UV. Between the wavelengths of 200 nanometre and 300 nanometre (nm), UV-C operates. As the use of UV light has become increasingly popular for disinfection purposes, ultraviolet germicidal irradiation (UVGI) systems are now much cheaper. UVGI systems can be installed in enclosed spaces where the constant flow of air or water ensures high levels of exposure. The effectiveness of this system relies on many factors, including the quality and type of equipment uses, the duration of exposure, wavelength and intensity of UV, the presence of protective particles and the microorganism's ability to withstand UV light. The effectiveness of UVGI systems can also be determined by something as simple as dust on the bulb; therefore, equipment must be regularly cleaned and replaced to ensure its efficacy for sterilisation procedures.

In this proposed scheme, the design is developed, including two IR sensors, aLCD display and Arduino UNO microcontroller board, based on ATmega328P microcontroller to make it an efficient automated system. IR sensors are used to detect the incoming and outgoing of human beings inside the room and based on that, the LCD display, fitted outside the classroom, will display the actual count of human beings inside the room. As UV ray is harmful to human health, this checking is very much essential. In this design scheme, the UV Sanitisation process will only start if and only if the human count inside the classroom is zero. This entire process is controlled by the above-mentioned microcontroller system. The microcontroller is used to command and process the sensor outcome data. A real-time clock (RTC) is also implemented in this design scheme for time-based monitoring of the entire process, given in Fig. 10.1.

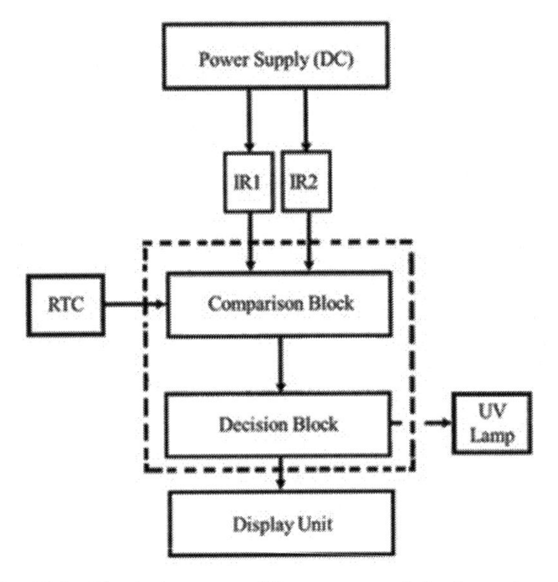

Fig. 10.1 Block diagram of the proposed design scheme

Hardware Implementation

Here, two IR sensors are used sequentially for detecting the human presence inside the room. The fitted LCD display shows the actual count of human being inside the room. When a person enters a room the IR1 sensor is first activated followed by the activation of IR2 sensor. The sequence of the activation of IR sensors will be just the opposite when a person exits the classroom. IR2 sensor is activated first followed by the activation of IR1 sensor. The counting system was implemented for real-time checking of human count inside the room. When the entire people inside the classroom leave the room, the count becomes zero and the UV-C lamp glows to start the process of UV sanitising. The microcontroller controls the entire process by receiving the data from IR sensors. The hardware model is given in Fig. 10.2.

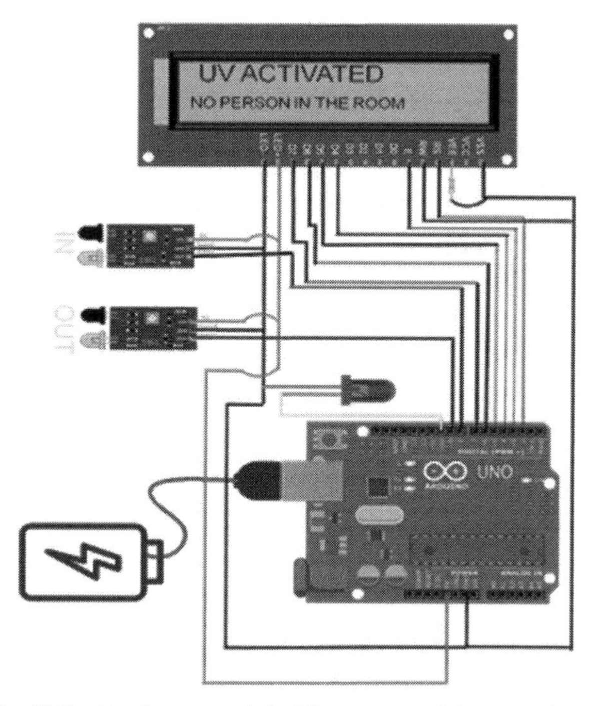

Fig. 10.2 Hardware model of the proposed design scheme

Observation

This design scheme is developed for the room where a single door entry and exit system is present. Here, the two IR sensors are placed in close gap to maintain the sequential operation of the sensor properly. When IR1 sensor is activated, first followed by the activation of IR2 sensor, it depicts the condition of a person who goes inside the room, which means the person count inside the room is increased by one. For the opposite sequence, that is, IR2 sensor activates first followed by the activation of IR1 sensor, it depicts the condition of a person who comes outside of the room, which means the person count inside the room is decreased by one. LCD display shows the count in real-time. The detailed observation table of IR sensor activation is given in Table 10.1. When person count becomes zero inside the classroom, the

Table 10.1 Observation table of IR sensor activation

Sequential Activation of IR Sensors			
Sl No	**Sequence of Activation for IR1 Sensor**	**Sequence of Activation for IR2 Sensor**	**Count in LCD Display**
1	1st	2nd	Count = Count + 1
2	2nd	1st	Count = Count – 1

UV-C lamp becomes ON condition and the sanitisation process starts. A detailed observation of UV Sanitisation process is given in Table 10.2. The RTC system is employed to regulate the time for which the sanitisation process is going on.

Table 10.2 Observation table of UV sanitisation

Status of UV Sanitisation		
Sl No	**Count in LCD Display**	**Status of UV Sanitisation Process**
1	Count > 0	OFF/Inactive Condition
2	Count = 0	ON/Active Condition

Conclusion

This design scheme is developed for an automated UV sanitisation system, which is very much essential nowadays to keep the campus free from bacterial attacks. Periodic sanitisation is required throughout the day in any office or institution campus for safety. But with sanitisation, a few other measures and procedures must be followed to keep the campus fully secured. In this design scheme, auto-temperature check for any person before entering a room can be included, as the temperature is a very important parameter for the probable detection of a virus affected. Above a certain body temperature, the persons will not be allowed to enter the room. It obviously will make the overall system more robust in terms of health protection. Also, a real-time attendance monitoring system may be developed for the classroom of any institution, which can give a real-time status of present and absent.

References

1. Bansal, D., Kumar, M., Dutta ,N., Rahman, H.,Singha,A., Alam,M.A. and Roy,A .(2017). Integration of Artificial & Day Light to Reduce Energy Demand, National Student Congress on Illumination, Kolkata, India.
2. Parise, F. Lamonaca, D.L. Carnì. (2015). Interior lighting control system: A practical case using daylight harvesting control strategy, IEEE 15th International Conference on Environment and Electrical Engineering (EEEIC), Rome, pp. 719–724.
3. Saha, A., Paul, D., Jana, R., Datta, S., Roy, A. (2018). Design andDevelopment of Energy Efficient Lighting Controller, International Journal of Computer Sciences and Engineering, pp. 297–300
4. Singh,R., Ahamed ,F.T., Ibrahim, Sani Irwan MD Salim, Wong Yan Chiew. (2009) Door Sensors for Automatic Light Switching System, Third UKSim European Symposium on Computer Modelling and Simulation, Athens, Greece, pp. 574–578.
5. Roy, A., Gangly,A., Roy, B. (2020)Real-time condition monitoring system for LED luminaires–An approach. IEEE Applied Signal Processing Conference, pp. 217–221.

CHAPTER 11

Energy Efficiency and Environmental Sustainability: A Multi-Criteria based Comparison of BRICS and G7 Countries

Sanjib Biswas*

Decision Sciences and Operations Management Area; Calcutta Business School (A Unit of Shikshayatan Foundation), Bishnupur, South 24 Paraganas, West Bengal, India

Debanshee Datta[1]

Business Analytics Area; Calcutta Business School (A Unit of Shikshayatan Foundation), Bishnupur, South 24 Paraganas, West Bengal, India

Samarjit Kar[2]

Department of Mathematics, National Institute of Technology, Durgapur, West Bengal, India

Abstract

Global warming and climate change have become urgent and important issues for governments, policymakers and corporates across the globe over the last few decades. Environmental sustainability (ES) is one of the prime concerns of the sustainable development goals (SDG) of the United Nation. The objective of ES is to protect the global ecosystem for safeguarding the lives and livelihoods on the planet. In this regard, energy efficiency (EE) has emerged as one of the key enablers for achieving ES and thereby contributing towards fulfilling SDGs. In this paper, we present a multi-criteria decision making (MCDM) framework to compare BRICS (developing nations) and G7 countries (developed nations) on the basis of their energy efficiency vis-à-vis ES. We consider six attributes such as energy security (ENS), energy equity (ENE), total renewable energy (TREN), ES, climate change performance index (CCPI) and CO_2 total emission (CTE) for comparative analysis. We modify the recently developed attribute weighting algorithm logarithmic percentage change-driven objective weighting (LOPCOW) and use the modified framework for comparison of BRICS and G7 countries. LOPCOW provides a number of benefits, such as uniformity in the criteria weights, efficient working with larger criteria set and withstanding negative values in the decision matrix. However, LOPCOW uses linear max-min normalisation. In this paper, we modify the LOPCOW algorithm by using a combination of additive and ratio-based normalisations with more flexibility in the aggregation process. Hence, the modified LOPCOW method provides more robustness.

We observe that ENS and ENE hold more significance based on objective information. In addition, the result shows that Brazil secures the top position along with France and Italy in

**Corresponding author:* sanjibb@acm.org

[1]debansheedatta2204@gmail.com; [2]dr.samarjitkar@gmail.com

the top bracket, while the Asian giants, such as China and India, remain in the lower group. We compare the results of the modified logarithmic percentage change-driven objective weighting (LOPCOW) method with another recently developed and popular algorithm (using objective information) such as preference selection index (PSI) for validation purposes and subsequently perform the sensitivity analysis. We notice that our method provides reasonably accurate and stable results.

Keywords: energy efficiency, environmental sustainability, logarithmic percentage change-driven objective weighting (LOPCOW), multi-criteria decision making (MCDM), normalisation, sustainable development goals (SDG)

Introduction

"Air pollution increases risk of premature death by 20%" (PTI, June 22, 2022), "Cement carbon dioxide emissions quietly double in 20 years" (TOI, June 22, 2022), "The amount of planet-warming carbon dioxide in the atmosphere broke a record in May, continuing its relentless climb….. It is now 50% higher than the preindustrial average…." (TOI, Jun 04, 2022), "Climate, big agriculture slashing insect populations "by half" (TOI, April 20, 2022), "Containing warming to 2 degreesC would require actions that actions that limit global economic growth by 1.3% to 2.7% by 2050" (TOI, April 05, 2022). From some of the above-mentioned headlines of the global news published recently in a leading daily in India, it is amply evident that environmental degradation is a major issue in the world. Climate change and environmental protection have become a concern for lives and livelihood over the last several decades alongside a surge of urbanisation, sociocultural transformation, technological development and rapid industrialisation (Wen et al., 2021). Reduction in the consumption of energy and water, optimum utilisation of natural resources and waste management have increasingly turned out as a necessity for the survival of the lives and livelihood on this planet (Woods, 2021). For justified reasons, environmental sustainability (ES) has been an area of paramount importance to the leaders of the nations, policymakers, corporate leaders and all decision-makers across the world. The movement for climate protection has been necessitated after the second world war.

In simpler words, ES can be viewed as an objective to optimally utilise the resources for meeting the needs of the present generation while ensuring the availability for future generations and maintaining the ecosystem (Morelli, 2011). The concept of ES goes beyond the conventional economic sustainability focus. Goodland (1995) portrayed the linkage of ES with economic sustainability. Economic sustainability requires optimum utilisation of natural resources as physical inputs for producing final goods, whereas ES ensures the preservation of environmental eco-system for the existence of mankind and industries. Hence, corporate sustainability significantly depends on ES. The organisations' active engagement in promoting ES through eco-friendly green marketing not only limns a favourable image of the firms in the mind of the consumers and helps the organisations in attracting a large number of potential customers (Nyame-Asiamah and Kawalek, 2021). Further, there is a need to reorient and redefine the entrepreneurial value for sustainable entrepreneurship considering ES as one of the key enablers (Gregori and Holzmann, 2022).

Therefore, embracing the practices to foster ES vis-à-vis the sustainable development goals (SDGs) of the United Nation is the call of the hour and is a recognised global need today. The counties need to reduce their ecological footprints through robust environmental regulation (Murshed et al., 2021). Ahmad et al. (2021) in their research have reflected on the urgent requirement of symbiosis of economic growth and ES and building a holistic and responsive ecosystem for eco-friendly innovation for the G7 countries.

In this context, the present paper aims at providing a multi-perspective analysis using MCDM models. The current study focuses on energy efficiency (EE), use of renewable energy, environmental performance and sustainability to compare a mix of developed and developing countries. The countries belonging to the groups of G7 and BRICS are considered as alternatives in this paper. Our best possible search reveals the scantiness of research related to multi-criteria-based cross-country comparisons. We enquire about the following research questions in this regard:

RQ1. To what extent do the countries differ from each other based on EE and ES criteria?

RQ2. Are the G7 countries superior in energy and environmental performance as compared with the BRICS nations?

RQ3. How can reliable and stable MCDM model be developed for carrying out a cross-country comparison on the basis of ES?

In order to carry out the comparative analysis, the present paper proposes a new modification of a recently developed MCDM algorithm, such as LOPCOW. The original LOPCOW method has been modified with respect to the normalisation scheme. Normalisation is an important aspect of any MCDM model. Normalisation converts the different units of measurement of various criteria into dimensionless numbers for providing a uniform platform for comparison. The efficacy of the MCDM models largely depends on four essential features of the normalisation, such as scaling, symmetry, ability to handle negative values in the decision matrix and susceptibility to the rank reversal issue (Jahan and Edwards, 2015; Vafaei et al., 2016; Jahan, 2018). The extant literature (for example, Liao and Wu, 2020; Aytekin, 2021) has argued for using a hybrid or multiple normalisation scheme instead of a single method for reducing the variations in the normalised values due to varying nature of the criteria and scale effect.

In place of the linear max-min normalisation scheme, the modified version of LOPCOW applies a mixed normalisation process that combines three types of normalisation, such as linear sum-based, linear ratio based and linear max-min type. Wen et al. (2020) have considered mixed aggregation normalisation in their original work. We adopt the normalisation scheme (MNS) from the work of Wen et al. (2020). Accordingly, three types of normalisation schemes are used in conjunction with the modified LOPCOW method, and finally, the normalised decision matrix is formulated after aggregating all schemes. The present paper has two major contributions to the expanding literature as under

- A multi-criterion based cross-country comparative assessment framework for EE and environmental performance.
- It proposes a modified version of LOPCOW method with a mixed normalisation scheme.

The rest of the paper is structured as follows. Section 2 provides a review of some of the recent work. In section 3, data description and research framework are presented. Section 4 exhibits the findings while in section 5, the outcome of the sensitivity analysis and validation test are included. Section 6 concludes the paper and mentions some of the future scope.

Literature Review

In this section, a brief overview of some of the recent past research work is presented. There has been a growing strand of literature in the stated field. For example, the impact of urban density on commuting behaviour and the consequences of CO_2 emissions was studied by Fabio et al. (2008). The paper intended to check the limit to which the urban form affects individual travel behaviour and, consequently, the transport-induced level of CO_2 emissions. Taylor et al. (2010) outlined the IEA indicator methodology and presents examples of how disaggregated indicators drive and restrain energy demand at the end-user level activity. The authors (Selvakkumaran and Limmeechokchai, 2013) felt the need to establish the importance of energy security (ENS) as an outcome of EE in the power sector. In this paper, ENS has been measured along three main themes oil security, gas security and sustainability. Comparing the policy measures revealed a powerful influence of financial subsidies and energy labels in Girod et al. (2017). Verma et al. (2018) compared the EE initiatives of the three top-performing members of OECD countries vis-a-vis renewable energy management, such as Iceland, Norway and New Zealand.

Roman et al. (2021) investigated circular efficiency using data envelopment analysis and slack-based models. The current research (Rabia et al., 2021) examined the heterogeneous impacts of EE, renewable energy consumption (REC), and other factors on EG. Perry (2021) used the logarithmic mean divisia index (LMDI) method for identifying the driving factors in wind energy consumption for developing and developed countries that are significant consumers of wind energy.

The researchers (Bo et al., 2021) also investigated the impact of globalization, financial development and energy utilisation on ES in the Gulf Cooperation Council (GCC) countries. Godil et al. (2021) examined the role of economic growth, technology innovation and renewable energy in reducing China's transport sector's CO_2 emissions. Gatto and Drago (2021) exploited a statistical approach to measure energy policy effectiveness. A statistical comparison of the different results is relevant to ensure robustness implications. Wei-Zheng et al. (2021) studied the general and heterogeneous long-run equilibrium relationships, short-run dynamic relationships, impact mechanism and lag effect between urbanisation and three carbon emission dimensions in OECD high-income countries. Dell'Anna (2021) investigated the potential of investments in Italy's energy sector through input-output analysis. Irfan et al. (2021) delved into the impact of EE and renewable energy (RE) on carbon emissions. The empirical investigation was carried out using the panel autoregressive distributed lag model. Popkova and Sergi (2021) elaborated on multiple actors' diverging interests in achieving and advancing EE. The paper studies the reasonable likelihood of the results in EE with insignificant mid-term changes. Tachega et al. (2021) assessed the EE, energy productivity improvement and the determinants of EE of 14 African oil-producing countries. The EE has been evaluated using the DEA-SMB approach. Brockway et al. (2021) showed the relationship between energy consumption and GDP. This study employed energy, population growth and financial development as moderator variables to examine the impact of the energy trilemma. Rusydiana (2021) measured the effectiveness of energy access based on the indicators of the seventh SDGs in 50 countries that are members of the OIC.

The study of Baloch et al. (2022) investigated the role of energy innovation in combating GHGs emissions by taking the environmental Kuznets curve for BRICS economies. Aguir Bargaoui (2022) explored the impact of adopting EE and RE strategies on the CO_2 emissions of 36 OECD countries from 2000–2019. The study of Luisa and Rosa (2022) aimed to access modern energy sources for achieving the decarbonisation of the energy system and reducing pollution of the soil, water and air. Sudharshan et al. (2022) investigated the role of technology on energy demand and efficiency for 28 OECD economies. Zhang et al. (2022) investigated how a set of economic factors determine energy consumption in organisations.

Data and Methodology

In this paper, we aim to carry out a comparative analysis of the BRICS and G7 enlisted countries that are our decision-making units (DMU) based on their relative performance in energy management and ES criteria. Figure 11.1 depicts the flow of the steps followed in the current study. A modified version of the original LOPCOW method has been used for comparative analysis purpose.

Sample

The DMUs are given in Table 11.1. The study has been conducted on the basis of the latest reports (2021–22).

Table 11.1 Description of the DMUs

S/L	Country	Group		S/L	Country	Group
A_1	Brazil	BRICS		A_7	USA	G7
A_2	Russia	BRICS		A_8	France	G7
A_3	India	BRICS		A_9	Canada	G7
A_4	China	BRICS		A_{10}	Germany	G7
A_5	South Africa	BRICS		A_{11}	Italy	G7
A_6	UK	G7		A_{12}	Japan	G7

Criteria Description

The present paper uses six (6) criteria for comparing the DMUs (see Table 11.2). The criteria are selected in line with the discussions made in the extant literature. The data source and unit of measurement (UOM) are also mentioned in Table 11.2. Accordingly, the decision matrix is given in the Table 11.3.

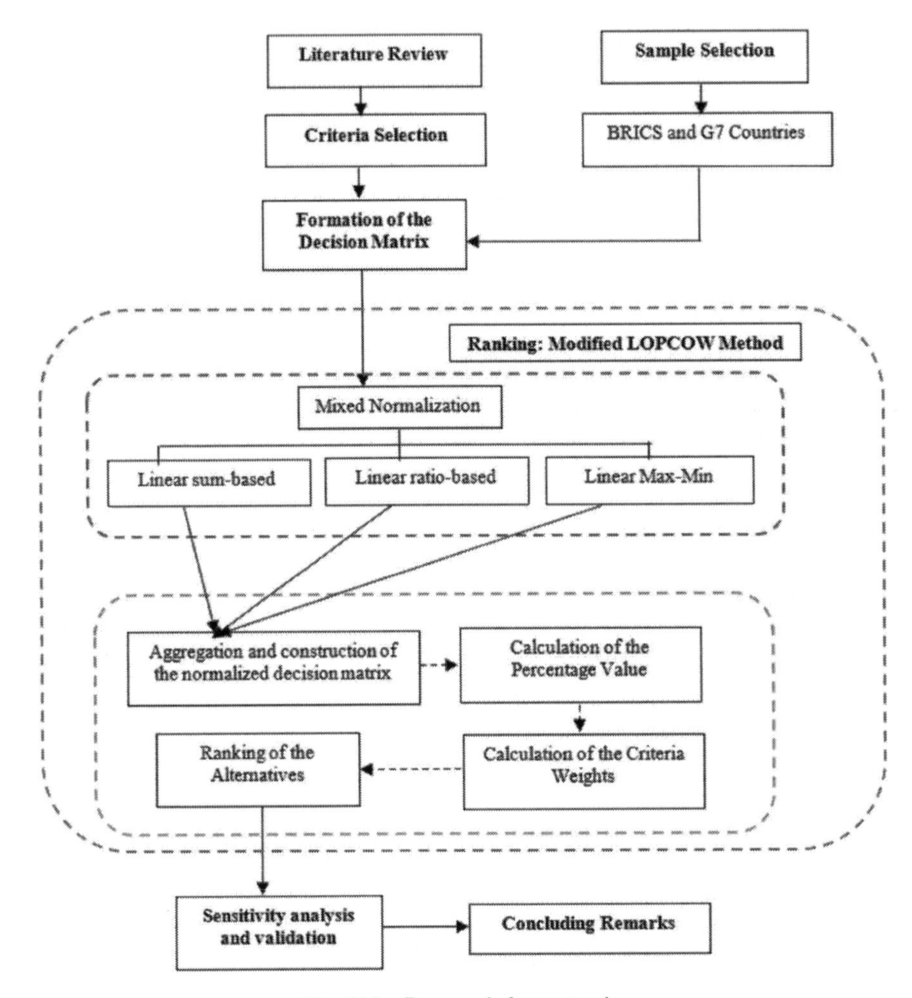

Fig. 11.1 Research framework

Table 11.2 Description of the criteria

S/L	Criteria	Effect	Data Source
C_1	Energy Security 2021	(+)	Marti and Puertas (2022); World Energy Council
C_2	Energy Equity 2021	(+)	Marti and Puertas (2022); World Energy Council
C_3	Total Renewable Energy (installed capacity in GW)	(+)	International Renewable Energy Agency (IRENA)
C_4	Environmental Sustainability 2021	(+)	Marti and Puertas (2022); World Energy Council
C_5	Climate Change Performance Index (CCPI) 2021	(+)	Germanwatch e.V
C_6	CO_2 Total Emission (Mton)	(-)	https://worldpopulationreview.com

Table 11.3 Decision matrix

Criteria	C_1	C_2	C_3	C_4	C_5	C_6
Country	(+)	(+)	(+)	(+)	(+)	(-)
A_1	72.6	77.8	159.943	83.4	54.86	478.15
A_2	69.1	97	56.217	62.5	34.73	1792.02
A_3	59.7	60	147.122	49.1	69.2	2597.36
A_4	64.5	80.8	1020.234	57.7	52.2	11535.2
A_5	52	77.4	10.193	58.6	51.13	494.86
A_6	68.4	96.3	50.293	82.5	73.09	364.91
A_7	72.2	96.7	325.391	71.6	37.39	5107.26
A_8	68.3	95.1	59.546	85.5	61.01	314.74
A_9	77.1	95.6	102.932	73.4	26.03	584.85
A_{10}	72	95	138.151	77.8	63.53	702.6
A_{11}	66.6	95.8	56.987	81.5	55.39	331.56
A_{12}	59.4	94.3	111.86	74.5	48.53	1153.72

Modified LOPCOW Method for Ranking

The original LOPCOW method has been developed to calculate the criteria weights based on objective information (Ecer and Pamucar, 2022). It provides the following advantages:

- Unlike the previous methods, LOPCOW distributes the criteria weights in a comparatively even manner.
- LOPCOW is capable to work fine with the negative performance values of the alternatives.
- A large number of criteria and alternatives can efficiently be dealt with by the LOPCOW method.

Let $X = [x_{ij}]_{m \times n}$ be the decision matrix where m is the number of alternatives (i.e. companies under comparison; $m = 30$) and r is the number of criteria (in our case, $n = 8$). In this paper, we use the modified LOPCOW method for criteria weight calculation as well as performance-based ranking of the alternatives (i.e. the countries). What follows are the computational steps of the modified version of the original LOPCOW method (Ecer and Pamucar, 2022).

Step 1. Normalisation of the decision matrix

Let, the normalised decision matrix be represented as $R = [r_{ij}]_{m \times n}$

As mentioned in section 1, in this paper, a mixed normalisation consisting of three schemes has been used as described below.

Normalisation scheme 1 (Linear sum-based)

$$r_{ij}^{1} = \frac{x_{ij}}{\sum_{i=1}^{m} x_{ij}} \quad \text{(when } j \in j^{+}\text{, desired effect: maximising)} \tag{1}$$

$$r_{ij}^{1} = \frac{\left(\frac{1}{x_{ij}}\right)}{\sum_{i=1}^{m} \frac{1}{x_{ij}}} \text{ (when } j \in j^{-}\text{, desired effect: minimising)} \tag{2}$$

Normalisation scheme 2 (Linear ratio-based)

$$r_{ij}^{2} = \frac{x_{ij}}{\max\limits_{i}(x_{ij})} \text{ (when } j \in j^{+}\text{, desired effect: maximising)} \tag{3}$$

$$r_{ij}^{2} = \frac{\min\limits_{i}(x_{ij})}{x_{ij}} \text{ (when } j \in j^{-}\text{, desired effect: minimising)} \tag{4}$$

Normalisation scheme 3 (Linear max-min type)

$$r_{ij}^{3} = \frac{x_{ij} - x_{\min}^{j}}{x_{\max}^{j} - x_{\min}^{j}} \text{ (when } j \in j^{+}\text{, desired effect: maximising)} \tag{5}$$

$$r_{ij}^{3} = \frac{x_{\max}^{j} - x_{ij}}{x_{\max}^{j} - x_{\min}^{j}} \text{ (when } j \in j^{-}\text{, desired effect: minimizing)} \tag{6}$$

The aggregation is done as described below:

$$r_{ij} = \lambda r_{ij}^{1} + \xi r_{ij}^{2} + (1 - \lambda - \xi) r_{ij}^{3} \tag{7}$$
$$0 \leq \lambda, \xi \leq 1; \lambda + \xi \leq 1$$

Here, λ and ξ are the adjustment coefficients that bring flexibility to the decision-makers in terms of their emphasis in the decision-making process. In what follows are the significance of the values of λ and ξ:

(a) Higher value of λ—If the decision maker wants to focus on specific alternatives out of all available options.
(b) Higher value of ξ—If the best performance of the alternatives is preferred
(c) Smaller value of λ and ξ—If the decision maker prefers to consider both best and worst possible performance while highlighting the former one.

In this way, the modified LOPCOW method provides wider options to the decision-makers depending upon the context and premise of the decision-making.

Step 2. Derive the percentage value (PV) for the criteria

The PV for each criterion is given by the natural log of the mean square value as a proportion of the standard deviation expressed in percentage. This step helps to reduce the uneven distribution of the weights. Accordingly, PV is calculated as:

$$P_j = \left| \ln \left(\frac{\sqrt{\frac{\sum_{i=1}^{m} r_{ij}^2}{m}}}{\sigma} \right) .100 \right| \quad (8)$$

σ denotes the standard deviation.

Step 3. Computation of criteria weights

The weight for the jth criterion is given by

$$w_j = \frac{P_{ij}}{\sum_{j=1}^{n} P_{ij}} \quad (9)$$

where $\sum_{j=1}^{n} w_j = 1$ (i.e. sum of the weights of all criteria = 1)

Step 4. Ranking of the alternatives

The alternatives are ranked as per their performance scores. Higher the score, the more preferred is the corresponding alternative. The performance score is expressed as:

$$S_i = \sum_{j=1}^{n} r_{ij} w_j \quad (10)$$

In the present paper, for calculation purposes, we have used MS Office (2016) and SPSS (version 25) software tools on a computer with Intel(R) Core(TM) i3-1005G1 CPU @ 1.20GHz 1.19 GHz, 8GB RAM.

Results

In this section, we briefly present the findings of the data analysis. The results of the normalisation of the decision matrix based on the three schemes as given in expressions (1) to (6) of the modified LOPCOW method are given in Appendix A (supplementary file attached to this paper). It may be noted that the normalised performance values of the alternatives under all schemes differ notably. We then utilise the aggregation (see expression (7)) to construct the final normalised decision matrix (see Table 11.4). For aggregation purposes, we set the values of λ and ξ as 0.3333 (i.e. 1/3) to give equal priority to all three normalisation schemes.

Table 11.4 Final normalised decision matrix (after aggregation)

Criteria	C_1	C_2	C_3	C_4	C_5	C_6
Country	(+)	(+)	(+)	(+)	(+)	(–)
A_1	0.6176	0.4521	0.1255	0.6716	0.4836	0.5860
A_2	0.5546	0.6971	0.0419	0.3907	0.2385	0.3582
A_3	0.3852	0.2250	0.1152	0.2105	0.6582	0.3129

Criteria	C_1	C_2	C_3	C_4	C_5	C_6
Country	(+)	(+)	(+)	(+)	(+)	(–)
A_4	0.4717	0.4904	0.8186	0.3261	0.4512	0.0107
A_5	0.2464	0.4470	0.0048	0.3382	0.4381	0.5769
A_6	0.5419	0.6882	0.0372	0.6595	0.7055	0.6693
A_7	0.6104	0.6933	0.2588	0.5130	0.2709	0.2151
A_8	0.5401	0.6729	0.0446	0.6999	0.5584	0.7246
A_9	0.6987	0.6793	0.0796	0.5372	0.1325	0.5359
A_{10}	0.6068	0.6716	0.1079	0.5964	0.5891	0.4971
A_{11}	0.5095	0.6818	0.0425	0.6461	0.4900	0.7043
A_{12}	0.3798	0.6627	0.0868	0.5520	0.4065	0.4152

Now we apply the definition given in the expression (8) to find out the percentage values of the alternatives subject to the influence of the criteria. Then we follow the expression (9) to calculate the criteria weights (see Table 11.5)

Table 11.5 Criteria weights

Criteria	C_1	C_2	C_3	C_4	C_5	C_6
Mean Square	0.2782	0.3671	0.0665	0.2853	0.2310	0.2604
Std. Dev.	0.1257	0.1506	0.2214	0.1596	0.1711	0.2144
PV	143.3971	139.1935	15.2645	120.7775	103.2849	86.7270
Wj	0.2356	0.2287	0.0251	0.1984	0.1697	0.1425

We note that $C_1 > C_2 > C_4 > C_5 > C_6 > C_3$ (based on the criteria weights)

It is reflected that energy security (C_1) obtains a higher weight than others while total renewable energy (TREN) (C_3) attains the least weight. We then use the criteria weights and the expression (10) to derive the performance-based ranking of the country (see Table 11.6). We observe that Brazil secures the top position along with France and Italy in the top bracket, while the Asian giants, such as China and India, remain in the lower group.

Table 11.6 Ranking of the alternatives

Country	Criteria						S_i	Rank
	C_1	C_2	C_3	C_4	C_5	C_6		
A_1	0.1455	0.1034	0.0031	0.1333	0.0821	0.0835	0.5509	1
A_2	0.0725	0.1111	0.0000	0.0303	0.0097	0.0183	0.2419	9
A_3	0.0350	0.0116	0.0003	0.0088	0.0735	0.0140	0.1431	12
A_4	0.0524	0.0550	0.0168	0.0211	0.0345	0.0000	0.1799	10
A_5	0.0143	0.0457	0.0000	0.0227	0.0326	0.0474	0.1627	11
A_6	0.0692	0.1083	0.0000	0.0863	0.0845	0.0638	0.4122	2

Country	Criteria						S_i	Rank
	C_1	C_2	C_3	C_4	C_5	C_6		
A_7	0.0878	0.1099	0.0017	0.0522	0.0124	0.0066	0.2707	7
A_8	0.0687	0.1035	0.0000	0.0972	0.0529	0.0748	0.3973	3
A_9	0.1150	0.1055	0.0002	0.0573	0.0030	0.0409	0.3219	6
A_{10}	0.0868	0.1032	0.0003	0.0706	0.0589	0.0352	0.3549	5
A_{11}	0.0612	0.1063	0.0000	0.0828	0.0407	0.0707	0.3618	4
A_{12}	0.0340	0.1004	0.0002	0.0605	0.0280	0.0246	0.2477	8

Validation and Sensitivity Analysis

MCDM models deal with complex real-life problems that are subject to the influence of a number of criteria conflicting with each other. Hence, given the problem, it is important to evaluate the problems using multiple lenses to confirm the validity of the results. The extant literature provides evidence of comparing the result obtained by using a specific MCDM method with the same derived with the use of other models (for instance, Gupta et al., 2019; Biswas, 2020; Pramanik et al., 2021; Biswas et al., 2022a). In this paper, we utilise the aggregated normalisation scheme (as used to modify the LOPCOW method) for performance-based ranking of the countries using Preference Selection Index (PSI) method (Maniya and Bhatt, 2010). The PSI method is used for deriving the overall preference of the criteria or attributes based on statistical concepts. In this method, there is no need to assess the relative importance of the attributes. Therefore, the PSI method is useful for complex problems dealing with extremely conflicting criteria (Biswas and Anand, 2020). Following the steps of the PSI algorithm (Maniya and Bhatt, 2010) with aggregated normalised decision matrix, we derive the ranking of the countries based on their PSI values and then a comparison with the modified LOPCOW is made (see Table 11.7). To examine the consistency between the results obtained by using both models, we calculate the coefficient of SRC. SRC has been used in a plethora of past work (e.g. Biswas et al., 2022b, 2022c) for comparing the MCDM results. Table 11.8 supports that there is a statistically significant and very high correlation between the modified LOPCOW and PSI in ranking the alternatives.

Table 11.7 Comparison of ranking by modified LOPCOW and PSI

Country	Ranking		Country	Ranking	
	Modified LOPCOW	PSI		Modified LOPCOW	PSI
A1	1	5	A7	7	7
A2	9	10	A8	3	2
A3	12	12	A9	6	6
A4	10	9	A10	5	3
A5	11	11	A11	4	4
A6	2	1	A12	8	8

Table 11.8 SRC test between modified LOPCOW and PSI

		Modified LOPCOW
Modified PSI	Spearman's rho	0.916***
	p-value	<.001

*Note. *p < .05, **p < .01, ***p < .001*

Next, it is also important for any MCDM-based analysis to test the stability of the outcome. Any change in the given conditions (e.g. change in the criteria weights) may result in variations in the outcome of a MCDM-based analysis (Biswas and Pamučar, 2021). Researchers, therefore, perform the sensitivity analysis which is carried out to examine the susceptibility of the MCDM method to changes in the criteria weights (Pamučar et al., 2021; 2022). The present paper varies the values of the adjustment coefficients (λ and ξ) used for aggregating three normalisation schemes in the modified LOPCOW framework. Accordingly, criteria weights get varied. Table 11.9 provides the different values of the adjustment coefficients under various experimental cases. As explained in the methodology section, the adjustment coefficients (λ and ξ) indicate the relative priorities assigned to the normalisation schemes. For the original case, the modified LOPCOW method treats equal priority (1/3) for schemes, i.e., λ(norm 1 importance) = ξ (norm 2 importance) = $(1 - \lambda - \xi)$ (norm 3 importance) = $\frac{1}{3}$ = 0.333. Now, for experimental cases 1 to 6, we set $\lambda = \frac{1}{2}, \frac{1}{4}, \frac{1}{5}, \frac{1}{6}, \frac{1}{7}, \frac{1}{8}$ while keeping ξ fixed at 0.333, which means that the importance of the normalisation schemes 1 and 3 get changed. Next, for experimental cases 7 to 12, we do the opposite, that is varying the values ξ while keeping λ fixed at 0.333. Therefore, the importance of normalisation schemes 2 and 3 get changed. Finally, for cases 13 to 15 either we increase λ and ξ decrease and/or decrease ξ while increasing λ. Hence, the relative importance of all normalisation schemes get varied. Accordingly, we record the calculated criteria weights (see Appendix B) and ranking orders of the alternatives (Table 11.10).

Table 11.9 Experimental cases for sensitivity analysis (with varying λ and ξ values)

	Priorities				Priorities		
	Norm 1	Norm 2	Norm 3		Norm 1	Norm 2	Norm 3
Cases	λ	ξ	$(1 - \lambda - \xi)$	Cases	λ	ξ	$(1 - \lambda - \xi)$
Original	0.333	0.333	0.333	Case 8	0.333	0.250	0.417
Case 1	0.500	0.333	0.167	Case 9	0.333	0.200	0.467
Case 2	0.250	0.333	0.417	Case 10	0.333	0.167	0.500
Case 3	0.200	0.333	0.467	Case 11	0.333	0.143	0.524
Case 4	0.167	0.333	0.500	Case 12	0.333	0.125	0.542
Case 5	0.143	0.333	0.524	Case 13	0.167	0.500	0.333
Case 6	0.125	0.333	0.542	Case 14	0.500	0.250	0.250
Case 7	0.333	0.500	0.167	Case 15	0.143	0.125	0.732

Table 11.10 Result of the sensitivity analysis

Country	Original	Case 1	Case 2	Case 3	Case 4	Case 5	Case 6	Case 7
A1	1	1	1	1	1	1	1	1
A2	9	9	9	9	9	9	9	9
A3	12	12	12	12	12	12	12	12
A4	10	10	11	11	11	11	11	10
A5	11	11	10	10	10	10	10	11
A6	2	2	2	2	2	2	2	2
A7	7	7	7	7	7	7	7	7
A8	3	3	3	3	3	3	3	3
A9	6	6	6	6	6	6	6	6
A10	5	5	5	5	5	5	5	5
A11	4	4	4	4	4	4	4	4
A12	8	8	8	8	8	8	8	8
Country	**Case 8**	**Case 9**	**Case 10**	**Case 11**	**Case 12**	**Case 13**	**Case 14**	**Case 15**
A1	1	1	1	1	1	1	1	1
A2	9	9	9	9	9	9	9	9
A3	12	12	11	11	11	12	12	11
A4	11	11	12	12	12	10	10	12
A5	10	10	10	10	10	11	11	10
A6	2	2	2	2	2	2	2	2
A7	7	7	7	7	7	7	7	8
A8	3	3	3	3	3	3	3	3
A9	6	6	6	6	6	6	6	6
A10	5	5	5	5	4	5	5	4
A11	4	4	4	4	5	4	4	5

We also plot the ranking of the alternatives (under various experimental cases) as depicted by Fig. 11.2 and examine for their consistencies by using SRC test (see Table 11.11). Figure 11.2 reflects that alternatives show consistency in their rankings. Table 11.11 exhibits high-rank correlation values with a statistical significance, which proves that our modified LOPCOW method provides a stable solution given the changes in the underlying conditions.

Conclusion

The current paper has provided a MCDM-based comparison of BRICS and G7 nations with respect to EE and ES. ES has been a matter of utmost importance for sustainable living, economic growth and achieving SDGs. The criteria considered are ENS, ENE, TREN,

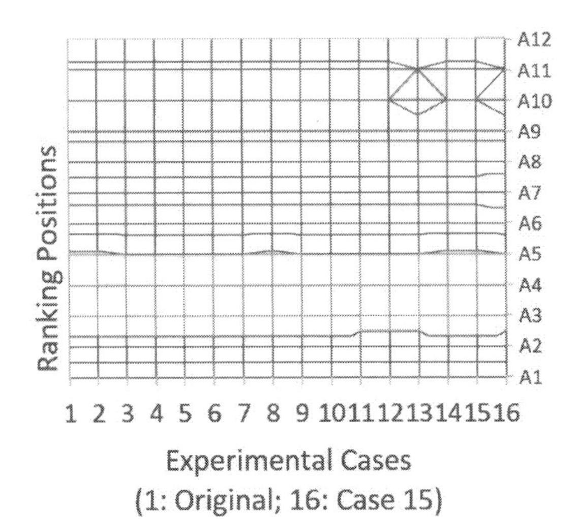

Fig. 11.2 Result of sensitivity analysis (ranking distribution patterns for various experimental cases)

ES, CCPI and CO_2 total emission CTE. A modified LOPCOW method with a mix of three normalisation schemes, such as linear sum-based, linear ratio based and linear max-min, has been proposed for comparison. It is observed that ENS and ENE hold more significance based on objective information. In addition, the result has shown that Brazil secures the top position along with France and Italy in the top bracket, while the Asian giants, such as China and India, remain in the lower group. Therefore, it is evident that G7 countries do not always show better performance as compared with the BRICS nations. The paper has two-fold implications. First, it provides a hybrid comparative assessment (considering energy and environmental aspects) of a group of developing and developed countries. Second, it provides a new extension of the original LOPCOW method with applications in ES area. However, the paper may be further extended with some of the following future scopes. Firstly, a future study may look at climate change parameters (related to global warming) to compare the G7 and BRICS countries. Secondly, the proposed model of this paper may be applied to compare OECD and other countries. Thirdly, an interesting study may be made to delve into the interrelationship between climate financing initiatives and the ES performance of the BRICS and G7 nations. Fourthly, a causal analysis may be done to explore how technological developments can improve ES. Nevertheless, the present paper is a kind of its own type that we believe may be of interest to researchers and policymakers.

Acknowledgement

The authors are grateful to all the anonymous reviewers for their valuable comments. Further, we are thankful to the conference organisers.

Table 11.11 Result of sensitivity analysis (SRC among the ranking under various cases)

Correlations		Original	Case1	Case2	Case3	Case4	Case5	Case6	Case7	Case8	Case9	Case10	Case11	Case12	Case13	Case14	Case15
Spearman's rho	Original	1															
	Case1	1.000**	1														
	Case2	.993**	.993**	1													
	Case3	.993**	.993**	1.000**	1												
	Case4	.993**	.993**	1.000**	1.000**	1											
	Case5	.993**	.993**	1.000**	1.000**	1.000**	1										
	Case6	.993**	.993**	1.000**	1.000**	1.000**	1.000**	1									
	Case7	1.000**	1.000**	.993**	.993**	.993**	.993**	.993**	1								
	Case8	.993**	.993**	1.000**	1.000**	1.000**	1.000**	1.000**	.993**	1							
	Case9	.993**	.993**	1.000**	1.000**	1.000**	1.000**	1.000**	.993**	1.000**	1						
	Case10	.979**	.979**	.993**	.993**	.993**	.993**	.993**	.979**	.993**	.993**	1					
	Case11	.979**	.979**	.993**	.993**	.993**	.993**	.993**	.979**	.993**	.993**	1.000**	1				
	Case12	.972**	.972**	.986**	.986**	.986**	.986**	.986**	.972**	.986**	.986**	.993**	.993**	1			
	Case13	1.000**	1.000**	.993**	.993**	.993**	.993**	.993**	1.000**	.993**	.993**	.979**	.979**	.972**	1		
	Case14	1.000**	1.000**	.993**	.993**	.993**	.993**	.993**	1.000**	.993**	.993**	.979**	.979**	.972**	1.000**	1	
	Case15	.965**	.965**	.979**	.979**	.979**	.979**	.979**	.965**	.979**	.979**	.986**	.986**	.993**	.965**	.965**	1

** Correlation is significant at the 0.01 level (2-tailed).

References

1. Aguir Bargaoui, S. (2022). The impact of energy efficiency and renewable energies on environmental quality in OECD countries. J. Knowl. Econ. 1-21. https://doi.org/10.1007/s13132-021-00864-0
2. Ahmad, M., Jiang, P., Murshed, M., Shehzad, K., Akram, R., Cui, L., & Khan, Z. (2021). Modelling the dynamic linkages between eco-innovation, urbanization, economic growth and ecological footprints for G7 countries: does financial globalization matter?
3. Aytekin, A. (2021). Comparative analysis of the normalization techniques in the context of MCDM problems. Decis. Mak. Appl. Manag. Eng. 4(2): 1–25.
4. Baloch, M. A., Danish and Qiu, Y. (2022). Does energy innovation play a role in achieving sustainable development goals in BRICS countries. Environ. Technol. 43(15): 2290–2299.
5. Biswas, S. (2020). Measuring performance of healthcare supply chains in India: A comparative analysis of multi-criteria decision making methods. Decis. Mak. Appl. Manag. Eng. 3(2): 162–189.
6. Biswas, S. and Anand, O. P. (2020). Logistics Competitiveness Index-Based Comparison of BRICS and G7 Countries: An Integrated PSI-PIV Approach. IUP J. Supply Chain Manag. 17(2): 32–57.
7. Biswas, S. and Pamučar, D. S. (2021). Combinative distance based assessment (CODAS) framework using logarithmic normalization for multi-criteria decision making. Serb. J. Manag. 16(2): 321–340.
8. Biswas, S., Majumder, S. and Dawn, S. K. (2022a). Comparing the socioeconomic development of G7 and BRICS countries and resilience to COVID-19: An entropy–MARCOS framework. Bus. Perspect. Res. 10(2): 286–303.
9. Biswas, S., Pamučar, D., Božanić, D. and Halder, B. (2022b). A New Spherical Fuzzy LBWA-MULTIMOOSRAL Framework: Application in Evaluation of Leanness of MSMEs in India. Math. Probl. Eng. 2022. https://doi.org/10.1155/2022/5480848
10. Biswas, S., Pamucar, D. and Kar, S. (2022c). A preference-based comparison of select over-the-top video streaming platforms with picture fuzzy information. Int. J. Commun. Netw. Distrib. Syst. 28(4): 414–458.
11. Bo Y., Atif J., Muhammad U. and Muhammad Atif K. (2021), The dynamic linkage between globalization, financial development, energy utilization, and environmental sustainability in GCC countries. Environ. Sci. Pollut. Res. 28: 16568–16588
12. Brockway, P. E., Sorrell, S., Semieniuk, G., Heun, M. K., and Court, V. (2021). Energy efficiency and economy-wide rebound effects: A review of the evidence and its implications. Renew. Sustain. Energy Rev. 141: 110781.
13. Dell'Anna, F. (2021). Green jobs and energy efficiency as strategies for economic growth and the reduction of environmental impacts. Energy Policy, 149: 112031.
14. Fabio G., Jeroen C.J.M.B. and Jos N. O. (2008), An Empirical Analysis of Urban Form, Transport, and Global Warming. The Energy Journal. 29(4): 97–122.
15. Gatto, A. and Drago, C. (2021). When renewable energy, empowerment, and entrepreneurship connect: Measuring energy policy effectiveness in 230 countries. Energy Res. Soc. Sci. 78: 101977.
16. Girod, B., Stucki, T., and Woerter, M. (2017). How do policies for efficient energy use in the household sector induce energy-efficiency innovation? An evaluation of European countries. Energy Pol. 103: 223–237.
17. Godil, D. I., Yu, Z., Sharif, A., Usman, R. and Khan, S. A. R. (2021). Investigate the role of technology innovation and renewable energy in reducing transport sector CO2 emission in China: A path toward sustainable development. Sustain. Dev. 29(4): 694–707.
18. Goodland, R. (1995). The concept of environmental sustainability. Annu Rev Ecol Evol Syst. 26: 1–24.

19. Gregori, P. and Holzmann, P. (2022). Entrepreneurial practices and the constitution of environmental value for sustainability. Bus Strat Env. 1–16. https://doi.org/10.1002/bse.3077
20. Gupta, S., Bandyopadhyay, G., Bhattacharjee, M. and Biswas, S. (2019). Portfolio Selection using DEA-COPRAS at risk–return interface based on NSE (India). Int. j. eng. explor. technol (IJITEE). 8(10): 4078–4086.
21. Irfan, M., Mahapatra, B., & Ojha, R. K. (2021). Examining the effectiveness of low-carbon strategies in South Asian countries: the case of energy efficiency and renewable energy. Environ. Dev. Sustain. 23(8): 11936–11952.
22. Jahan, A. and Edwards, K. L. (2015). A state-of-the-art survey on the influence of normalization techniques in ranking: Improving the materials selection process in engineering design. Mater. Des. 65: 335–342.
23. Jahan, A. (2018). Developing WASPAS-RTB method for range target-based criteria: toward selection for robust design. Technol. Econ. Dev. Econ. 24(4): 1362–1387.
24. Liao, H. and Wu, X. (2020). DNMA: A double normalization-based multiple aggregation method for multi-expert multi-criteria decision making. Omega, 94: 102058.
25. Luisa M. and Rosa P. (2022), Sustainable energy development analysis: Energy Trilemma, Sustain. Technol Entrepreneurship. 1: 1–10.
26. Maniya, K. and Bhatt, M. G. (2010). A selection of material using a novel type decision-making method: Preference selection index method. Mater. Des. 31(4): 1785–1789.
27. Marti, L. and Puertas, R. (2022). Sustainable energy development analysis: Energy Trilemma. Sustain. Technol Entrepreneurship. 1(1): 100007. https://doi.org/10.1016/j.stae.2022.100007
28. Morelli, J. (2011). Environmental sustainability: A definition for environmental professionals. J. environ. Sustain. 1(1): 1–9. DOI: 10.14448/jes.01.0002
29. Murshed, M., Rahman, M., Alam, M. S., Ahmad, P. and Dagar, V. (2021). The nexus between environmental regulations, economic growth, and environmental sustainability: linking environmental patents to ecological footprint reduction in South Asia. Environ. Sci. Pollut. Res. 28(36): 49967–49988.
30. Nyame-Asiamah, F. and Kawalek, P. (2021). Sustainability and Consumer Behaviour: Towards a Cohered Emergent Theory. The Palgrave Handbook of Corporate Social Responsibility, 1177–1194.
31. Pamucar, D., Torkayesh, A. E. and Biswas, S. (2022). Supplier selection in healthcare supply chain management during the COVID-19 pandemic: a novel fuzzy rough decision-making approach. Ann. Oper. Res. 1–43. https://doi.org/10.1007/s10479-022-04529-2
32. Pamucar, D., Žižović, M., Biswas, S. and Božanić, D. (2021). A new logarithm methodology of additive weights (LMAW) for multi-criteria decision-making: Application in logistics. Facta Univ. Ser.: Mech. Eng. 19(3): 361–380.
33. Perry S. (2021), Wind energy for sustainable development: Driving factors and future outlook. J. Clean. Prod. 289: 1–15.
34. Popkova, E. G. and Sergi, B. S. (2021). Energy efficiency in leading emerging and developed countries. Energy, 221, 119730.
35. Pramanik, P. K. D., Biswas, S., Pal, S., Marinković, D. and Choudhury, P. (2021). A comparative analysis of multi-criteria decision-making methods for resource selection in mobile crowd computing. Symmetry. 13(9): 1713. https://doi.org/10.3390/sym13091713
36. Rabia A., Fuzhong C., Fahad K., Guanhua H. and Muhammad I. (2021), Heterogeneous effects of energy efficiency and renewable energy on economic growth of BRICS countries: A fixed effect panel quantile regression analysis. Energy. 215: 1–11.
37. Roman L., Zuzana H. and Marcin Z. (2021), The Efficiency of Circular Economies: A Comparison of Visegrád Group Countries. Energies. 14: 1680.

38. Rusydiana, A. S. (2021). Energy efficiency in OIC countries: SDG 7 output. 670216917.
39. Selvakkumaran, S. and Limmeechokchai, B. (2013). Energy security and co-benefits of energy efficiency improvement in three Asian countries. Renew. Sustain. Energy Rev. 20: 491–503.
40. Sudharshan R., Paramati, Umer S. and Buhari D. (2022), The role of environmental technology for energy demand and energy efficiency: Evidence from OECD countries, Renew. Sustain. Energy Rev.153: 1–9.
41. Tachega, M. A., Yao, X., Liu, Y., Ahmed, D., Li, H. and Mintah, C. (2021). Energy efficiency evaluation of oil producing economies in Africa: DEA, malmquist and multiple regression approaches. Clen. Environ. Sys. 2: 100025.
42. Taylor, P. G., d'Ortigue, O. L., Francoeur, M. and Trudeau, N. (2010). Final energy use in IEA countries: The role of energy efficiency. Energy Pol. 38(11): 6463–6474.
43. Vafaei, N., Ribeiro, R. A. and Camarinha-Matos, L. M. (2016, April). Normalization techniques for multi-criteria decision making: analytical hierarchy process case study. In doctoral conference on computing, electrical and industrial systems (pp. 261-269). Springer, Cham.
44. Verma, P., Patel, N., Nair, N. K. C. and Brent, A. C. (2018). Improving the energy efficiency of the New Zealand economy: A policy comparison with other renewable-rich countries. Energy Pol. 122: 506–517.
45. Wei-Zheng W., Lan-Cui L., Hua L. and Yi-Ming W. (2021), Impacts of urbanization on carbon emissions: An empirical analysis from OECD countries. Energy Pol.151: 1–15.
46. Wen, Z., Liao, H. and Zavadskas, E. K. (2020). MACONT: Mixed aggregation by comprehensive normalization technique for multi-criteria analysis. Informatica. 31(4): 857–880.
47. Wen, J., Mughal, N., Zhao, J., Shabbir, M. S., Niedbała, G., Jain, V. and Anwar, A. (2021). Does globalization matter for environmental degradation? Nexus among energy consumption, economic growth, and carbon dioxide emission. Energy Pol. 153: 112230.
48. Woods, L. (2021). Why environmental sustainability is important. BDJ In Practice, 34(3), 40–41.
49. Zhang Y., Pablo P., Ateeq Ur Rehman I., Muhammad T., Katerine P. and Abdul Rehman K. (2022), Energy efciency and Jevons' paradox in OECD countries: policy implications leading toward sustainable development. J Pet Explor Prod Technol. 1–14. https://doi.org/10.1007/s13202-022-01478-1

CHAPTER 12

Detection of Trace Level Heavy Metal Ions by an Aluminium Complex: Preparation, Crystal Structure, Emission and Lifetime Studies

Supriya Dutta*
Department of Chemistry, Nistarini College,
Purulia, West Bengal

Abstract

With a view to detect trace amounts of heavy metal ions as well as other toxic ions, the mononuclear complex [AlHL] (1) has been prepared using $AlCl_3$, H_4L and triethylamine in 1:1:3 ratio (H_4L = 1, 1, 1, 1-tetrakis [(2-salicylaldiminomethyl)] methane). The objective of preparing the said complex is to detect cations through the photoluminescence method at as low as nano-order or near nano-order levels. The complex [Al(HL)] crystallises in the space group monoclinic *P2(1)/c*. Although H_4L does not emit at room temperature, the remarkable chelation enhanced fluorescence (CHEF) effect is observed by the formation of complex 1. When excited at a wavelength of 350 nm, complex 1 emits at 433 nm. Emission titration of the solution of complex 1 in acetonitrile with Zn^{2+}, Cd^{2+}, Hg^{2+} and Pb^{2+} shows selectivity towards Zn^{2+} ion. Unlike Cd^{2+}, Hg^{2+} and Pb^{2+}, which quench the emission, Zn^{2+} enhances the fluorescence intensity by 1.64 folds. The quenching process follows the Stern-Volmer equation and, in order to get an insight into whether the quenching was static or dynamic in nature, lifetime measurements were carried out. Interestingly, zinc and mercury were found to follow static quenching, while cadmium and lead adopt dynamic quenching pathways.

Keywords: aluminium, quenching, emission spectra, lifetime

Introduction

Over the years, there has been a paradigm shift in the consciousness of humankind and environmentalists about the protection of the environment. Global warming has led to the breakage of the polar ice caps and scientists have predicted a global average rise in sea level of 3–4 metres by 2050. Extensive industrialisation has been a major factor behind the temperature rise and it has also contributed vastly towards the dispersion of toxic wastes in the drinking water, soil and consequently food products. Hence, the detection of heavy metal ions, as well as toxic ions, has become necessary. In addition to the heavy metal ions, even the lighter

*supriyadutta78@gmail.com
All the data provided in the tables as well as the figures are experimental data carried out by the author

p-block elements are known to possess a significant role in biological systems. For example, the coordination chemistry of the group 13 (IIIA) metal ions is of biomedical interest because of the involvement of aluminium with neurological dysfunctions and has been suspected1 of inducing neurological dysfunctions such as Alzheimer's disease. The contributions of neurotoxicity of aluminium (III) in animals were first reported in 1897 by Dollken. The modern understanding of the effects of aluminium(III) in animal was initiated by the extraordinary discovery of Klatzo et al. who reported that injections of Al salts into the rabbit brain lead to the formation of NFTs (intraneuronal neurofibrillary tangles). But the subsequent research has produced conflicting results and overwhelming medical evidence still does not convincingly demonstrate the relationship between aluminium and the disease. In addition to Al(III), Zn(II) is also thought to take a vital role in certain neurological disorders. Disruption in the uptake of zinc may lead to the generation of amyloid plaques in Alzheimer's disease. Thus, due to the presence of zinc as an essential component of many biological enzymes and substrates, such as carbonic anhydrase as well as its use in the area of cancer, neuroscience, sensing of zinc in solution has become a prime area of research during the last decade. Macrocyclic zinc sensing fluorophores developed by Kimura et. al. showed 8 folds increase in fluorescence intensity, while Nagano et. al. enhanced the fluorescence intensity to 14 folds by irradiating with visible light. The Zinpyr family developed by Lippard6 and co-workers still serves as one of the best models for the fluorescence-based detection of zinc. Apart from biological importance, aluminium complexes, mainly aluminium tris(quinolin-8-olate) (AlQ3) also finds a special place in the field of organic light-emitting diodes (OLEDs) by virtue of their excellent emission properties, high quantum yields and good colour purity. So far as flat full-colour displays are concerned, luminophores with colour tunability encompassing the entire visible range (red, green and blue) are desired. Although efficient green and red emitting OLEDs have been designed, the scarcity of aluminium complexes emitting in the indigo-blue (430–500 nm) region restricts its extensive use. Designed fabrications of ligands have enabled a few aluminium complexes to emit at around 475 nm, while only one complex to the best of our knowledge emits at 438 nm.

In our effort to explore the metal complex chemistry of tetrapodal ligands with a carbon bridgehead, we have prepared the ligand 1,1,1,1-tetrakis(2-salicylaldiminomethyl)methane. Herein, we report the synthesis, structure and photophysical properties of an aluminium complex that blue emits at 433 nm, and serves as an excellent sensor for zinc.

Experimental Materials

Reagent-grade chemicals obtained from commercial sources were used as received. Solvents were purified and dried according to standard methods. Tetrapodal ligand H_4L (Fig. 12.1) has been synthesised according to the procedure previously reported by us.[7]

Preparation of the Complex $[Al(HL)]\cdot CH_3CN$ (1)

To a boiling acetonitrile solution (25 mL) containing the ligand H_4L (0.55 g, 1.00 mmol) and anhydrous $AlCl_3$ (0.14 g, 1.04 mmol) was added dropwise a solution of triethylamine (0.30 g, 2.97 mmol) in acetonitrile. During the addition of the base, the colour of the solution changed from yellow to almost colourless. The solution was heated under reflux for 1 h, after which

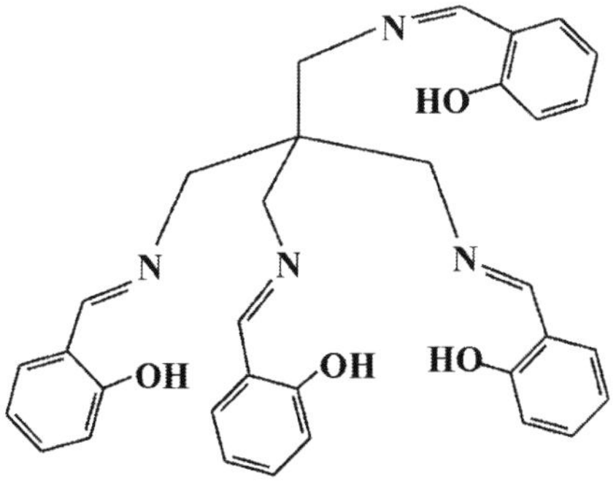

H_4L

Fig. 12.1 Ligand used for detection of heavy metal ions

it was evaporated to dryness on a rotary evaporator. The pale-yellow solid thus obtained was recrystallised from 1:1 acetonitrile-ethanol. Single crystals for X-ray analysis were obtained by slow diffusion of diethyl ether into a concentrated acetonitrile solution of the complex.

Yield: 0.46 g (75%). Anal. Calcd for $C_{35}H_{32}N_5O_4Al$: C, 68.51; H, 5.22; N, 11.42. Found: C, 68.45; H, 5.17; N, 11.48. ESI-MS (positive) in CH_3CN: m/z = 572.86 $[Al(HL^1) + H]^+$ (100%), 594.83 $[Al(HL^1) + Na]^+$ (95%). FT-IR (KBr, υ/cm^{-1}): 3433 (w, br), 2914 (w), 1629 (s), 1602 (m, sh), 1545 (s), 1473 (s), 1454 (s), 1402 (m), 1342 (m), 1323 (m), 1276 (w), 1209 (m), 1149 (m), 1031 (w), 904 (w), 808 (w), 761 (s), 740 (w), 626 (m), 507 (w), 460 (w), 426 (w). ^1H NMR (500 MHz, CD_2Cl_2): δ 12.64 (s, 1H, Ar-OH); 8.52 (s, 1H, CH(e′)=N); 8.15 (s, 3H, CH(e)=N); 7.40 (d, 2H, J = 7.50 Hz, Ar-H(a′+d′)); 7.33 (t, 3H, J_1 = 8.06 Hz, J_2 = 7.78 Hz, Ar-H(b)); 7.23 (d, 3H, J = 7.78 Hz, Ar-H(a)); 6.99 (t, 2H, J_1 = 6.46 Hz, J_2 = 7.13 Hz, Ar-H(b′+c′)); 6.71 (d, 3H, J = 8.15 Hz, Ar-H(d)); 6.68 (t, 3H, J_1 = 8.00 Hz, J_2 = 7.50 Hz, Ar-H(c)); 4.13 (d, 3H, J = 13.25 Hz, CH_2 (g/g′)); 3.71 (dd, 2H, $J_1 = J_2 = J_3$ = 12.91 Hz, CH_2 (h)); 3.50 (d, 3H, J = 12.91 Hz, CH_2 (g′/g)).

Physical Measurements

C, H and N analyses were performed on a Perkin-Elmer 2400 II elemental analyser. IR spectra were recorded using KBr disks on a Shimadzu FTIR 8400S spectrometer. The electronic spectra were recorded using a Perkin-Elmer 950 UV/VIS/NIR spectrophotometer. The luminescence spectra were recorded on a Perkin-Elmer LS 55 luminescence spectrophotometer at room temperature. Lifetime measurement for photo-emission was carried out with a Horiba Jobin YVON steady-state spectrometer. ESI-MS in acetonitrile was obtained on a Micromass Qtof YA 263 mass spectrometer. ^1H and ^1H-^1H COSY NMR spectra were obtained on a Bruker Advance DPX spectrometer at 500 MHz.

X-ray Crystallography

Crystals suitable for structure determinations of [Al(HL)] (**1**) were obtained by slow diffusion of diethyl ether into a concentrated acetonitrile solution of the complex. The crystal was mounted on a glass fibre using perfluoropolyether oil. Intensity data were collected on a Bruker-AXS SMART APEX II diffractometer at 120(2) K using graphite-monochromated Mo-Ka radiation

($\lambda = 0.71073$ Å). The data were processed with SAINT[8] and absorption corrections were made with SADABS. The structures were solved by direct and Fourier methods and refined by full-matrix least-squares methods based on F^2 using SHELX-97.[9] For the structure solutions and refinements, the SHELX-TL software package was used. A highly disorder solvent molecule (acetonitrile) in compound **1** was deleted using SQUEEZE facility of PLATON[10]. The non-hydrogen atoms were refined anisotropically, while the hydrogen atoms were placed at geometrically calculated positions with fixed thermal parameters. Crystal data and details of data collection are listed in Table 12.1.

Table 12.1 Crystallographic data of [Al(HL)] (1)

	1
Empirical formula	C33H29N4O4Al
Fw	572.58
T, K	120(2)
crystal syst, space group	Monoclinic, P2(1)/c
a, Å	12.5964(6)
b, Å	14.7675(7)
c, Å	16.8111(8)
α, deg	90
β, deg	90.144(2)
γ, deg	90
V, Å3	3127.1(3)
Z, ρcalcd, Mg/m^3	4, 1.216
μ, mm-1	0.107
F(000)	1200
Crystal size, mm^3	0.44 × 0.32 × 0.28
No. of data/restraints/params	5720/0/379
No. of reflns [I > 2σ (I)]	5720
GOF on F2	1.092
Final R indices[I > 2σ (I)]	R1a = 0.0595, wR2b = 0.1216
R indices (all data)	R1a = 0.0756, wR2b = 0.1268

Results and Discussion

ESI-MS

The electrospray ionisation mass spectra (ESI-MS positive) of the complexes **1** have been measured in acetonitrile. Three peaks observed for **1** are $[Al^{III}(HL)+H]^+$ and $[Al^{III}(HL)+Na]^+$ at m/z = 572.86 (100%) and m/z = 594.84 (95%), respectively.

X-ray Crystal Structure

The ORTEP representation of [Al(HL)] (**1**) is shown in Fig. 12.2. Selected bond distances and bond angles for **1** is given in Table 12.2. Compound **1** crystallises in monoclinic P2(1)/c space group. The coordination environment around the central metal ion is a slightly distorted octahedral in which the Al(III) centre is coordinated by three imine nitrogens and three phenolate oxygens provided by three pendant salicylaldimines. Each N3 and O3 donor set coordinates to the metal ion in a facial way to form a distorted octahedral geometry. Figure 12.2 shows that in compound **1** one of the salicylaldimine moiety remains uncoordinated and the phenolic OH and the imine nitrogen are intramolecularly hydrogen-bonded. The average Al(III)–N(imine) and Al(III)–O(phenolate) distances are 2.043(2) and 1.831(2)Å, respectively. The transoid angles in **1** vary between 172.18(10)° and 173.78(10)°, while the cisoid angles vary between 83.91(9)° and 95.14(10) °. The packing diagram of compound 1 (Fig. 12.3) reveals several short contacts between either a methylene hydrogen atom or an aromatic hydrogen atom and the centroid of the π cloud of another aromatic ring. These C–H····π interactions are intermolecular in nature. Generally, the upper acceptable limit for the distance of a C-H hydrogen atom from the centroid of an aromatic ring in a C–H····π interaction is 3.5 Å. The metrical parameters for C–H····π interactions in compounds **1** indicate that the C–H····π distances range from 2.564 to 3.335 Å, albeit a majority of them lie between 2.6 and 2.9 Å, indicating strong intermolecular C–H····π interactions.

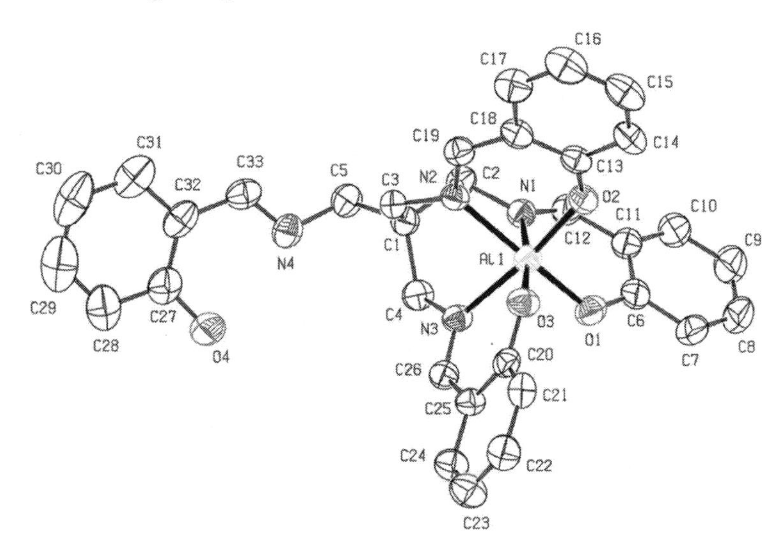

Fig. 12.2 ORTEP representation of [Al(HL)] (1) at 50% probability ellipsoids. Hydrogen atoms have been removed for clarity

Proton NMR Spectra

^1H NMR spectroscopic studies were carried out for complex **1** in CD_2Cl_2. The observed chemical shifts along with the spectral assignments are given in the experimental section and the proton NMR spectra with incremental addition of Zn^{2+}, is shown in Fig. 12.4. The NMR

Table 12.2 Selected bond distance (Å) and angle (1)

1			
Al–N(1)	2.037(2)	O(1)–Al–O(2)	92.62(9)
Al–N(2)	2.050(2)	O(1)–Al–O(3)	95.14(10)
Al–N(3)	2.042(2)	O(2)–Al–O(3)	91.53(9)
Al–O(1)	1.837(2)	N(1)–Al–O(1)	88.37(10)
Al–O(2)	1.831(2)	N(1)–Al–O(2)	94.40(10)
Al–O(3)	1.826(2)	N(1)–Al–O(3)	172.97(10)
		N(2)–Al–O(1)	172.18(10)
		N(2)–Al–O(2)	89.25(9)
		N(2)–Al–O(3)	92.39(10)
		N(3)–Al–O(1)	93.52(9)
		N(3)–Al–O(2)	173.78(10)
		N(3)–Al–O(3)	88.94(9)
		N(1)–Al–N(2)	83.91(9)
		N(1)–Al–N(3)	84.76(9)
		N(2)–Al–N(3)	84.53(9)

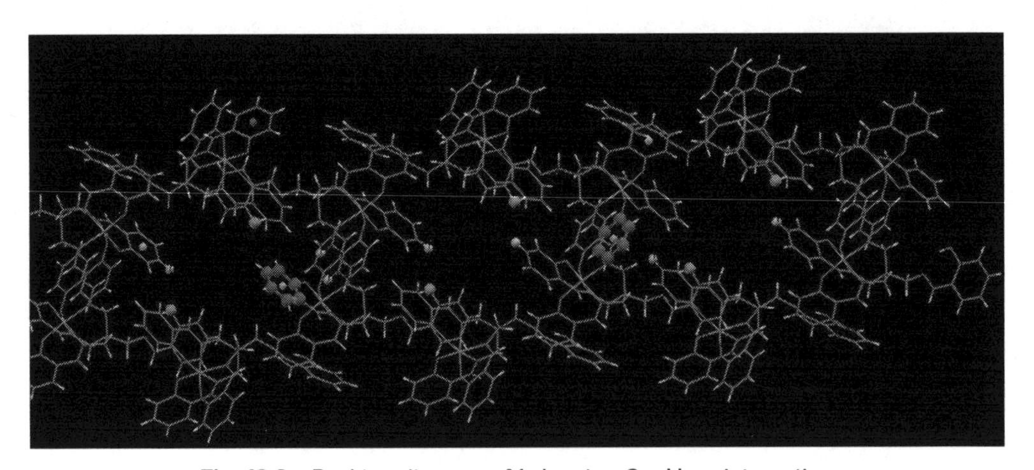

Fig. 12.3 Packing diagram of 1 showing C – H····π interaction

spectrum at the bottom refers to the complex [Al(HL)], while the upper-most spectrum refers to the complex **1** saturated with Zn^{2+}. The hydrogen-bonded phenolate O–H····N is observed as a sharp singlet at 12.64 ppm. Compound **1** shows 12 signals, including a pair of singlets, due to CH=N at 8.52 and 8.15 ppm with the proton intensity ratio of 1:3, in coherence with the structure with one pendant arm lying uncoordinated to the metal center.

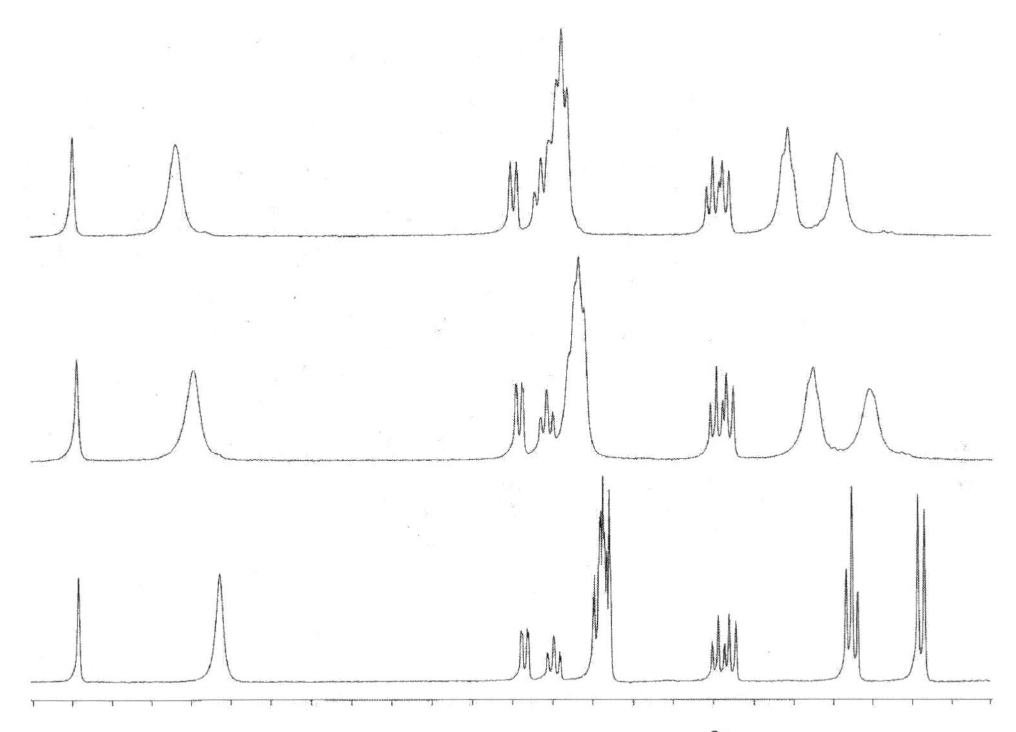

Fig. 12.4 NMR spectra of complex 1 with incremental addition of Zn^{2+} ion. Bottom most spectrum depicts the NMR spectrum of pure Complex 1, while the uppermost spectra depict the spectrum of complex 1 saturated with Zn^{2+} ion

The 16 aromatic protons of each ligand occur in between 7.40 to 6.68 ppm with their intensity ratio being 2:3:3:2:3:3. Further, the eight methylene protons appear as two doublets and one quartet in the range 4.13 to 3.50 ppm with intensity ratio being 2:1:2. It seems that methylene protons of three metal chelated pendant salicylaldiminomethyl arms are anisochronous and are observed as two doublets but they are isochronous for metal-free fourth arm. ^1H NMR titration of Al(HL) (**1**) with $Zn(ClO_4)_2{\cdot}4H_2O$ in CD_3CN showed a slight perturbation of the aromatic as well as the aliphatic protons of the uncoordinated pendant arm. With addition of 0.31 equivalent of $Zn(ClO_4)_2{\cdot}4H_2O$, the originally sharp, well-defined aromatic/aliphatic peaks broadened, accompanied by a small shift towards the low field region as shown in Fig. 12.4. The addition of 1 equivalent of Zn^{2+} led to further deformation of the peaks, ultimately merging to form a broad multiplet. The unperturbed aromatic signals (as shown in Fig. 12. 4) corresponding to the metal-bound pendant arms signify no interaction of Zn^{2+} with the coordinated salicylaldimine moieties.

Absorption Spectral Studies

Absorption spectra of complex **1** recorded in acetonitrile shows two peaks at 350 nm (ε = 20,000 $M^{-1}cm^{-1}$) and 263 nm (ε = 45,000 $M^{-1}cm^{-1}$), respectively, accompanied by a shoulder at 275 nm. Both the peaks are likely to be ligand-based $\pi \rightarrow \pi^*$ transition. Titration of complex **1** with d^{10} metal ions (Zn^{2+}, Cd^{2+}, Hg^{2+} and Pb^{2+}) yields an almost unchanged spectra for Cd^{2+} and Pb^{2+}, while Zn^{2+} and Hg^{2+} show a small but distinct change in the ground state absorption spectra. In case of titration of complex **1** with Zn^{2+}, the small absorption spectral change observed is in coherence with the ^1H NMR titration, which also shows a small perturbation of the signals due to weak interaction of Zn^{2+} with the aromatic protons of the uncoordinated salicylaldimine moiety. Upon titration of **1** with Hg^{2+} ion, the absorption maxima at 350 nm undergo a redshift of 5 nm, while in the UV region the intensity of the shoulder at 275 nm is increased at the expense of the diminishing peak at 263 nm. Cd^{2+} somehow does not alter the absorption spectra of **1**, although it has a small quenching effect upon the emission intensity of **1** as discussed later. The unavailability of sufficient space to accommodate the larger Pb^{2+} in the uncoordinated pendant arm is likely to be the reason that the metal ion cannot interact with the CH=N as well as the phenolic-OH of the free salicylaldimine moiety, thereby keeping the ground state absorption spectra unperturbed.

Emission Spectral Studies

The ligand H_4L is non-emissive at room temperature. Coordination of Al(III) with the subsequent formation of [Al(HL)] (**1**) leads to a CHEF effect with a remarkable increase of emission intensity at λ_{em} = 433 nm when complex **1** is excited (λ_{ex}) at 350 nm in acetonitrile at room temperature. To the best of our knowledge, blue-emitting aluminium complexes have been reported with λ_{em} as low as 438 nm, while complex **1** is a better blue emitter by 5 nm. The addition of Zn^{2+} increases the intensity of emission as shown in Fig. 12.5.

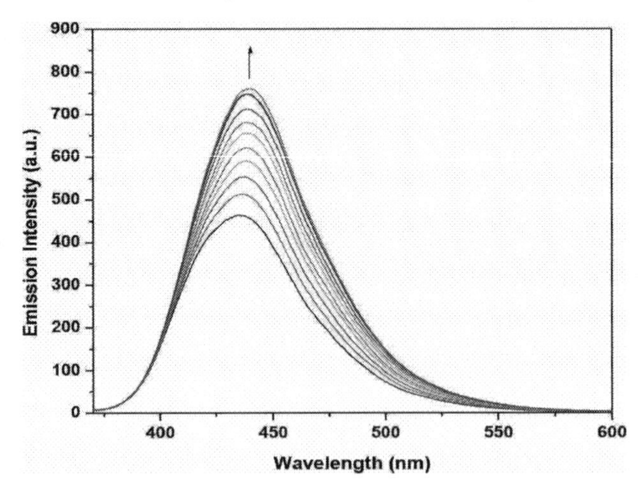

Fig. 12.5 Increase of emission intensity of [Al(HL)] (1) with addition of 0.3 equivalents of Zn^{2+} ion

Interestingly, complex **1** when titrated with the d^{10} metal ions (Zn^{2+}, Cd^{2+}, Hg^{2+} and Pb^{2+}) shows an ability to sense Zn^{2+}. As shown in Fig. 12.6, Cd^{2+}, Hg^{2+} and Pb^{2+} simply quench the emission intensity without practically altering the emission wavelength (λ_{em} = 433 nm). The extent of quenching increases with heavier ions with Cd^{2+}, Hg^{2+} and Pb^{2+} quenching the emission intensity by 1.35, 2.2 and 2.5 folds, respectively. This quenching of emission intensity may be ascribed to the heavy atom effect, especially in the case of Hg^{2+} and Pb^{2+}.

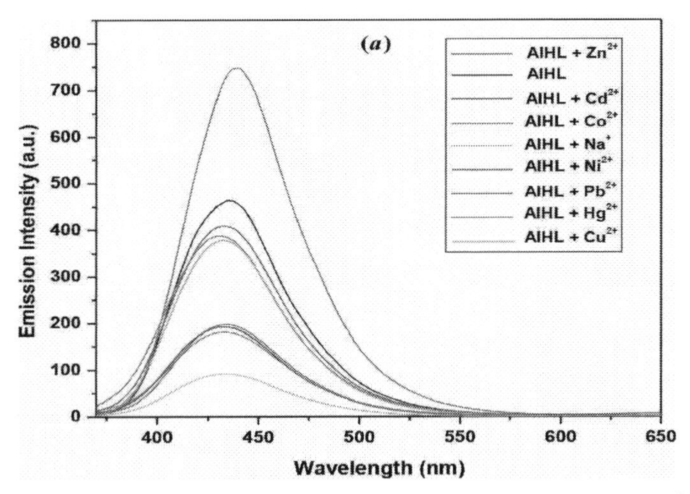

Fig. 12.6 Extent of quenching of emission intensity of [Al(HL)] (1) with various M^{2+} and Na^{+} ions

On the other hand, when complex **1** is titrated with Zn^{2+}, unlike Cd^{2+}, Hg^{2+} and Pb^{2+}, initially enhances the emission intensity by 1.64 folds (0.31 equivalents of Zn^{2+} with respect to complex **1**, accompanied by a red-shift of emission wavelength by 5 nm, and then starts quenching until 1 equivalent of Zn^{2+} is added, as shown in Figure 5. The increase of emission intensity up to the addition of 0.31 equivalent of Zn^{2+}, may probably indicate a Zn^{2+} : [AlHL] = 1:3 weak interaction, where three uncoordinated salicylaldimine moieties from three molecules of complex **1** weakly coordinate with Zn^{2+} to enhance the emission intensity.

Thus, complex **1** itself acts as a Zn^{2+} sensor, which can sense Zn^{2+} in the concentration range as low as 6.6×10^{-8} (M). A comparative study is shown in Fig. 12.6 where the sensing of Zn^{2+} ions by means of fluorescence enhancement, is clearly depicted against the quenching effect of the other d^{10} metal ions. Figure 12.7 shows a bar diagram depicting the same observation quantitatively in terms of $(F - F_0)$ [F_0 = emission intensity for AlHL (**1**) in absence of quencher;

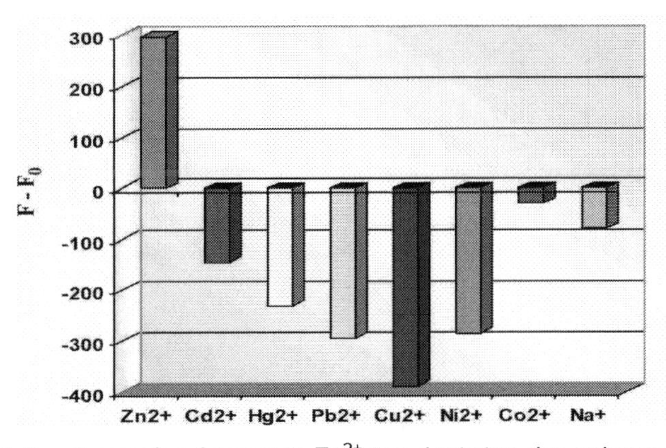

Fig. 12.7 Selectivity of complex 1 towards Zn^{2+} ion depicting the enhancement of emission intensity as compared to other metal ion quenchers

F = emission intensity of **1** saturated with the quencher] for the metal ions in a more perceptible manner. In quest of having an insight into the nature of quenching, Stern-Volmer plots (F_0/F versus [Q] [Q = quencher concentration]) were studied. In all cases, F_0/F vs [Q] gave linear plots. Figures 12.8 shows a representative Stern-Volmer Plot of quenching of emission intensity of complex **1** by Pb^{2+} ion. Lifetime measurements were carried out to ascertain whether the quenching of emission intensities was static or dynamic in nature. The absorption, emission and lifetime data for all the quencher ions are listed in Table 12.3.

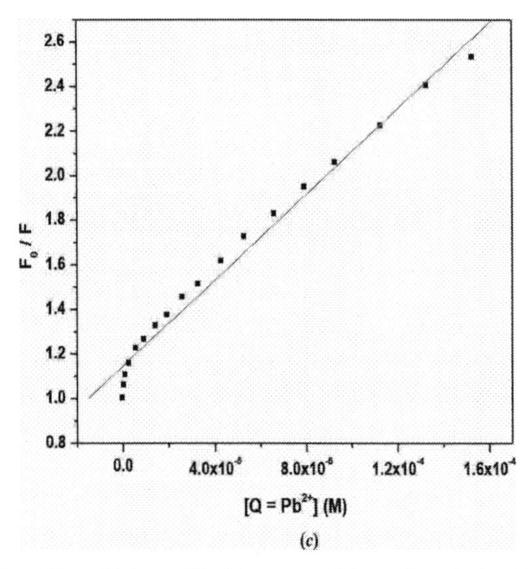

Fig. 12.8 Representative Stern-Volmer Plot for quenching of emission intensity of complex 1 by Pb^{2+} ion

Lifetime Measurements

Figure 12.9 shows the lifetime profile of AlHL (**1**) at room temperature in acetonitrile. Lifetime titrations of complex **1** with Zn^{2+}, Cd^{2+}, Hg^{2+} and Pb^{2+} were carried out in the same way as emission titrations using acetonitrile solutions at room temperature. Table 12.3 lists the room temperature lifetimes of AlHL (**1**) in the absence of quencher, and saturated with quencher. While titration of **1** with Zn^{2+} and Hg^{2+} ions gave a (τ_0 / τ) value around 1 implying a static quenching as already indicated by the absorption spectral profile in the ground state, titration of **1** with Cd^{2+} and Pb^{2+} showed dynamic quenching, with (F_0 / F) » (τ_0 / τ), as expected from their ground state absorption spectra. Table 3 shows the corresponding (F_0 / F) and (τ_0 / τ) values for the titration of **1** with Zn^{2+}, Cd^{2+}, Hg^{2+} and Pb^{2+}, respectively.

Conclusion

In view of making a blue emitting aluminium complex, the mononuclear complex [Al(HL)] (**1**) has been prepared using the ligand 1,1,1,1-tetrakis[(2-salicylaldiminoaminomethyl)] methane, $AlCl_3$ and triethylamine in the ratio 1:1:3. Although the ligand in non-emissive at room temperature, **1** shows remarkable CHEF effect with λ_{em} = 433 nm, one of the bluest

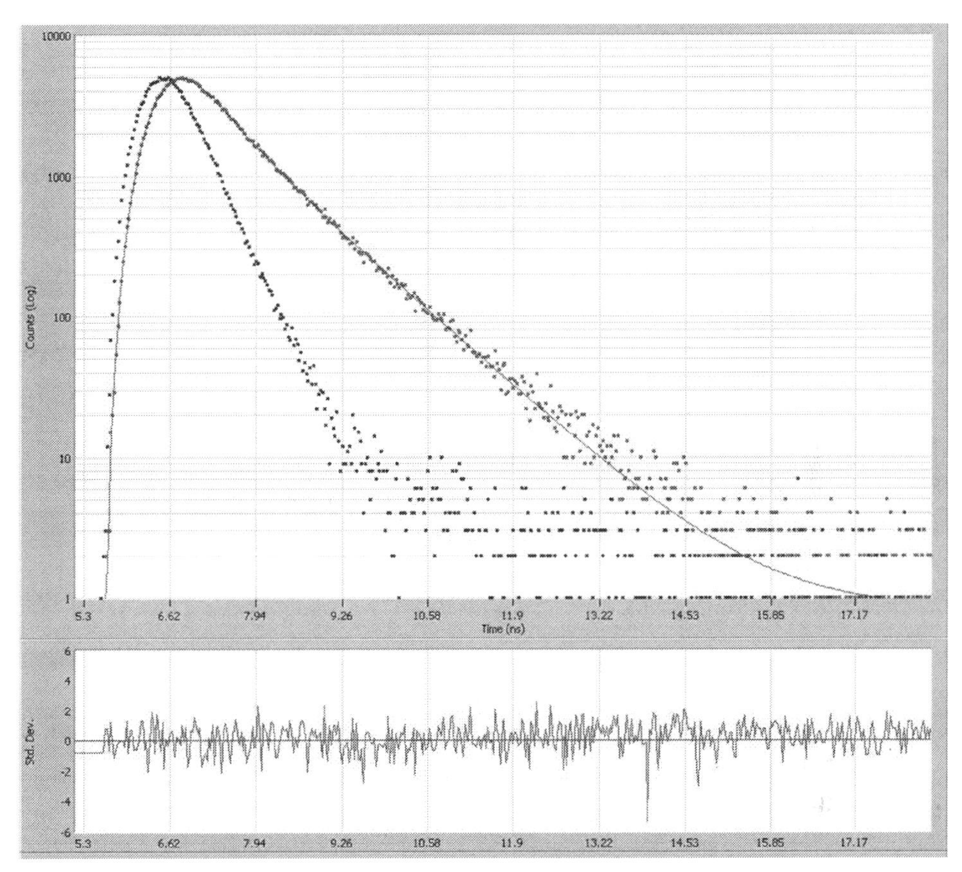

Fig. 12.9 Lifetime profile of [Al(HL)] (1) at room temperature in acetonitrile. Red points represent the experimental data, the green line shows the best fit line and the blue points represent the data of the promt

Table 12.3 Absorption, emission and lifetime data of [Al(HL)] (1) with Zn^{2+}, Cd^{2+}, Hg^{2+} and Pb^{2+} metal ions

Compound	λ_{abs} (nm)	λ_{em} (nm)^a	ϕ	$\tau_0 \times 10^{-10}$ (s)	$\tau_0 \times 10^{-10}$ (s)^c	F_0/F^c	τ_0/τ
[Al(HL)] (**1**)	350 (20,000)	433	0.002	5.70			
	263 (45,000)						
(**1**) + Zn^{2+}	345 (16,700)	438	0.003	6.70^b	6.85	0.61	-
(**1**) + Cd^{2+}	350 (17,200)	433	0.0018	5.67^b	4.35	1.31	1.35
	264 (39,500)						
(**1**) + Hg^{2+}	354 (33,400)	440	0.0012	5.37^b	5.65	2.20	1 ± 0.2
	275 (70,000)						
(**1**) + Pb^{2+}	350 (19,000)	433	0.0008	5.73^b	2.57	2.50	2.23
	264 (47,500)						

$^a\lambda_{ex}$ = 350 nm. ^{b}Lifetime of (**1**) without addition of the corresponding M^{2+}. ^{c}Value of emission intensity and lifetime at saturation.

emitting aluminium complexes known to us. Moreover, an acetonitrile solution of complex **1** when titrated with zinc perchlorate (0.3 equivalents with respect to **1**) enhanced the emission intensity of **1** by 1.64 folds and then quenches the emission till **1** equivalent of Zn^{2+} is added. All other d^{10} metal ions ($M^{2+} = Cd^{2+}$, Hg^{2+} and Pb^{2+}) only quench the emission of [AlHL] (**1**) to varying extents. Thus, [AlHL] (**1**) acts as an excellent sensor for zinc, showing completely opposite emission behavior as compared to other d^{10} metal ions. In order to understand the nature of quenching, lifetime experiments were carried out that revealed a dynamic quenching of emission intensity of complex **1** for Cd^{2+} and Pb^{2+} ion, while Hg^{2+} followed a static quenching pathway.

References

1. Alzheimer's Society Information Sheet (June 2, 2002).
2. Burdette, S.C., Frederickson, C.J., Bu, W. and Lippard, S.J. (2003). J. Am. Chem. Soc. 125: 1778–1787.
3. Dutta, S., Biswas, P., Flörke, U. and Nag, K. (2010). Inorg. Chem. 49: 7382–7400.
4. Exley, C. (2001). Aluminium and Alzheimer's Disease. Amsterdam.
5. Green, M. A. and Welch, M.J. (1989). Nucl. Med. Biol. 16:435.
6. Kimura, E., Aoki, S., Kikuta, E. and Koike, T. (2003). Proc. Natl. Acad. Sci. U.S.A. 100: 3731–3736.
7. Klatzo, I., Wisniewski, H. M., and Streicher, E. (1965). Experimental production of neurofibrillary degeneration. J. Neuropath. Exp. Neur. 24(1): 87–99.
8. SAINT (version 6.02), SADABS (version 2.03), (2002). Bruker AXS Inc., Madison, Wisconsin.
9. Sheldrick, G.M. (1997). SHELXL-97, Program for the Refinement of Crystal Structures, University of Götingen, Götingen, Germany.
10. Spek, A.L. (2010). PLATON, A Multipurpose Crystallographic Tool, Utrecht University, Utrecht, The Netherlands.

CHAPTER 13

Recognising Mosquito Breeding Zones Applying Machine Learning Over UAV Images: A Comparative Study between Dense Urban Area Urban Slum Area

Joydip Datta
Esri India Technologies Ltd

Souvik Chakraborty*
Assistant Professor, CE department,
Dr. Sudhir Chandra Sur Institute of Technology & Sports Complex, Kolkata, India

Prathama Paul
Department of Geography, University of Calcutta, Kolkata, India

Sudipta Bhomick and Saumyadip Kar
U.G. Student of C.E. Department,
Dr. Sudhir Chandra Sur Institute of Technology & Sports Complex, Kolkata, India

Abstract

The transmission of diseases carried by mosquitoes has mostly been influenced by human ecology. Manufactured containers that have been thrown off as waste as well as seem to be capable of keeping standing water offer mosquito larvae, the ideal settings for growth because stagnant water is a crucial factor in insect reproduction. The growth of all these larvae increases the potential for indigenous mosquito-borne illnesses, such as Zika and dengue virus, to spread among humans in populated areas. Utilising unmanned aerial vehicles, which are currently gradually being used more frequently in the geospatial intelligence sector, becomes a possible approach to solving this problem. India gets affected by malaria, dengue, and other mosquito-related diseases every year, which is quite unfortunate. Records show that the proper availability of necessary medicines is not enough to prevent such incidents. The major challenge also lies in the identification of the distribution of breeding habitats of these diseases. Processor architectures, which are physical objects that link to data through wireless as well as satellite systems, have been getting more sophisticated and effective. For instance, devices that use geospatial information from sensors, such as GPS units, do so by using a satellite-based system to obtain positional data. By calculating the separations between receiving antennas as well as satellites with known locations, positioning data is obtained and converted into coordinates. In central focus, coordinates have been solely calculated using information

**Corresponding author:* schakraborty@dsec.ac.in

collected from sensors. Triangulation allows for centimetre-level positional accuracy, giving the derived coordinates base truth information. Superior accuracy, a broad measuring range, real-time possibilities, a shared reference system, as well as closeness to the assessment site are just a few of GPS' benefits. A GPS-accessed set of static as well as kinematic testing reveals a high standard of precision. A tiny antenna was placed at a known location in 2002, as well as global positioning data was gathered for 14.5 hours. The standard variances of the latitudinal, as well as longitudinal values, have both been less than 2 m, indicating that the relocation of the coordinates was reliable. Superior accuracy, a broad measuring range, real-time possibilities, a shared reference system, as well as closeness to the assessment site are just a few of GPS' benefits. A GPS-accessed set of static as well as kinematic testing reveals a high standard of precision. A tiny antenna was placed in a known place in 2002, as well as global positioning data were gathered for 14.5 hours. The standard variances of the latitudinal, as well as longitudinal values, have been just under 2 m, indicating that the relocation of the coordinates was reliable. Often, government organisations responsible for destroying such breeding habitats face problems regarding proper detection and as a result, these areas cannot be accessed. Such areas include roof gutters, rooftops, overhead water reservoirs and cement matters, which are efficient in retaining water. Apart from these, government organisations also fail to detect several other depressed areas which play vital roles in water accumulation and hence provide mosquitoes with an ideal environment for breeding. The major objective of this study is to design an elementary and novel approach for the identification of suitable habitats for mosquitoes using unmanned aerial vehicle (UAV) technology applying machine learning using the Mahalanobis distance classifier and digital terrain models. Mahalanobis distance classifier based supervised classification provide 78% and 81% accuracy for dense urban area of Khanjarpur, Roorkee and urban slum area of Chingrajpara slum, respectively. The research focuses on the identification and comparison of mosquito breeding habitats using remotely-sensed images obtained from UAV for two different areas: urban area of Khanjarpur, Roorkee and urban slum area of Chingrajpara slum.

Keywords: recognising mosquito breeding zones, machine learning (ML), unmanned aerial vehicle (UAV), Mahalanobis distance classifier, dense urban area of Khanjarpur, urban slum area of Chingrajpara slum

Introduction

The landscape ecology of India is very rich and that supports its rapidly growing population. Alongside, India has fairly a large spatial extent of variables and breeding zones, which enhances the potential for dipterous insect breeding and other vector-borne epidemics throughout each and every state of India. Mosquito-related problems and the wide-ranging mosquito-incurred disorder are challenging issues for this nation. It has turned out to be a vital warning to the public well-being, mainly in the fast-growing cities (El-Zeiny, El-Hefni, and Sowilem 2017; Nguyen et al. 2011; Palaniyandi 2014; Palaniyandi Vector et al. 2016; Tran et al. 2008). India emerged as a base for various types of diseases, such as malaria, chikungunya and dengue, and is also recognised as the regional host of the filariasis parasites (Palaniyandi Vector et al. 2016).

The most well-known strategies to overcome mosquito-related diseases are long-lasting insecticide-treated nets (LLINs) (Pryce, Richardson, and Lengeler 2018) and use of indoor residual spray (IRS) (Pluess et al. 2010), which aim to destroy endophilic and endophagic dipterous insect vectors. However, the decrease in their effectiveness is associated mainly with the evasion of pesticide contact by fast-departing nature of mosquitoes, which sustain interiors (Killeen 2014), growing exterior fuelling and transmitting nature, zoophilic response and insecticide tolerance (Lyimo and Ferguson 2009).The urgent requirement to re-plan vector supervision devices for dipterous insects, resistive to present-day interference, has guided to focus on key ecological assets and rising importance of larval source management (LSM) (Tusting 2014; Tusting et al. 2013). Gravid female mosquitoes obtain the ability for distinguishing wetlands and search appropriate zones for laying eggs by utilising optical as well as olfactory signals (Bentley and Day 1989). As a result, awareness of the categorisation and recognition of the extremely fruitful and favourable wetlands should assist in extending the effectiveness of intended larval mosquito management.

The conventional survey approach on the habitats of the larva, in general, accomplishes small-scale space-based study, restricting investigation about mosquito breeding zones. Larger aquatic zones over vast areas are not possible for ground surveying because of not only the composite earth structure, but also the dynamic character of those aquatic zones. Multiple pieces of research revealed the ability of satellite imagery for identifying big mosquito breeding zones across many nations (Achee et al. 2006; Caldas De Castro et al. 2006; Vittor et al. 2009). Freely available satellite images (e.g. ~10 metres/pixel of Sentinel - 2) is a fairly low spatial resolution and hence insufficient for the study of mosquito breeding zones. Moreover, the quality of images due to climatic conditions, particularly throughout the rainy season, is also an issue for satellite images. However, there are implementations for unmanned serial vehicles (UAV) across many sectors, such as crop monitoring (Yang et al. 2017) and monitoring of forest (Dash et al. 2017). Few researchers have applied this technology to scrutinise mosquito breeding zones associated with transmitting sectors. A couple of contemporary researchers have utilized UAVs to make the layout of land use and breeding zones of mosquitoes (Fornace et al. 2014; Hardy et al. 2017) and to relate landscape ecology with malaria epidemiology (Kaewwaen and Bhumiratana 2015). Although, very less similar research was done in India, which is functionally identifying potential mosquito breeding grounds with a comparative approach betwixt dense urban areas and urban slum zone. GIS is utilised for mapping the determinants of the environment and ecology of mosquito dominance (Leonardo et al. 2005; Palaniyandi 2014). This space-related analysis is important for highlighting the malady diffusion, as well as to identify important zones for mosquito restraint schemes and managing practices (Ahmad et al. 2018; Gluskin et al. 2014; Hay et al. n.d.; Nguyen et al. 2011; Palaniyandi, Palaniyandi, et al. 2014; Palaniyandi, Anand, and R.Maniyosai 2014; Palaniyandi and Palaniyandi 2014b, 2014a; Pam et al. 2017; Sharma et al. 1996). These issues have been a major setback for economical consequence of the nation. They include freight of disability-adjusted life years (DALY), disparity within the social sector, physical distress and individual's economic loss. The year-wise expense for controlling, management and treatment of current situation is huge.

The present research is focused on the use of GIS and remote sensing technology to map mosquito-genic breeding grounds and to design maps that can be used for recommending mosquito control. This study also compares the situation betwixt dense urban areas and urban slum zones in the India context (El-Zeiny et al. 2017; Leonardo et al. 2005; Palaniyandi 2014; Palaniyandi, Palaniyandi, et al. 2014; Palaniyandi Vector et al. 2016; Pryce et al. 2018).

Study Area

The study has compared the identification of mosquito breeding zones betwixt a dense urban area and an urban slum area. For the dense urban area, the study is concentrated at the Khanjarpur area of Roorkee (Uttrakhand). Khanjarpur area ranges from 29°51′55″N to 29°52′6″N and 83°54′14″E to 83°54′28″E covering an area of 0.061 km^2. For urban slum area, the study considered the Chingrajpara slum area of Bilaspur (Chhattisgarh). Chingrajpara slum area ranges from 22°5′20″N to 22°5′36″N and 82°9′59″E to 82°10′14″E covering an area of 0.130 km^2. In both areas, mosquito-related diseases have remained a significant issue for a long time and every year a large number of people in these areas get affected and die due to mosquito-related diseases.

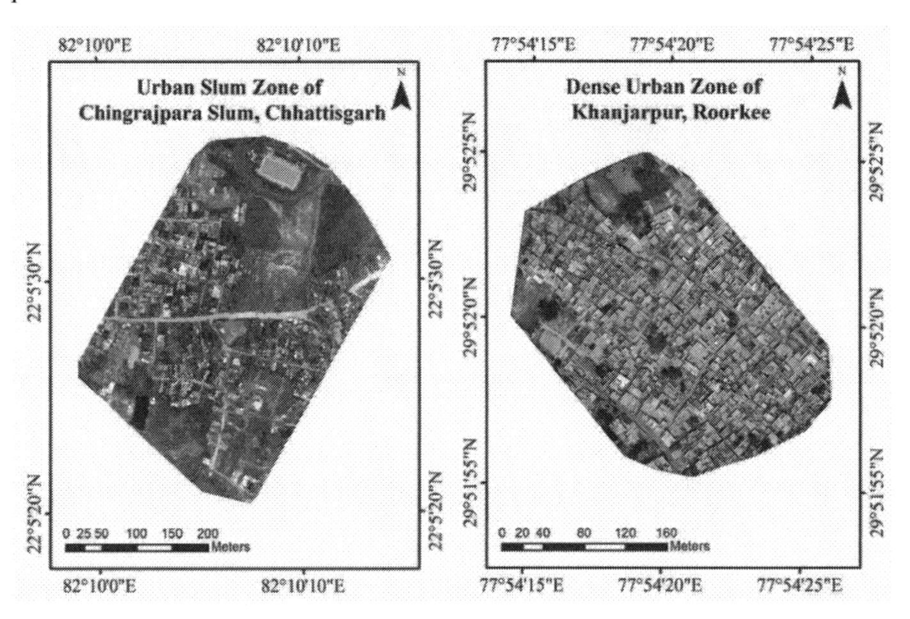

Fig. 13.1 Location map of the study areas

Data Collection and Processing

The study has been carried out using the UAV dataset for the dense urban area of the Khanjarpur area in Uttarakhand of India and for the urban slum area of Chingrajpara area in Chhattisgarh of India. Sample UAV dataset available for the dense urban area of Khanjarpur and urban slum area of Chingrajpara were used. DJI Phantom 4pro and DJI Inspire 2 are used to collect the UAV image datasets. For pre-processing of these images, Pix4Dmapper Pro version 4.3.31 was used. For the dense urban area of Khanjarpur, the flying height of the drone was set at 150 metres and the average ground sampling distance was specified to be 1.79 cm.

This dataset contains 102 UAV images of a highly dense urban area. For the urban slum area of Chingrajpara, flying height of the drone was fixed at 100 metres and the average ground sampling distance was set to be 2.19 cm. This dataset contains 85 UAV images of the slum area with sub-urban features

Methodology and Model Specifications

To find the mosquito breeding grounds, first, we need to classify the UAV images into different classes. In our study, both images have been classified into four major classes; vegetation, open land, roads and urban structures. For classification, we have considered the machine learning technique Mahalanobis distance classifier for image classification (Perumal and Bhaskaran 2010). For supervised classification, we classify the datasets into two subgroups for training and testing purposes. We preferred 70:30 partitions of study area data in favour of training and testing, respectively. It is advocated that the majority of data train properly at the ratio of 70:30 ensuing high accuracy (Ehlers et al. n.d.). ENVI software was used for image classification using the training dataset.

Mahalanobis Distance Classifier

Mahalanobis distance is the space measurement betwixt two nodes in the area specified with two or more than two interrelated parameters. It accepts the interrelations betwixt the variables within the data set. For the case of non-correlated two variables in a two-dimensional scatter space, Mahalanobis distance of the variables betwixt the points is similar to Euclidean distance (Panuju, Paull, and Griffin 2020; Xu n.d.).

If the covariance matrix is the unit matrix, then the Mahalanobis distance is same as the Euclidean distance. An exact case forms when the standardised data matrix having the two columns is orthogonal (Madhura and Venkatachalam 2013). Mahalanobis distance relies upon on the attribute's dispersion matrix, which is also responsible to depict the interrelations. Dispersion matrices are used between two components of random variables to correct the effects of cross-covariance. The distance betwixt a node and the midpoint for every cluster in multidimensional space specified by multi-variable and their covariance is known as Mahalanobis distance. So, a small value of Mahalanobis distance enhances the chances of a node to become close to centre of the cluster and more presumably being assigned to that cluster. (Al-Ahmadi and Hames 2009) Mahalanobis distance towards every class means, for each feature vector is calculated by an equation. This includes the calculation of the variance-dispersion matrix of each class.

$$dk2 = (X - \mu k)\ TV\ (X - \mu k)$$

Here, X is vector of image data, μk is the mean vector of the kth class and V is the variance-covariance matrix (Khan et al. 2015).

Accuracy Assessment

Classification accuracy can be measured by comparing a few training data, rejecting the ground truth with the classified pixels. In any mapping project, accuracy assessment is recognised as a critical component. Every classifier's accuracy is done on the foundation of the testing pixels using an error matrix and two further renowned criteria.

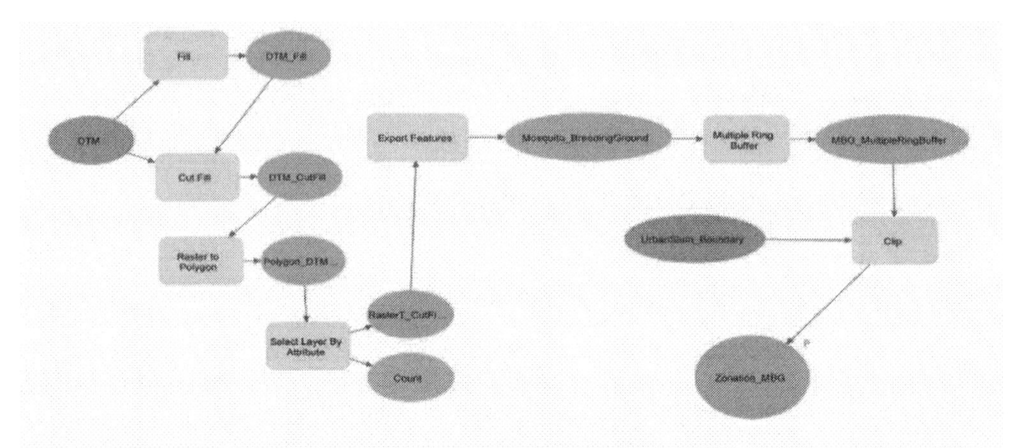

Python Script For The Model

```
import arcpy
def Model():  # Model
    arcpy.env.overwriteOutput = False
    arcpy.ImportToolbox(r"c:\program files\arcgis\pro\Resources\ArcToolbox\toolboxes\Analysis Tools.tbx")
    DTM = arcpy.Raster("DTM")
    UrbanSlum_Boundary = "UrbanSlum_Boundary"
    DTM_Fill = "C:\\Users\\joydip.datta\\Documents\\ArcGIS\\Projects\\Howrah\\Howrah.gdb\\DTM_Fill"
    Fill = DTM_Fill
    DTM_Fill = arcpy.sa.Fill(in_surface_raster=DTM, z_limit=None)
    DTM_Fill.save(Fill)
    DTM_CutFill = "C:\\Users\\joydip.datta\\Documents\\ArcGIS\\Projects\\Howrah\\Howrah.gdb\\DTM_CutFill"
    Cut_Fill = DTM_CutFill
    DTM_CutFill = arcpy.sa.CutFill(in_before_surface=DTM, in_after_surface=DTM_Fill, z_factor=1)
    DTM_CutFill.save(Cut_Fill)
    Polygon_DTM_CutFill =
"C:\\Users\\joydip.datta\\Documents\\ArcGIS\\Projects\\Howrah\\Howrah.gdb\\Polygon_DTM_CutFill"
    with arcpy.EnvManager(outputMFlag="Disabled", outputZFlag="Disabled"):
        arcpy.conversion.RasterToPolygon(in_raster=DTM_CutFill, out_polygon_features=Polygon_DTM_CutFill,
simplify="SIMPLIFY", raster_field="VALUE", create_multipart_features="SINGLE_OUTER_PART",
max_vertices_per_feature=None)
    RasterT_CutFill3_Layer, Count = arcpy.management.SelectLayerByAttribute(in_layer_or_view=Polygon_DTM_CutFill,
selection_type="NEW_SELECTION", where_clause="GRIDCODE > 1 And Shape_Area > 1",
invert_where_clause="NON_INVERT")
    Mosquito_BreedingGround =
"C:\\Users\\joydip.datta\\Documents\\ArcGIS\\Projects\\Howrah\\Howrah.gdb\\Mosquito_BreedingGround"
    arcpy.conversion.ExportFeatures(in_features=RasterT_CutFill3_Layer, out_features=Mosquito_BreedingGround,
where_clause="", use_field_alias_as_name="NOT_USE_ALIAS", field_mapping="", sort_field=[])
    MBG_MultipleRingBuffer =
"C:\\Users\\joydip.datta\\Documents\\ArcGIS\\Projects\\Howrah\\Howrah.gdb\\MBG_MultipleRingBuffer"
    arcpy.analysis.MultipleRingBuffer(Input_Features=Mosquito_BreedingGround,
Output_Feature_class=MBG_MultipleRingBuffer, Distances=[10, 20, 30, 40, 50], Buffer_Unit="Meters",
Field_Name="distance", Dissolve_Option="ALL", Outside_Polygons_Only="FULL", Method="PLANAR")
    Zonation_MBG = "C:\\Users\\joydip.datta\\Documents\\ArcGIS\\Projects\\Howrah\\Howrah.gdb\\Zonation_MBG"
    arcpy.analysis.Clip(in_features=MBG_MultipleRingBuffer, clip_features=UrbanSlum_Boundary,
out_feature_class=Zonation_MBG, cluster_tolerance="")
if _name_ == '_main_':
    # Global Environment settings
    with arcpy.EnvManager(scratchWorkspace=r"C:\Users\joydip.datta\Documents\ArcGIS\Projects\Howrah\Howrah.gdb",
workspace=r"C:\Users\joydip.datta\Documents\ArcGIS\Projects\Howrah\Howrah.gdb"):
        Model()
```

Fig. 13.2 Python script of the model

A confusion matrix is composed of spatial comparison of evaluating pixels of classifications. The main diagonal represents the number of pixels that are given the same identification by test data classifiers, these are the correctly classified number of pixels. The classification accuracy is high if the cell value is large. The misclassified pixels are represented by the off-main diagonal in the form of errors or omissions and errors of commissions. The overall accuracy is calculated by dividing the total of main diagonal entries of the confusion matrix by the total number of samples (Foody 2002; Ibrahim, Zakariya, and Wahab 2018; Viera and Garrett 2005).

Omission error = (Off diagonal elements/Total no. of columns)

Commission error = (Off diagonal elements/Total no. of rows)

User and producer accuracy have direct relationships to errors of commission and errors of omission, respectively. The relationships are presented as

User's accuracy (reliability) = 100% – Commission error (%)

Producer's accuracy = 100% – Error of omission (%)

Overall accuracy is a measurement of the ratio of the count of pixels classified similarly in the satellite image and on the ground to the total count of pixels.

Overall accuracy = (Count of classified pixels/Sum of number of pixels)

As the term specifies, it depicts the overall accuracy of the classification instead of the accuracy of each class being categorised independently.

Kappa Coefficient

It is helpful to measure the observed agreement betwixt two classifiers which can classify N items into C classes. Performances of classifiers statistically evaluated by Kappa co-efficient and with respect to a random classifier denote their accuracy.

$$K = (N\Sigma ji = 1\ xii - \Sigma ji = 1\ xi + .x + i)/(N^2 - \Sigma ji = 1\ xi + .x + i)$$

where N = total amount of pixels in all classes

j = total amount of classes

Xii = pixels on diagonal of confusion matrix

$X + i$ = sum of all rows on column i

$Xi +$ = sum of all columns on row i

After classifying both UAV images, we identified the water retention zones in our study areas to identify the mosquito breeding grounds. Mosquito lay their eggs in the aquatic environment and during rainy season, the precipitation creates habitats for mosquitoes to lay eggs and their larvae to grow. For that alarming reason, identifying depressed elevation zones from the digital terrain models (DTM) is very necessary to spot major mosquito breeding grounds. In the DTM, if a pixel (or region of pixels) is at a lesser height than its neighbouring pixels, the pixel behaves like a 'sink', and surface flow will be blocked and water will be assembled there. To identify such pixels, fill tool under the spatial analyst tools in ArcGIS Pro is applied. Reconditioning DTM with this tool produces a new elevation surface with all sinks filled to their spill elevation.

After that, the output raster developed through the fill tool was subtracted from the actual DTM. The results in the new raster only represent the depressed elevation zones or sinks and their deepness value. By providing a threshold of 0.05 m to the new raster map containing the depressed elevation zones, potential water retention zones were identified. A threshold of 10 m^2 was provided on the depressed elevation zone map to identify major potential mosquito breeding zones. Buffer analysis under the analysis tools in ArcGIS Pro was used in the major potential mosquito breeding zones at 10-m interval, and the whole study area was classified into five classes (on 10-m interval) to represent mosquito-prone areas. Thus, zones that are more or less prone to mosquito-induced diseases were identified for both study areas.

Results and Discussions

Using Mahalanobis distance classification method, ortho-mosaic UAV images of dense urban area of Khanjarpur, Roorkee and urban slum area of Chingrajpara slum, Chhattisgarh were classified into four major land cover classes: vegetation, open land, roads and urban structures. The map below shows the land use land cover zones of the study areas. The accuracy assessment mentioning producers and user accuracy as well as omission and commission errors of the classification for both areas is discussed in Table 13.1 with the help of the confusion matrix.

From the analysis of Table 13.1 and 13.2, we can state that for the dense urban area of Khanjarpur, Roorkee, we can forecast further analysis with a 79.4% confidence and for the urban slum area of Chingrajpara slum, Chhattisgarh, we can forecast further analysis with an 80.1%

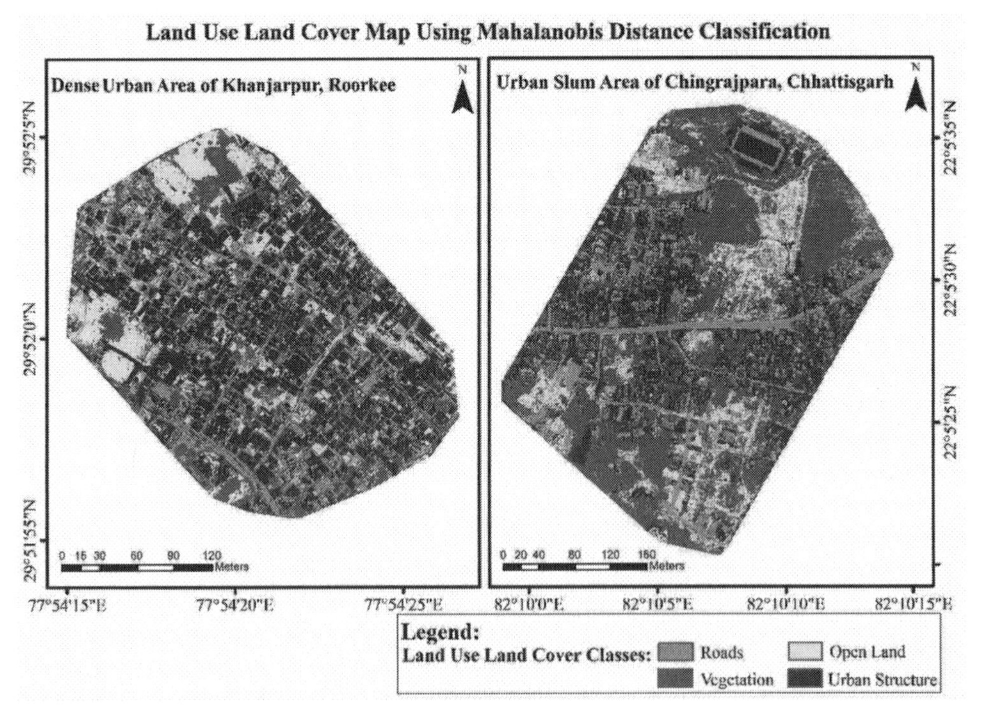

Fig. 13.3 Land use land cover map using Mahalanobis distance classification

Table 13.1 Confusion matrix of dense urban area of Khanjarpur, Roorkee

Dense urban	Road	Vegetation	Open land	Urban structure	Row total
Road	147	21	45	37	250
Vegetation	8	228	14	0	250
Open land	28	6	208	8	250
Urban Structure	21	2	16	211	250
Column total	204	257	283	256	1000
	Producers Accuracy	**Omission Errors**	**User Accuracy**	**Commission Errors**	
Road	72.06	27.94	58.80	41.20	
Vegetation	88.72	11.28	91.20	8.80	
Open Land	73.50	26.50	83.20	16.80	
Urban Structure	82.42	17.58	84.40	15.60	
Overall classification accuracy				79.4	
Overall Kappa statistics				0.7253	

Table 13.2 Confusion matrix of urban slum area of Chingrajpara slum, Chhattisgarh

Urban Slum	Road	Vegetation	Open Land	Urban Structure	Row Total
Road	207	1	9	33	250
Vegetation	2	236	9	3	250
Open land	31	18	189	12	250
Urban structure	68	2	11	169	250
Column total	308	257	218	217	1000
	Producers Accuracy	**Omission Errors**	**User Accuracy**	**Commission Errors**	
Road	67.21	32.79	82.80	17.20	
Vegetation	91.83	8.17	94.40	5.60	
Open land	86.70	13.30	75.60	24.40	
Urban structure	77.88	22.12	67.60	32.40	
Overall classification accuracy				80.1	
Overall Kappa statistics				0.7347	

confidence. After the land cover analysis, the digital terrain model maps for both study areas were considered. It is observed that the elevation of the Khanjarpur area varies from 267 m to 277 m, and the elevation of the Chingrajpara varies from 318 m to 327 m. Both DTM maps have been classified into five classes for better visual representations.

To determine the depressed elevation zones, fill maps from the respective DTM maps were subtracted and zones were classified into more than and less than 0.05 m. The mosquito can lay

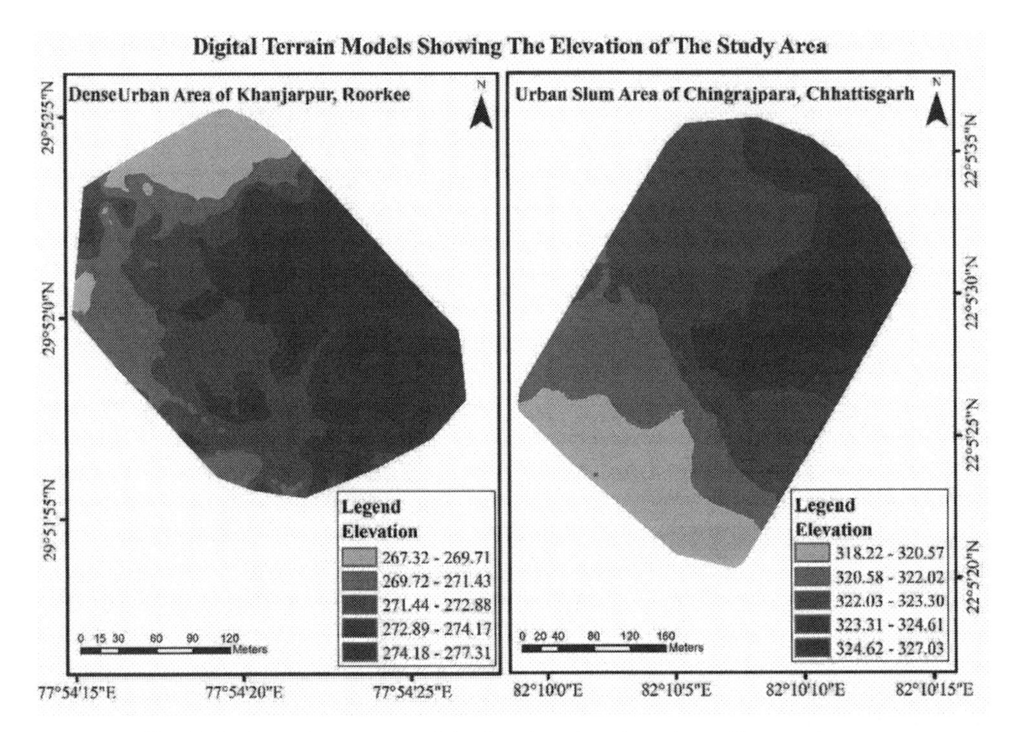

Fig. 13.4 Digital terrain models showing the elevation of the study area

eggs even in wet soil. Therefore, we considered 0.05 m as a good threshold, and zones having more depth than 0.05 m are considered as major depressed zones for both the study areas. It is noticed that in the Khanjarpur area, major depressed zones are majorly in the northeast, south and in western part covering nearly 4678.93 m^2 which is around 7.67% of the total area, whereas, in the Chingrajpara area, the major depressed zones are majorly in the northern, central and western part covering nearly 8348.45 m^2 which is around 6.42% of the total area.

Generally, mosquitoes breed in stagnant water, and to identify such areas from the map of depressed elevation zones, zones having areas 10 m^2 and above are identified and labelled as major mosquito breeding grounds. In the above map, we showed the zones having major mosquito breeding grounds and applied buffer techniques of 10-m distance over those zones repeatedly for four times to identify the degree of mosquito-prone zones. For the dense urban area of Khanjarpur, Roorkee, it is observed that 55.37% of the total area comes in the 10-m zone, 31.68% of the total area comes in the 10-m–20-m zone, 10.13% of the total area comes in the 20-m–30-m zone, 2.43% of the total area comes in the 30-m–40-m zone and only 0.38% comes in more than 40-m zone. But for the case of the urban slum area of Chingrajpara, the scenario is slightly different. In the urban slum area of Chingrajpara, 34% of the total area comes in the 10-m zone, 29.81% of the total area comes in the 10-m–20-m zone, 17.02% of the total area comes in the 20-m–30-m zone, 10.89% of the total area comes in the 30-m–40-m zone and 8.28% comes in the more than 40-m zone.

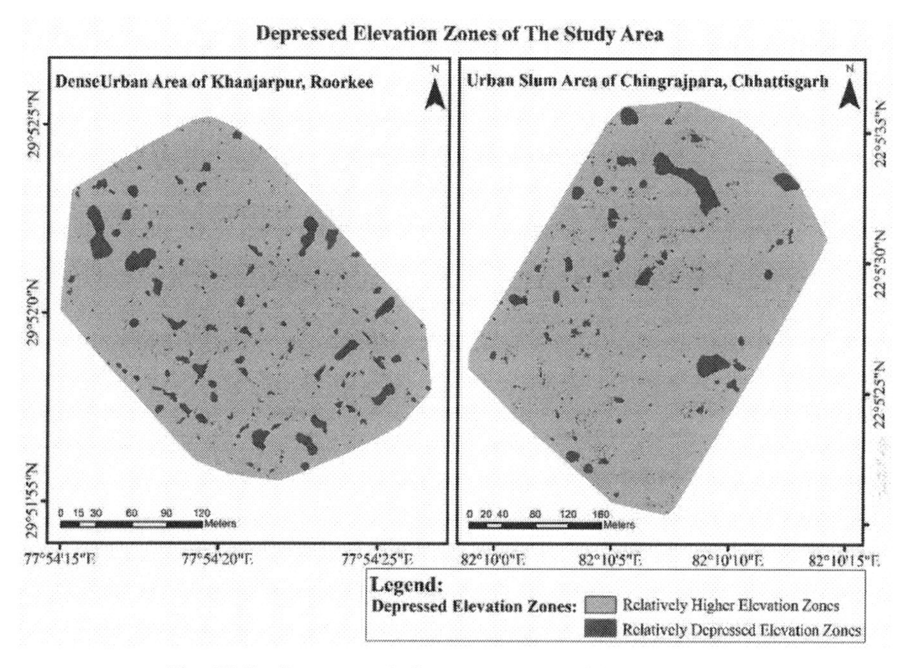

Fig. 13.5 Depressed elevation zones of the study area

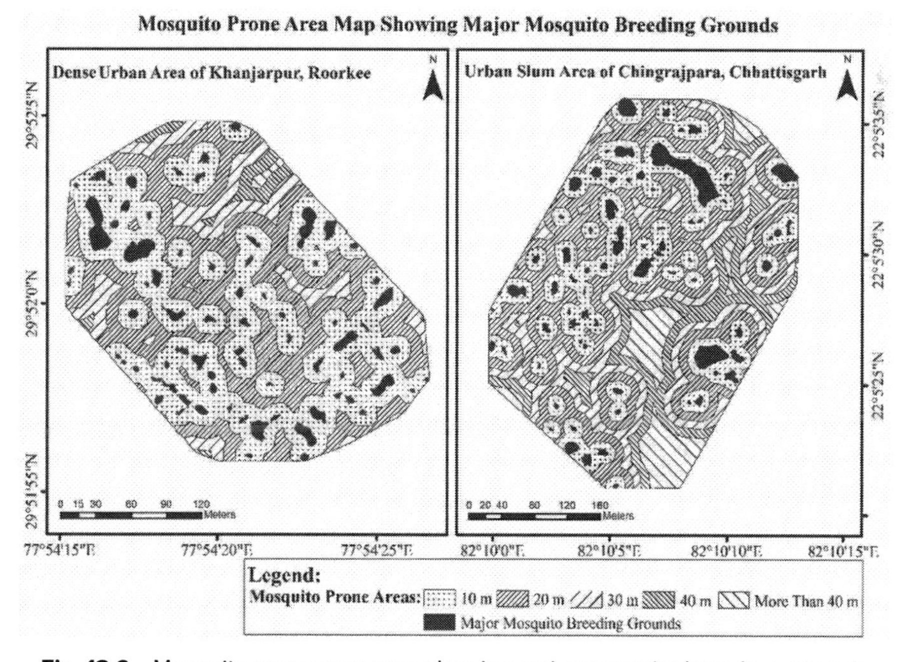

Fig. 13.6 Mosquito prone area map showing major mosquito breeding grounds

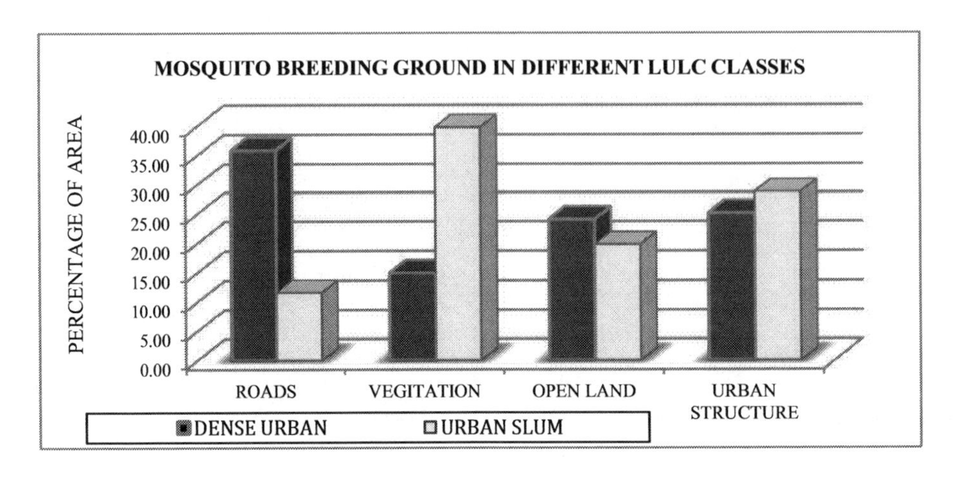

Fig. 13.7 Mosquito breeding ground in different LULC classes

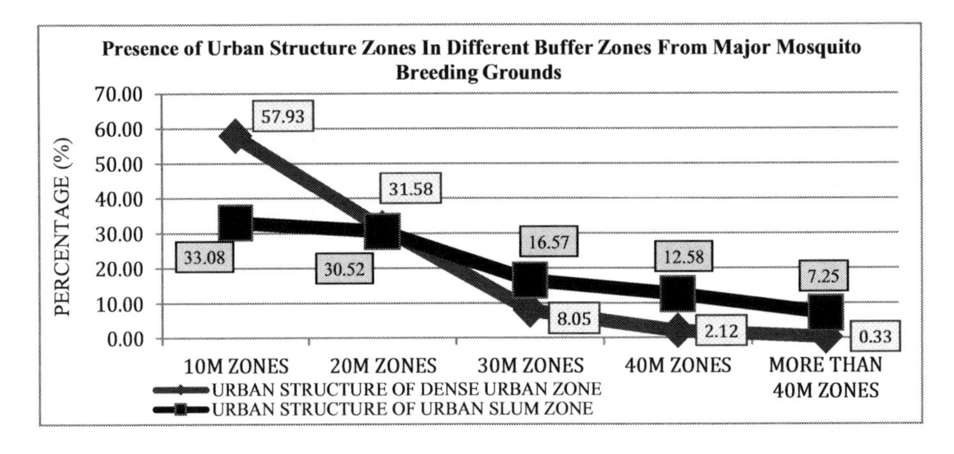

Fig. 13.8 Presence of urban structure zones in different buffer zones from major mosquito breeding grounds

After identifying the mosquito breeding grounds, land cover of those places was analysed. It is observed that mosquitoes generally breed in stagnant water. Hence, roads cannot be a place that can be considered as potential mosquito breeding grounds, as it is a place of regular motion. From that perspective, it can be said that among the potential mosquito breeding zones 35.91% of the dense urban area of Khanjarpur and 11.59% of the urban slum area of Chingrajpara are not the places where mosquito can breed due to the regular turbulence in water on that zone as they are part of roads. Apart from roads in the potential mosquito breeding ground, if vegetation and open land area are included in the dense urban area of Khanjarpur, it is covering 38.97% and in the urban slum area of Chingrajpara; it is 59.61%. These statistics are alarming for the slum zone as these areas are the potential zones for mosquito breedings. These vegetation and open land zones need special care from the local government. In the case of urban structures, potential mosquito breeding grounds in the dense urban zone of Khanjarpur is 25.12% and the urban slum zone of Chingrajpara is 28.80%. The urban structures in these zones should be on high alert. Apart from the urban structures in the potential mosquito breeding grounds, if conditions of the urban structure in different buffer zones are considered, we will come to know how settlement patterns also play a role. From the approach of settlement geography, it can be said that the settlement pattern in the dense urban zone is compact and in the urban slum area is relatively dispersed. That is the reason why the curve of the dense urban area falls on a rapid slope and the curve of the urban slum area shows a steady slope from 10-m zone to more than 40-m zone.

Conclusion

The comparative analysis betwixt the dense urban zone of Khanjarpur and the urban slum zone of Chingrajpara shows some interesting facts. In the dense urban zone of Khanjarpur, it was noticed that due to the compact settlement patterns, 55% of the total urban structures are in the 10-m buffer zone. Similarly, around 36% of the depressed zone comes under roads which cannot be a potential mosquito breeding ground. About 25% comes under urban structure, where the problems of mosquito breeding can be solved with daily surveillance. In the case of urban slum zone of Chingrajpara, due to its less compact nature of the urban structure zone, there are around 33% urban structure pixels in the 10-m buffer zone, but alarmingly there are around 40% pixel of vegetation and 20% pixel of open land in the depressed zones. This 60% area has a high potential to convert into mosquito breeding grounds after rainy days. So, comparatively, the urban slum area of Chingrajpara is in a highly sensitive mode during and after the rainy days than the dense urban area of Khanjarpur, Roorkee. In both images, many things like cycle tyres, bamboos, dumped building materials and others are observed within the urban structures. These can also turn into potential mosquito breeding grounds and need special care from the local people to eradicate these problems during and after rainy days.

Similar studies, when carried out using a large number of UAV images, field surveys and analysing tools, such as combination of machine learning and GIS techniques, can be very useful to municipal authorities to eradicate the problem due to mosquito and other vector-borne diseases in India.

Acknowledgements

We are highly thankful to Dr. Kamal Jain, Professor, Indian Institute of Technology-Roorkee for providing the processed UAV images as well as the digital surface models (DSM) and DTM for the study areas.

References

1. Achee, N. L., Grieco, J. P., Masuoka, P., Andre, R. G., Roberts, D. R., Thomas, J., Briceno, I., King, R., & Rejmankova, E. (2006). Use of remote sensing and geographic information systems to predict locations of Anopheles darlingi-positive breeding sites within the Sibun River in Belize,

Central America. Journal of Medical Entomology, 43(2), 382–392. https://doi.org/10.1603/0022-2585(2006)043[0382:UORSAG]2.0.CO;2

2. Ahmad, S., Asif, M., Talib, R., Adeel, M., Yasir, M., & Chaudary, M. H. (2018). Surveillance of intensity level and geographical spreading of dengue outbreak among males and females in Punjab, Pakistan: A case study of 2011. Journal of Infection and Public Health, 11(4), 472–485. https://doi.org/10.1016/j.jiph.2017.10.002
3. Al-Ahmadi, F. S., & Hames, A. S. (2009). Comparison of four classification methods to extract land use and land cover from raw satellite images for some remote arid areas, Kingdom of Saudi Arabia. Journal of King Abdulaziz University, Earth Sciences, 20(1), 167–191. https://doi.org/10.4197/Ear.20-1.9
4. Bentley, M. D., & Day, J. F. (1989). CHEMICAL ECOLOGY AND BEHAVIORAL ASPECTS OF MOSQUITO OVIPOSITION Further ANNUAL REVIEWS. In Ann. Rev. Entomol (Vol. 34). www.annualreviews.org
5. Caldas De Castro, M., Monte-Mó, R. L., Sawyer, D. O., & Singer, B. H. (2006). Malaria risk on the Amazon frontier. In PNAS February (Vol. 14, Issue 7). www.pnas.orgcgidoi10.1073pnas.0510576103
6. Dash, J. P., Watt, M. S., Pearse, G. D., Heaphy, M., & Dungey, H. S. (2017). Assessing very high resolution UAV imagery for monitoring forest health during a simulated disease outbreak. ISPRS Journal of Photogrammetry and Remote Sensing, 131, 1–14. https://doi.org/10.1016/j.isprsjprs.2017.07.007
7. Ehlers, A., Baumann, F., Spindler, R., Glasmacher, B., & Rosenhahn, B. (n.d.). PCA Enhanced Training Data for Adaboost.
8. El-Zeiny, A., El-Hefni, A., & Sowilem, M. (2017). Geospatial techniques for environmental modeling of mosquito breeding habitats at Suez Canal Zone, Egypt. Egyptian Journal of Remote Sensing and Space Science, 20(2), 283–293. https://doi.org/10.1016/j.ejrs.2016.11.009
9. Foody, G. M. (2002). Status of land cover classification accuracy assessment. Remote Sensing of Environment, 80(1), 185–201. https://doi.org/https://doi.org/10.1016/S0034-4257(01)00295-4
10. Fornace, K. M., Drakeley, C. J., William, T., Espino, F., & Cox, J. (2014). Mapping infectious disease landscapes: Unmanned aerial vehicles and epidemiology. In Trends in Parasitology (Vol. 30, Issue 11, pp. 514–519). Elsevier Ltd. https://doi.org/10.1016/j.pt.2014.09.001
11. Gluskin, R. T., Johansson, M. A., Santillana, M., & Brownstein, J. S. (2014). Evaluation of Internet-Based Dengue Query Data: Google Dengue Trends. PLoS Neglected Tropical Diseases, 8(2). https://doi.org/10.1371/journal.pntd.0002713
12. Hardy, A., Makame, M., Cross, D., Majambere, S., & Msellem, M. (2017). Using low-cost drones to map malaria vector habitats. Parasites and Vectors, 10(1). https://doi.org/10.1186/s13071-017-1973-3
13. Hay, S. I., Omumbo, J. A., Craig, M. H., & Snow, R. W. (n.d.). Earth Observation, Geographic Information Systems and Plasmodium falciparum Malaria in Sub-Saharan Africa.
14. Ibrahim, I., Zakariya, K., & Wahab, N. A. (2018). Satellite Image Analysis along the Kuala Selangor to Sabak Bernam Rural Tourism Routes. IOP Conference Series: Earth and Environmental Science, 117(1). https://doi.org/10.1088/1755-1315/117/1/012013
15. Kaewwaen, W., & Bhumiratana, A. (2015). Landscape ecology and epidemiology of malaria associated with rubber plantations in Thailand: Integrated approaches to malaria ecotoping. In Interdisciplinary Perspectives on Infectious Diseases (Vol. 2015). Hindawi Publishing Corporation. https://doi.org/10.1155/2015/909106
16. Khan, U., Minallah, N., Junaid, A., Gul, K., & Ahmad, N. (2015). Parallelepiped and Mahalanobis Distance based Classification for forestry identification in Pakistan. 2015 International Conference on Emerging Technologies (ICET), 1–6. https://doi.org/10.1109/ICET.2015.7389199

17. Killeen, G. F. (2014). Characterizing, controlling and eliminating residual malaria transmission. In Malaria Journal (Vol. 13, Issue 1). BioMed Central Ltd. https://doi.org/10.1186/1475-2875-13-330
18. Leonardo, L., Rivera, P., Crisostomo, B., Sarol Jr, J., Bantayan, N., Tiu, W., & Bergquist, R. (2005). A study of the environmental determinants of malaria and schistosomiasis in the Philippines using Remote Sensing and Geographic Information Systems. Parassitologia, 47, 105–114.
19. Lyimo, I. N., & Ferguson, H. M. (2009). Ecological and evolutionary determinants of host species choice in mosquito vectors. In Trends in Parasitology (Vol. 25, Issue 4, pp. 189–196). https://doi.org/10.1016/j.pt.2009.01.005
20. Madhura, M., & Venkatachalam, S. (2013). Comparison of Supervised Classification Methods On Remote Sensed Satellite Data: An Application In Chennai, South India. In International Journal of Science and Research (Vol. 4) www.ijsr.net
21. Nguyen, L. A. P., Clements, A. C. A., Jeffery, J. A. L., Yen, N. T., Nam, V. S., Vaughan, G., Shinkfield, R., Kutcher, S. C., Gatton, M. L., Kay, B. H., & Ryan, P. A. (2011). Abundance and prevalence of Aedes aegypti immatures and relationships with household water storage in rural areas in southern Viet Nam. International Health, 3(2), 115–125. https://doi.org/10.1016/j.inhe.2010.11.002
22. Palaniyandi, M. (2014). Gis Based Site Selection For Fixing Uv Light Adult Mosquito Trap And Gravid Adult Mosquito Trap For Epidemic Control In The Urban Settlements. INTERNATIONAL JOURNAL OF SCIENTIFIC & TECHNOLOGY RESEARCH, 3(8). www.ijstr.org
23. Palaniyandi, M., Anand, P., & R.Maniyosai. (2014). GIS based community survey and systematic grid sampling for dengue epidemic surveillance, control, and management: a case study of Pondicherry Municipality. International Journal of Mosquito Research, 1, 72–80.
24. Palaniyandi, M., & Palaniyandi, M. (2014a). Epidemic Control in the Urban Settlements Article in. International Journal of Scientific & Technology Research, 3(8). www.ijstr.org
25. Palaniyandi, M., & Palaniyandi, M. (2014b). Web mapping GIS: GPS under the GIS umbrella for Aedes species dengue and chikungunya vector mosquito surveillance and control Rapid Epidemiological Mapping of Lymphatic Filariasis in Southern India View project Application of Remote Sensing and Geographical Information Systems (GIS) for Epidemiology and Control of Lymphatic Filariasis View project Web mapping GIS: GPS under the GIS umbrella for Aedes species dengue and chikungunya vector mosquito surveillance and control. International Journal of Mosquito Research, 18(3), 18–25. https://www.researchgate.net/publication/265420844
26. Palaniyandi, M., Palaniyandi, M., Maniyosai, R., & Anand, P. H. (2014). Spatial cognition: A geospatial analysis of vector borne disease transmission and the environment, using remote sensing and GIS Rapid Epidemiological Mapping of Lymphatic Filariasis in Southern India View project Environmental Modeling for Prediction of Dengue Epidemics in India View project Spatial cognition: a geospatial analysis of vector borne disease transmission and the environment, using remote sensing and GIS. Article in International Journal of Mosquito Research, 1(3), 39–54. https://www.researchgate.net/publication/265555358
27. Palaniyandi Vector, M., Palaniyandi, M., Anand, P. H., Maniyosai, R., Mariappan, T., & Das, P. K. (2016). The integrated remote sensing and GIS for mapping of potential vector breeding habitats, and the Internet GIS surveillance for epidemic transmission control, and management. ~ 310 ~ Journal of Entomology and Zoology Studies, 4(2).
28. Pam, C., Omalu, D. D., Akintola, I., Azumi, D., Kalesanwo, Y., Babagana, A. O., Muhammad, M., Ocha, S. A., & Adeniyi, I. M. (2017). The Role of GIS And Remote Sensing in the Control of Malaria. In Online J Health Allied Scs (Vol. 16, Issue 3).
29. Panuju, D. R., Paull, D. J., & Griffin, A. L. (2020). Change detection techniques based on multispectral images for investigating land cover dynamics. In Remote Sensing (Vol. 12, Issue 11). MDPI AG. https://doi.org/10.3390/rs12111781

30. Perumal, K., & Bhaskaran, R. (2010). SUPERVISED CLASSIFICATION PERFORMANCE OF MULTISPECTRAL IMAGES (Vol. 2).
31. Pluess, B., Tanser, F. C., Lengeler, C., & Sharp, B. L. (2010). Indoor residual spraying for preventing malaria. Cochrane Database of Systematic Reviews. https://doi.org/10.1002/14651858.cd006657.pub2
32. Pryce, J., Richardson, M., & Lengeler, C. (2018). Insecticide-treated nets for preventing malaria. In Cochrane Database of Systematic Reviews (Vol. 2018, Issue 11). John Wiley and Sons Ltd. https://doi.org/10.1002/14651858.CD000363.pub3
33. Sharma, V. P., Dhiman, R. C., Ansari, M. A., Nagpal, B. N., Srivastava, A., Manavalan, P., Adiga, S., Radhakrishnan, K., & Chandrasekhar, M. G. (1996). Study on the feasibility of delineating mosquitogenic conditions in and around Delhi using Indian Remote Sensing Satellite data. Indian Journal of Malariology, 33(3), 107—125. http://europepmc.org/abstract/MED/9014394
34. Tran, A., Ponçon, N., Toty, C., Linard, C., Guis, H., Ferré, J. B., Lo Seen, D., Roger, F., de la Rocque, S., Fontenille, D., & Baldet, T. (2008). Using remote sensing to map larval and adult populations of Anopheles hyrcanus (Diptera: Culicidae) a potential malaria vector in Southern France. International Journal of Health Geographics, 7, 9. https://doi.org/10.1186/1476-072X-7-9
35. Tusting, L. S. (2014). Larval source management: A supplementary measure for malaria control. Outlooks on Pest Management, 25(1), 41–43. https://doi.org/10.1564/v25_feb_13
36. Tusting, L. S., Thwing, J., Sinclair, D., Fillinger, U., Gimnig, J., Bonner, K. E., Bottomley, C., & Lindsay, S. W. (2013). Mosquito larval source management for controlling malaria. In Cochrane Database of Systematic Reviews (Vol. 2013, Issue 8). John Wiley and Sons Ltd. https://doi.org/10.1002/14651858.CD008923.pub2
37. Viera, A. J., & Garrett, J. M. (2005). Understanding interobserver agreement: the kappa statistic. Family Medicine, 37 5, 360–363.
38. Vittor, A. Y., Pan, W., Gilman, R. H., Tielsch, J., Glass, G., Shields, T., Sánchez-Lozano, W., Pinedo, V. V, Salas-Cobos, E., Flores, S., & Patz, J. A. (2009). Linking Deforestation to Malaria in the Amazon: Characterization of the Breeding Habitat of the Principal Malaria Vector, Anopheles darlingi. In Am. J. Trop. Med. Hyg (Vol. 81, Issue 1). www.ajtmh.org
39. Xu, C. (n.d.). Obtaining Forest Description for Small-scale Forests Using an Integrated Remote Sensing Approach.
40. Yang, G., Liu, J., Zhao, C., Li, Z., Huang, Y., Yu, H., Xu, B., Yang, X., Zhu, D., Zhang, X., Zhang, R., Feng, H., Zhao, X., Li, Z., Li, H., & Yang, H. (2017). Unmanned aerial vehicle remote sensing for field-based crop phenotyping: Current status and perspectives. In Frontiers in Plant Science (Vol. 8). Frontiers Media S.A. https://doi.org/10.3389/fpls.2017.01111

CHAPTER 14

Treatment of Anionic Surfactant Contaminated Wastewater by Combined Advanced Oxidation and Biological Processes

Bijoli Mondal*
Associate Professor, Department of Civil Engineering,
Haldia Institute of Technology, Haldia, Purba Medinipur, India

Asok Adak[1]
Associate Professor, Department of Civil Engineering,
Indian Institute of Engineering Science and Technology, Shibpur, Howrah, India

Pallab Datta[2]
Assistant Professor Center for Healthcare Science and Technology, Indian Institute of Engineering Science and Technology, Shibpur, Howrah, India

Abstract

Surfactants are widely used for industrial as well as domestic purposes, and these chemicals are potentially harmful to the environment. This study reports the degradation of sodium dodecyl sulphate (SDS), a representative of anionic surfactants, present in wastewater using a combined approach of UV-H_2O_2 process and biological treatment. The toxicity of raw SDS and partially (50%) UV-H_2O_2 treated SDS was analysed using zebrafish as a model organism and indicating that partially degraded SDS samples were less toxic. Five batch reactors with a working volume of 500 ml each were used for the treatability study. The reactors operated with a hydraulic retention time (HRT) of 24 h. Each reactor was operated at different solid (biomass) retention time, that is, 2, 4, 6, 8 and 10 days. Bio-kinetic parameters (k_S, k, Y of and K_d) for raw SDS and partially degraded (50%) samples were evaluated using a modified form of the Monod equation.

Keywords: bio-kinetics parameters, sodium dodecyl sulfphte, toxicity, UV-H_2O_2

Introduction

Continuing industrialisation, urbanisation, population growth, deforestation and pollution are increasing the number of refractory organics, such as surfactants, herbicides, dyes and phenolic compounds, classified as emerging contaminants (EC) present in the water environment and

*Corresponding author: bijolimondal15@gmail.com
[1]asok@civil.iiests.ac.in; [2]contactpallab@gmail.com

has recently become concern of environmental engineering. They are responsible for causing enormous ecological and human health problems. Among these contaminants, surfactants or surface-active agents are commonly used in household as detergents for cleaning, textile industries, dying industries and so on. After use, the remaining concentration of surfactants and their degraded products are directly discarded into the sewage treatment plants and surface water from where they pollute the different environmental components. Surfactants bearing wastewater can be treated by various physicochemical methods, such as coagulation (Mahmoud and Ahmed, 2014), adsorption (Kyzas et al., 2013), reverse osmosis (Kowalska, 2012) and chemical oxidation (Lin et al., 1999). Each of the above methods has advantages and disadvantages; whereas, in general, treatment of wastewater with high surfactant concentration is very difficult to be treated using conventional physio-chemical treatment methods. Surfactant-contaminated wastewater may be treated by different advanced oxidation processes (AOPs). AOPs are generally recommended as alternative treatment method for the removal of refractory organics in wastewater. AOPs are based on the generation of hydroxyl radicals (HO·). These are highly reactive species with an oxidation potential of +2.8 eV and react with electron-rich organic or inorganic compounds (Buxton et al., 1988). AOPs are employed for the treatment of various types of surfactants by photochemical processes, photocatalytic processes and photo-Fenton processes. Application of AOPs in wastewater plants still remains limited due to high operational costs.

In the present study, an attractive treatment chain comprising short pre-treatment using one AOP followed by a biological treatment process is proposed. The short-term AOP is expected to convert recalcitrant organics into biodegradable intermediates, which will then be treated by the biological process without causing any toxicity to the microorganisms. Among AOP, photocatalytic and photo-Fenton processes have certain limitations. Therefore, we have selected homogeneous UV-H_2O_2 AOP followed by the biological process to provide a more efficient removal method to treat the most widely used anionic (sodium dodecyl sulphate, SDS).

Earlier, Alaton et al. (2007) investigated an integrated photochemical (H_2O_2-UV) and biological (activated sludge) treatment of an anionic and a non-ionic surfactant that was widely used in the textile industries. They established and optimised a photochemical treatment system that was able to degrade the slowly biodegradable surfactant and improve its ultimate biodegradability. Rivera-Utrilla et al. (2012) explored the performance of coupled treatment of ozonation and biological process to remove the anionic SDBS (Sodium Dodecylbenzenesulfonate) from an aqueous environment. The degradation efficiency of SDBS and total organic carbon increased due to the combined use of AOP and biological processes. However, one area of these integrated treatments is still limited, such as kinetic modelling for the combined processes and evaluation of the toxicity effect of intermediates after chemical oxidation. Therefore, the present research work aims towards the removal of surfactant from surfactant-bearing wastewater by a combined pre-treatment with UV-H_2O_2-AOP followed by a biological process. In particular, the specific objectives of this study are as follows: (1) to study the toxicity of the effluent generated in UV-H_2O_2-AOP treatment of surfactants, and (2) to investigate the influence of the AOPs on the biodegradation kinetics.

Materials and Methods

Reagents

Acridine orange II, glacial acetic acid and sodium dodecyle sulphate and toluene were obtained from Hi-Media, India, while chloroform, 30% H_2O_2 was obtained from Merck, Germany. All other chemicals used in this study were of more than 99% pure and analytical grade used without further purification.

Instrumentation

A high-precision electrical balance (Afcoset ER-182) was used for weighing. A digital pH metre (Orion 420A+, Thermo) was used to measure pH. A spectrophotometer (single beam spectrophotometer 117, Systronics) was used to measure absorbance. UV Reactor (M/s. Lab Tree, India) was used for the degradation of anionic surfactant (SDS) which has eight low-pressure mercury lamps, and emits monochromatic light at 253.7 nm. The important parameter of the UV reactor is photon flux, which is measured by the potassium ferrioxalate actinometry method (Hatchard and Parker, 1956). The value of photon flux of the reactor was $1.90 \times 10^{-05} \pm 9.66 \times 10^{-07}$ Einstein/min/100 mL.

Analytical Method for Determination of SDS Concentration

The concentration of SDS was determined spectrophotometrically using an ion-pairing agent acridine orange (ACO) chemically known as 3, 6-bis (dimythylamino), which shows absorbance at 467 nm after pairing with SDS (Adak et al., 2005b). SDS solutions containing 10 mL (0.1 to 6.0 mg L^{-1}) was taken in a 25 mL separating funnel. Then ACO (5×10^{-3} M) and glacial acetic acid 100 μL were added and followed by the addition of 5 mL toluene and agitated for 1 min. The aqueous layer was discarded and the absorbance of toluene layer at wavelength of 467 nm was measured. The corelation coefficient was found to be 0.9976. The relation between absorbance (abs.) and concentration (conc.) was determined as Absorbance = 0.1121 × Concentration (mg/L) + 0.0213. From this equation, the unknown concentration of SDS can be found after measuring its absorbance.

Advanced Oxidation of Surfactant

All advanced oxidation experiments were carried batch wise and illuminated by UV lamp at a regular time interval from 1 to 10 mins using 100 mL of SDS-bearing wastewater at 27 (±2)°C at pH 7 (10 mM phosphate buffer). The hydrogen peroxide concentration and initial concentration of SDS were taken as 2 mol/mol and 100 mg L^{-1} at pH 7, respectively. Samples were taken after every 1 min of irradiation. The percentage degradation of SDS was calculated and the experimental data were analysed by different kinetic models. In the present work, the degradation of SDS followed pseudo-first-order kinetic model, the linearised form of kinetic model is shown below (Adak et al., 2015).

$$ln\frac{[SDS]}{[SDS]_o} = -k'_{app}\ H' \quad (1)$$

where [SDS] is SDS concentration at any time *t*, $[SDS]_0$ is initial SDS concentration, *k'* is fluence-based pseudo-first-order reaction rate constant ($cm^2\ mJ^{-1}$) and *H¢* is the fluence of the UV lamp ($cm^2\ mJ^{-1}$).

Toxicity Evaluation of By-products of AOP

The intermediate products formed during the advanced oxidation of surfactants may be more toxic compared to the parent compounds (Olmez-hanci et al., 2014). Hence, it is important to evaluate the toxicity of the by-products formed during the advanced oxidation of anionic surfactant (SDS). Toxicity studies were carried out for and their degraded products on the zebrafish (*Danio rerio*) as a model organism. Zebrafish is a very good biosensor of aquatic pollutants or contaminants, which respond with great sensitivity to changes in the aquatic environment (Lal et al., 1983; Ogundiran et al., 2009). Healthy 300 zebrafishes were collected from the fish farms near the IIEST, Shibpur, India. These were acclimatised in a large aquarium containing tap water under laboratory conditions for 30 days and fed with commercial fish food (Fig. 14.1). Care was taken to avoid any sudden change in pH, alkalinity and temperature.

Fig. 14.1 Experimental setup for toxicity bio-assay

For toxicity study, 10 numbers of zebrafishes of the same size were exposed to different concentration of surfactant and its degraded samples. The mortality of the fish was observed after 24, 48, 72 and 96 hours of exposure. The mortality percentage is shown in Table 14.1–14.2 for both surfactants. Probits of mortality were obtained through probit analysis (Finney, 1971). Then, log dose (concentration of surfactant) and probit of mortality were plotted and LC_{50} values were obtained corresponding to probit value of 5.

Table 14.1 Mortality response of *zebrafish* to raw SDS sample

Sl. No.	Concentration ($mg\ L^{-1}$)	Mortality (%) in			
		24 h	48 h	72 h	96 h
1	14	-	-	-	-
2	15	-	-	-	10
3	16	-	10	20	30
4	17	10	20	50	70
5	18	10	30	50	80
6	19	30	50	60	90
7	20	50	60	70	100
8	21	70	80	100	-

Table 14.2 Mortality response of *zebrafish* to partially degraded SDS sample

Sl. No.	Concentration (mg L^{-1})	Mortality (%) in			
		24 h	48 h	72 h	96 h
1	22	-	-	-	-
2	24	-	-	-	10
3	26	-	-	10	30
4	28	-	10	30	60
5	30	20	30	50	80
6	32	40	50	80	100
7	34	60	80	100	-
8	36	90	100	-	-
9	38	90	-	-	-

Biodegradability and Treatability Study of AOP By-products using Aerobic Batch Reactor

Biodegradability and treatability studies were conducted for both the raw and degraded surfactant-bearing wastewater. The 5 days biochemical oxygen demand (BOD_5) and chemical oxygen demand (COD) of the raw and degraded surfactant-bearing wastewater were determined (Greenberg et al., 1989) and the ratio of BOD_5 to COD was used to indicate the biodegradability of the effluent. Residual H_2O_2 in the treated samples was quenched by catalase before the measurement of COD and BOD_5.

Treatability studies of SDS and its degraded products were conducted using an aerobic batch reactor (ABR). Initially, one bioreactor (10 L capacity) was started by adding activated sludge (collected from lab scale bioreactor treating kitchen waste) and fed with glucose solution, ammonia nitrate and phosphoric acid, such as carbon, nitrogen and phosphorus sources (Fig. 14.2). After 30 days of growth, the active sludge was thickened and 250 mL of thickened sludge was added to each of the two reactors of 5 L capacity. The reactors were then fed with gradually increasing SDS concentration up to 10 mg L^{-1} and partially degraded SDS sample for acclimatisation for another 4 weeks.

Five aerobic ABRs (each capacity of 500 mL) were then started with the acclimatised SDS bearing active sludge for each of the surfactants separately (Fig. 14.3). The ABRs were operated for a 24 hours cycle and were fed with raw and 50% degraded SDS sample. The volume of supernatant (effluent) was withdrawn after 24 hours and the volume was made-up with adding wastewater at each cycle. Each reactor was operated at different solid (biomass) retention time, that is, 2, 4, 6, 8 and 10 days. The performance of the reactors was monitored by measuring COD, mixed liquor suspended solids (MLSS) and mixed liquor volatile suspended solids (MLVSS).

The above experimental data were analysed to determine bio-kinetics parameters, such as growth yield coefficient, decay rate, maximum specific rate of substrate utilisation and

Fig. 14.2 Experimental setup for aerobic batch reactor

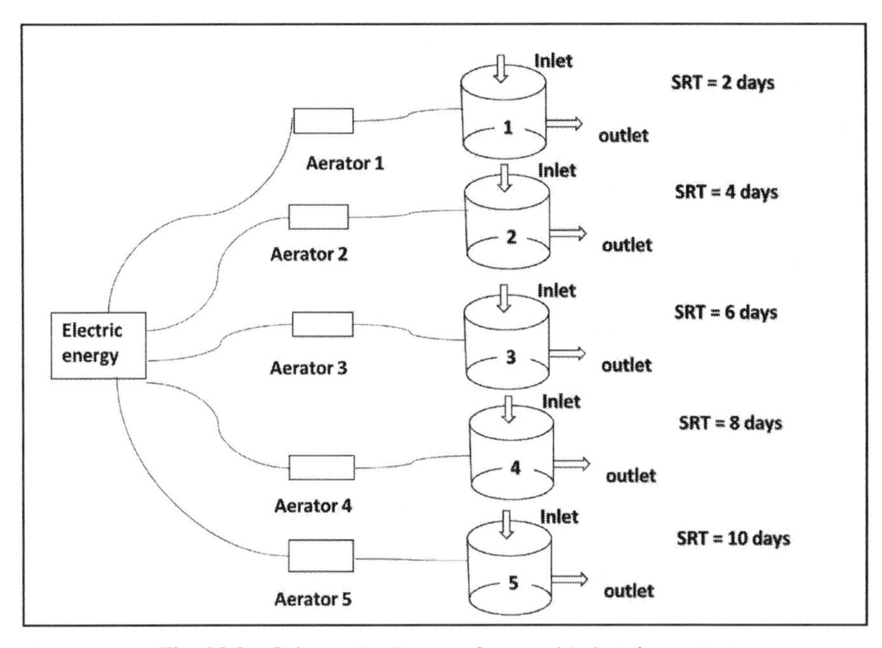

Fig. 14.3 Schematic diagram for aerobic batch reactor

maximum specific growth rate. For this purpose, the Monod equation (Eddy and Metcalf, 2003) was used. The rate of substrate utilisation can be represented by:

$$r_{SU} = \frac{kXS}{k_s + S} = \frac{S_0 - S}{\theta} \tag{2}$$

where

k = Maximum sp. rate of substrate utilisation (mg/mg-day).

k_S = Half-saturation coefficient or shape factor of the Monod equation (mg L^{-1})

θ = Hydraulic retention time (day)

S = Substrate concentration as COD (mg L^{-1})
S_o = Initial substrate concentration as COD (mg L^{-1})
X = Biomass concentration (mg L^{-1})

Dividing the above equation by X results

$$\frac{kS}{k_s + S} = \frac{S_0 - S}{\theta X} \tag{3}$$

The linearised form of this equation is obtained by taking its inverse and shown below:

$$\frac{X\theta}{S_0 - S} = \frac{k_S}{k}\frac{1}{S} + \frac{1}{k} \tag{4}$$

The values of k_S and k can be determined by plotting the term $\left(\frac{X\theta}{S_0 - S}\right)$ versus $\left(\frac{1}{S}\right)$. Similarly, in order to determine the values of Y and K_d following equation can be used:

$$\frac{1}{\theta_c} = Y\frac{r_{SU}}{X} - K_d \tag{5}$$

where

Y = Growth yield coefficient, which represents the mass of cells produced per unit mass of substrate utilised (mg VSS/mg of COD)
K_d = Decay rate, which represents the fraction of cells oxidised by endogenous respiration per time unit.
θ_c = Biomass retention time (day)

Substituting the expression for r_{su} given above, the resultant expression is given as

$$\frac{1}{\theta_c} = Y\frac{S_0 - S}{X\theta} - K_d \tag{6}$$

By plotting $\frac{1}{\theta_c}$ versus $\frac{S_0 - S}{X\theta}$, Y and K_d can be found from the slope and intercept of the best fit line. Further, the maximum specific growth rate (μ_m) was determined by multiplication of k and Y.

Results and Discussion

Optimisation of Photochemical Pre-treatment Conditions Study

The effectiveness of UV-H_2O_2-AOP on SDS degradation was presented in our previous research (Mondal et al., 2017). The summary of the research is presented here. Samples were withdrawn at specified time intervals and analysed for the kinetic model. It was observed that full mineralisation of SDS was completed within 4 min. To optimise the peroxide dose in UV-H_2O_2, 100 mg L^{-1} SDS solution was degraded with the fluence of 0.490 J cm^{-2} at pH 7±0.1 and varied the peroxide dose as follows concentration: 0.125, 0.25, 0.5, 1, 2, 5 and

7.5, mol H_2O_2/ mol SDS. The degradation efficiency as well as rate constants significantly increase with the increase of H_2O_2 dose when the H_2O_2 dose is below 2 mol H_2O_2/ mol SDS. When the H_2O_2 dose is over 2 mol H_2O_2/ mol SDS, the degradation efficiency showed a significant decrease as the H_2O_2 dose was increased. So, the optimum H_2O_2 concentration in this case is selected as 2 mol H_2O_2/ mol SDS.

The experiments of UV-H_2O_2 process were conducted at different pH in the range of 7 to12 with an initial SDS concentration of 100 mg L^{-1} and a peroxide dose of 1 mol H_2O_2/mol SDS. It was observed that the degradation of SDS was effective near pH 7–8 with no significant difference in these two pH levels. Further increase in pH beyond 10 resulted in a significant decrease in removal efficiency. Therefore, all experiments were conducted at pH 7.

From our previous research work (Mondal et al., 2017), it was observed that BOD to COD ratio increased with time. The initial BOD_5/COD ratio for 100 mg L^{-1} of SDS solution was found to be 0.11 (±0.02), indicating that SDS is not biodegradable. BOD_5/COD fraction is increasing from 0.11 to 0.67 fraction within 8 min of degradation for 100 mg L^{-1} of SDS.

Toxicity Study of Anionic Surfactant and its Degraded By-product

Toxicity study of the raw and partially degraded sample was conducted to check the toxicity level of intermediates before going to the biological treatment. After 24 h exposure to raw SDS sample, 10% mortality of zebrafish was observed at a concentration of 17 mg L^{-1}. 50% and 70% mortalities were found at 20 and 21 mg L^{-1}, respectively. Probit analysis of the toxicity responses of zebrafish to raw SDS samples was used to find out LC_{50} values. The LC_{50} value for 24 h exposure was 20.30 mg L^{-1} (Fig. 14.4). Similarly, the LC_{50} value for 48, 72 and 96 hrs exposures were calculated to be 19.18, 18.08 and 16.76 mg L^{-1}, respectively. It was observed that the LC_{50} value decreased with an increase in exposure time.

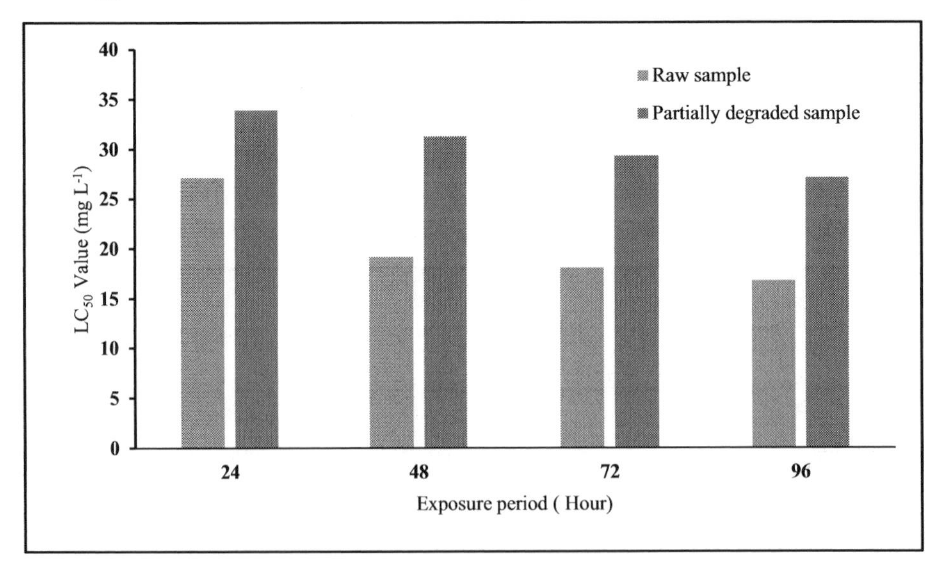

Fig. 14.4 LC_{50} values of raw SDS and 50% degraded sample

Similarly, the LC_{50} values for 24, 48, 72 and 96 h exposures were found to be 33.89, 31.28, 29.30 and 27.10 mg L^{-1}, respectively for 50% degraded samples (Fig. 14.4). Therefore, it can be concluded that partially degraded surfactant-bearing wastewater is less toxic than parent surfactant-bearing wastewater.

Biodegradability and Treatability Study of SDS and its Degraded Products

Biodegradability of SDS and its Degraded Sample

Due to the high operating costs of AOPs, the application of short-term AOP is followed by biological treatment and is being pursued for recalcitrant pollutants, such as SDS. Thus, it is required to examine the biodegradability of the SDS transformation products. Biodegradability can be represented by the ratio of the BOD and COD. Figure 14.5 shows the BOD_5/COD of the UV-H_2O_2 transformation products. It was observed that BOD to COD ratio increased significantly with time. In particular, 100 mg L^{-1} SDS solutions were treated by UV-H_2O_2 for 2, 4, 6 and 8 min. Residual H_2O_2 in the treated samples was quenched by catalase before the measurement of COD and BOD_5. The initial BOD_5/COD ratio for 100 mg L^{-1} of SDS solution was found to be 0.12 (±0.02), indicating that SDS is not biodegradable. BOD_5/COD fraction increased from 0.12 to 0.6 fraction within 8 min of degradation for 100 mg L^{-1} of SDS. Further, it is to be noted that the initial BOD_5 of 100 mg L^{-1} SDS solution was 30 mg L^{-1} and it was increased to 90 mg L^{-1} after 8 min of UV-H_2O_2 treatment. These results demonstrate that the transformation products were more biodegradable than the SDS parent compound. Since the BOD_5/COD ratio in UV-H_2O_2 effluent was 0.602, the effluent can be effectively treated by biological processes.

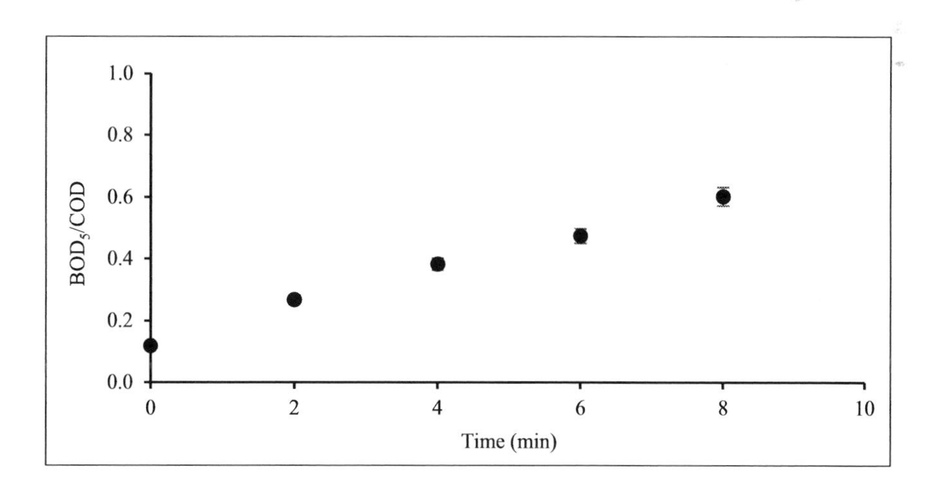

Fig. 14.5 Increase in biodegradability for UV-H_2O_2 degraded 100 mg L^{-1} SDS at [H_2O_2]/[SDS] = 1 at pH 7.0(±0.1) for 0 to 8 min

Treatability Study of SDS and its Degraded Products using Aerobic Reactor

Aerobic SBRs were used to examine the treatability of surfactant-bearing wastewater. COD, pH, MLSS and MLVSS were monitored on a daily basis for raw SDS solution and partially UV-H_2O_2 degraded samples. The growth of biomass with time is presented in Fig. 14.6. It was observed that MLSS reached 3300 mg L^{-1} and 4200 mg L^{-1} after 10 days of activity and maintained the steady-state conditions thereafter for raw SDS and partially UV-H_2O_2 degraded

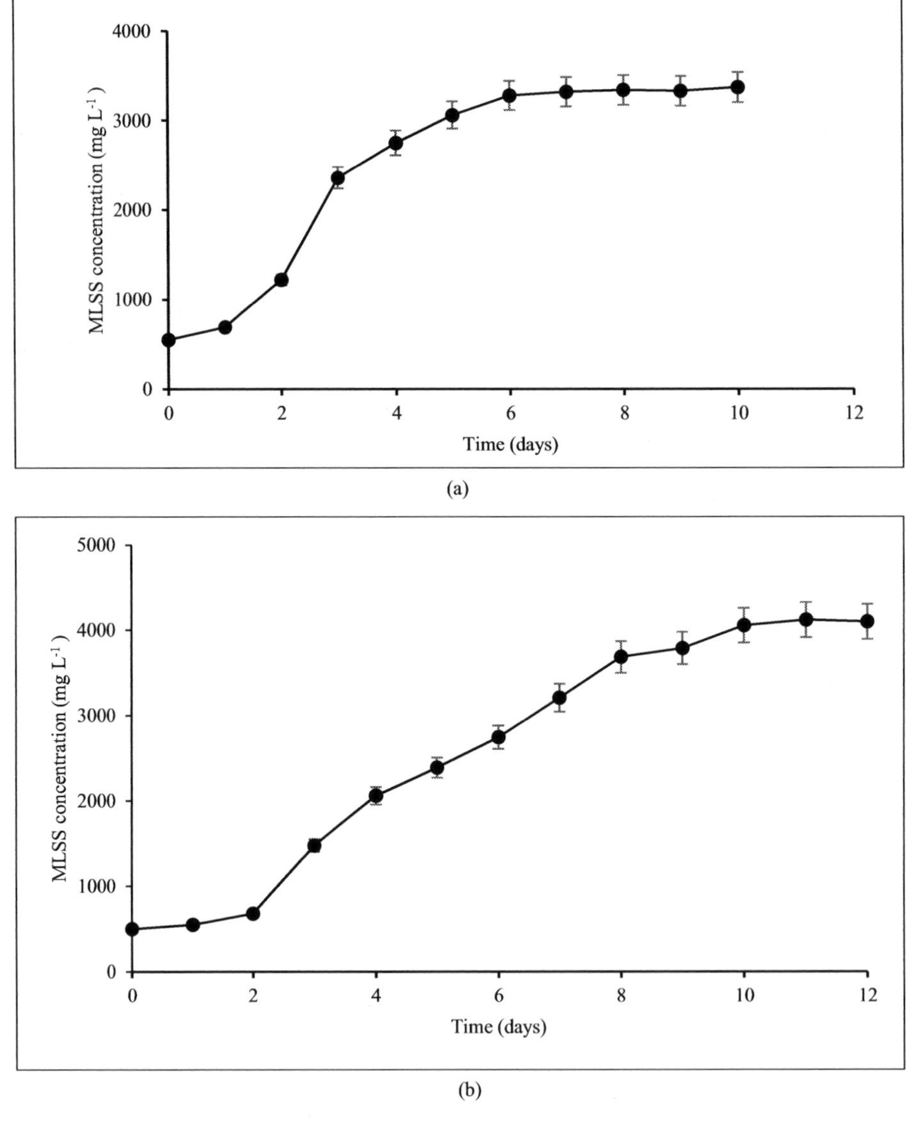

Fig. 14.6 Variation of MLSS concentration with time for (a) raw SDS sample and (b) partially degraded SDS sample

samples, respectively. Initially, the growth rate of biomass in the raw sample was rapid. Then, after 4 days, the growth of biomass slowly increased and maintained a steady state condition. Similarly, in the degraded sample, the growth of biomass rapidly increased up to 8 days. Then, it maintained a steady state condition. The overall removal of SDS (for partially AOP-treated samples) was found to be ~ 89% by coupled advanced oxidation process and biological process. Here, the growth rate of biomass in the degraded sample was found to be more than in the raw sample due to its more toxicity. The intermediates formed during UV- H_2O_2-AOP were more degradable than parent compounds.

For evaluation of bio-kinetic coefficients, values of the parameters $(S_0 - S)$, $\frac{X\theta}{S_0 - S}, \frac{1}{S}$ and $\frac{1}{\theta_C}$ were determined and plotted as per Eqs. 4 and 5 (Figs 14.7–14.10). The values of and were 130.20 mg L^{-1} and 0.2935 day^{-1} for raw SDS, respectively and 120.66 mg L^{-1} and 0.3331 day^{-1} for partially degraded (50%) sample, respectively. The values of and were 0.18 mg VSS/mg COD and 0.0019 day^{-1} for raw SDS, respectively, and 0.342 mg VSS/mg COD and 0.0002 day^{-1} for partially degraded (50%) sample, respectively.

A comparison of the values of bio-kinetic coefficients of surfactant-contaminated wastewater was made between the values obtained in this present work and those reported in the literature. Table 4.8 presents a summary of these values and indicates that the bio-kinetic coefficient values of this investigation are comparable to those reported by other researchers. The values of maximum specific growth rate (μ_m) were found to be 0.05 and 0.11 day^{-1} for raw SDS and partially degraded sample, respectively, and this indicates that partially degraded SDS samples can be more easily treated by biological process.

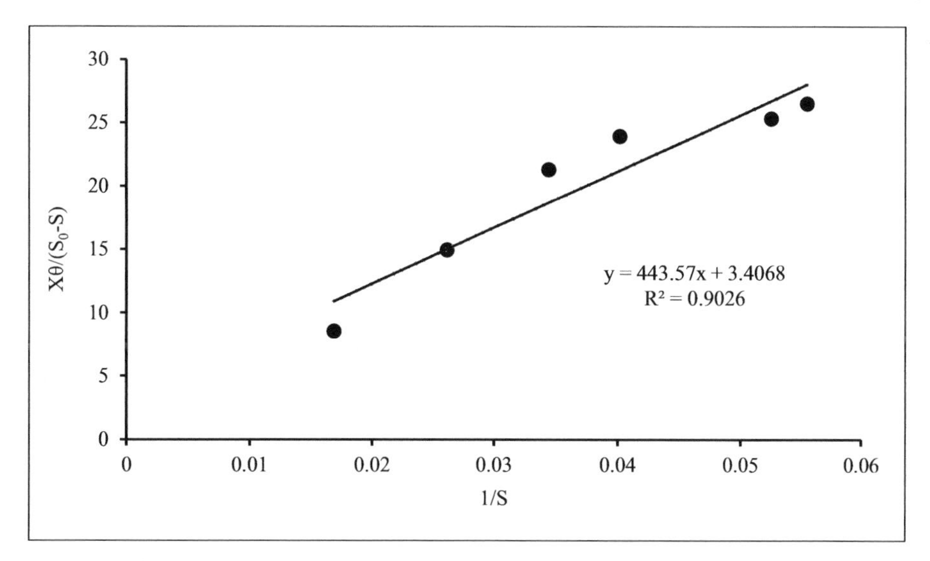

Fig. 14.7 Plot of 1/S versus 1/U on COD basis for raw SDS sample

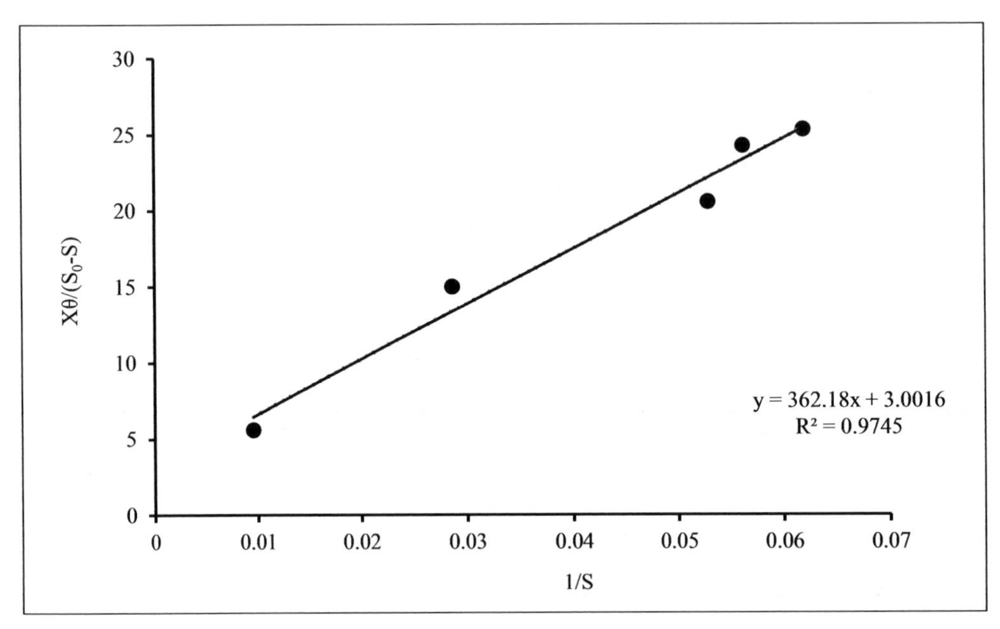

Fig. 14.8 Plot of 1/S versus 1/U on COD basis for partially degraded SDS sample

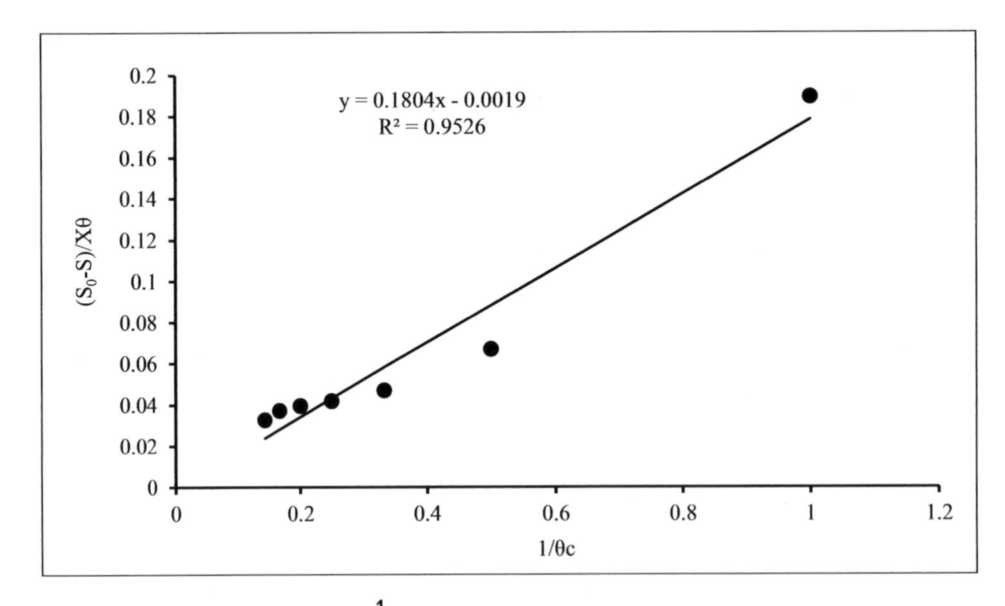

Fig. 14.9 Plot of $\frac{1}{\theta_C}$ versus on COD basis for raw SDS sample

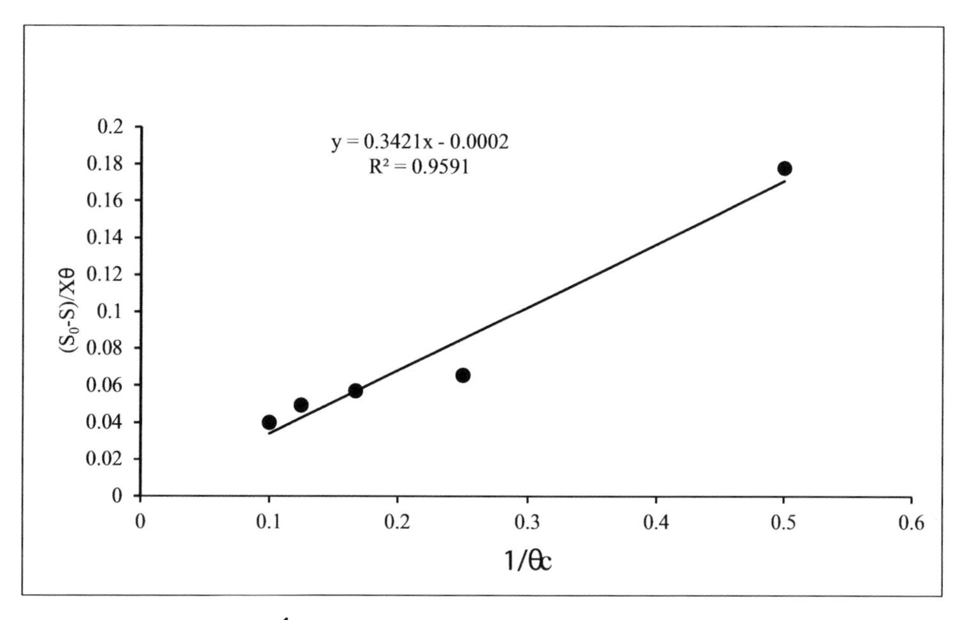

Fig. 14.10 Plot of $\frac{1}{\theta_C}$ versus on COD basis for partially degraded SDS sample

Table 14.3 Bio-kinetics coefficients of different surfactant wastewater

	K Day^{-1}	**K_s mg/COD**	**Y mg VSS/mg COD**	**K_d Day^{-1}**	
SDS raw	0.2935	130.20	0.18	0.0019	Present work
SDS degraded	0.3331	120.66	0.342	0.0002	
Branched alkylbenzene sulphonates*	8.8	111	0.54	0.048	(Hashim et al., 1989)
Sodium dodecyl sulphate	9	95	0.73	0.25	(Liwarska-Bizukojc et al., 2008)
Sodium alkylbenzene sulphonates	2.74	245	0.70	0.25	
Sodium alkyltrioxyethylene Sulphate	4.37	135	0.69	0.25	
Decaoxyethylene alkyl ether	4.23	120	0.46	0.25	
Nonylphenylheptaoxyethylene glycol ether	3.1	140	0.44	0.25	
Rhamnolipid	0.12	20	0.41	0.05	(Mohan et al., 2006)
TritonX-100	0.14	37	0.70	0.10	

**Note:* on BOD basis

Conclusions

The UV-H_2O_2-AOP pre-treatment of the surfactant-bearing wastewater is an effective method for detoxification. The toxicity of the intermediates formed during UV-H_2O_2 process reduces. The LC_{50} value for 24, 48, 72 and 96 h exposures were calculated to be 20.30, 19.18, 18.08 and 16.76 mg L^{-1}, respectively. Similarly, the LC_{50} values for 24, 48, 72 and 96 h exposures were found to be 33.89, 31.28, 29.30 and 27.10 mg L^{-1}, respectively, for 50% degraded sample. Therefore, it can be concluded that partially degraded surfactant-bearing wastewater is less toxic than parent surfactant-bearing wastewater. This positive impact is reported by a significant decrease in the inhibitory action of the growth of microbes in aerobic batch reactor. Partially degraded surfactant-bearing wastewater can be easily degraded by aerobic biological processes. The increase in biodegradability index and maximum specific growth rate of partially degraded cationic surfactants by UV-H_2O_2 process indicate that combined advanced oxidation-biological processes are a viable solution for the treatment of such refractory organic compounds.

References

1. Adak A., Pal A., Bandyopadhyay M. (2005b) Spectrophotometric determination of anionic surfactants in wastewater using acridine orange. Indian J. Chem. Technol. 12: 145–148.
2. Adak A., Mangalgiri K.P., Lee J., Blaney L. (2015) UV irradiation and UV-H_2O_2 advanced oxidation of the roxarsone and nitarsone organoarsenicals. Water Res. 70:74–85.
3. Arslan-Alaton I. and Olmez Hanci T. (2007) Effect of Photochemical Advanced Oxidation Processes on the Bioamenability and Acute Toxicity of an Anionic Textile Surfactant and a Textile Dye Precursor Adv. Treat. Text. Effl. 73–90.
4. Buxton G. V., Greenstock C.L., Helman W.P. and A.B. Ross (1988) Critical review of rate constants for reactions of hydrated electrons, hydrogen atoms and hydroxyl radicals (OH./O.) in aqueous solution, J. Phys. Chem. 17: 513–886.
5. Eddy, Metcalf (2003) Wastewater engineering treatment and reuse," 4th edition, McGraw Hill, New York.
6. Mondal B., Adak A. and Datta P. (2019) UV-H_2O_2 advanced oxidation of anionic surfactant: Reaction kinetics, effects of interfering Substances and operating conditions. Environment Eng. Manag. 18(6): 1245–1255.
7. Finney D. (1971) Probit analysis Cambridge University Press, New York
8. Greenberg A.E., Clesceri L.S. and Eaton A.D. (1989) Standard Methods for the Examination of Water and Wastewater.
9. Hashim M.A., Kulandai J. and Hassan R.S. (1989) Performance and kinetics of an activated sludge system treating wastewater containing branched alkylbenzene sulphonates. Environ. Technol. Lett. 10: 645–652.
10. Kowalska I. (2012) Separation of anionic surfactants in a sequential ultrafiltration – ion exchange purification system. J. Environ. Stud.21 (3): 677–684.
11. Kyzas G. Z., Peleka E. N. and Deliyanni E. A. (2013) Nanocrystalline akaganeite as adsorbent for surfactant removal from aqueous solutions. Materials (Basel). 6:184–197.
12. Lin S. H., Lin C. M. and Leu H. G. (1999) Operating characteristics and kinetics studies of surfactant wastewater treatment by Fenton oxidation. Wat. Res. 33 (7): 1735–1741.

13. Liwarska-Bizukojc E. and M. Bizukojc (2005) Digital image analysis to estimate the influence of sodium dodecyl sulphate on activated sludge flocs. Process Biochem. 40: 2067–2074.
14. Mahmoud S. S. and Ahmed M. M. (2014) Removal of surfactants in wastewater electrocoagulation method using iron electrodes. Phys. Sci. Res. Int. 2: 28–34.
15. Mishra L., Vishwanathan V., Krishnamurthi P.N C. (1983) Comparative studies on ecotoxicology of synthetic detergents," Ecotoxicol Env. Saf 8: 447–450.
16. Mohan P.K., Nakhla G. and Yanful E.K. (2006) Biokinetics of biodegradation of surfactants under aerobic, anoxic and anaerobic conditions. Water Res 40: 533–540.
17. Narkis N. and Schneider-Rotel M. (1980) Evaluation of ozone induced biodegradability of wastewater treatment effluents.Water. Res.: 19: 929–939.
18. Ogundiran M., Fawole O., Adewoye S. and Ayandiran T. (2009) Pathologic Lesions in the Gills Structures of Clariasgariepinus on exposure to sub lethal concentrations of soap and detergent effluents. J Cell and Animal. Biol. 3(5): 78–82.
19. Olmez-hanci T., Dursun D. Aydin E., I Arslan-alaton, et al, (2014) UV-C and H_2O_2/UV-C treatment of Bisphenol A : Assessment of toxicity, estrogenic activity, degradation products and results in real water. Chemosphere 119: 1–9.
20. Rivera-Utrilla J., Bautista-Toledo M. I., Sanchez-Polo M. and Mendez- Diaz J.D. (2012) Removal of surfactant dodecylbenzenesulfonate by consecutive use of ozonation and biodegradation, Eng. Life Sci. 12: 113–116.

CHAPTER 15

Investigation into the Mechanical Properties of Hybrid Bamboo-Carbon Reinforced Composite

Omprakash Wadekar[1], Surbhi Razdan[2] and Arun Mali[3]
School of Mechanical Engineering, Dr. Vishwanath Karad MIT-World Peace University, Pune, Maharashtra, India

Abstract

Due to increasing awareness of the environment, a large number of synthetic materials are replaced by natural fibre composite. Studies have shown that enforcing synthetic fibres in natural composite enhances the properties of the composite. As bamboo has properties such as biodegradable and higher tensile strength with low density, it has gained the attention of many researchers. Hybrid composites are being studied to enhance the properties of composite using synthetic fibres. The hybrid composition of bamboo with various synthetic and natural are yet to be fully explored and studied, which can prove its vast applications. In this research, the properties of bamboo/carbon/epoxy composite with varying carbon percentage were studied. The woven bamboo fabric and UD carbon mat were taken as the reinforcement with different percentages. The fibre-matrix percentage was 50/50. Eight different compositions of the composite were manufactured by using the hand lay-up method and then compressed and dried in the oven. These four specimens were then tested for tensile and impact strength and then they were compared. The effect of carbon percentage in bamboo fibre was examined.

Keywords: bamboo fabric, UD carbon fibre, composites

Introduction

The replacement of metal with plastics and composites in the automotive industry has increased the service life of the components and also increased the noise and vibration absorption capacity. The capacity to absorb the kinetic energy in case of shocks and accidents in an eco-friendly way. In recent decades, the use of composite materials in the automotive industry has increased significantly. Depending on the role of the composite in an automobile, the composite materials are chosen and the technology to obtain the composite also plays a crucial role in deciding the composite. The most challenging part in selecting the composite is that it should meet the requirements, such as low weight and low cost. The optimum structure strategies, such as limits of compromise in design, materials and geometry of design, are

[1]omprakashwadekar@gmail.com; [2]surbhi.razdan@mitwpu.edu.in; [3]arun.mali@mitwpu.edu.in

used to select the composite for the particular component. The selection of material for the automotive application depends upon basic factors, such as its durability, impact on energy absorption and ease in manufacturing. Nearly 7.5% of the secondary automobile components are manufactured from polymer-reinforced composite. Both the thermoplastic and thermoset are used to manufacture the composite with number of varying reinforcements from synthetic fibre to natural fiber. The most used synthetic fibre is carbon, aramid fibre and glass, whereas natural fibre contains hemp, jute, sisal, bamboo and flax fibre. In many applications, synthetic fibres were used most commonly. But, the main concern about synthetic fibres are that it is not eco-friendly and economical. Therefore, the need for replacing synthetic fibre was needed on an urgent basis. Hence, researchers began to study natural fibre for its biodegradability, weight-to-strength ratio and low energy consumption at low cost. Bamboo fibre comes with advantages such as its rate of growth with a high weight-to-strength ratio, therefore, it is called natural glass fibre.

A composite is a combination of two or more than two materials, such that the materials can be distinguished as separate phases after manufacturing. The basic requirements of any material in industrial and commercial applications are its durability, ease in manufacturing and greater impact strength. These requirements can be achieved with composite with the reduction in part weight of the component. Composite can reduce part weight by nearly 30%–40% with greater mechanical properties. Among various engineering materials for many applications, the fibre reinforced composites are the most advance materials. The load-bearing capacity of the composite can be optimized for a specific application by enforcing appropriate matrix and fibre orientation by specifically selecting the fibre.

The natural fibre-reinforced polymer composite has great potential for automotive application, construction applications and other products. There are many types of natural fibre sources, such as wood, wheat, hemp, kenaf, banana, sisal, pineapple leaf, kapok, coir, mulberry paper, rice husk, barley, sugarcane, grass, reeds, water hyacinth, empty fruit bunches, papyrus and raphia. In which bamboo fibre has the most promising properties to substitute synthetic fibres with the addition of strength and stiffness with low density and also it is less expensive. As bamboo fibre has properties such as sustainability, lightweight with low density and high growth rate it is been used as the reinforcement in a natural composite. It contains cellulose and lignin 61% and 32%, respectively, which gives strength and stiffness to the fibre. In the previous studies, the BFs are enhanced by altering their surface treatment or chemically treating with polypropylene to gain the polarity of polymer.

Bamboo estimating for having more than 1250 species with a production rate of $10 \text{x} 10^6$ tons worldwide. After wood and cotton, it is the third-largest in production with USD 0.5–1/kg. The second-largest exporter is Indonesia having a value of USD 269 million and with a value of USD 1 billion and 34 million chains it is the third-largest country. Among all 65% of the total production is in Asian countries now then followed by the USA with 28% and then 7% Africa. Japan is the major importer of bamboo with a value of USD 194 million followed by the USA with USD 254 million and Europe with USD 230 million. The bamboo has greater mechanical properties, such as high tensile strength with 140–800 MPa, modulus of elasticity 33GPa, density of 0.6–0.8 g/cm^3 and also having low specific gravity, specific strength and stiffness compared to glass fibre.

The woven fabric composite is also known as woven fabric reinforced composites (WRF) and is manufactured by stacking the fabric and the polymer in the composite. Plain, twill, diamond, herringbone and zig-zag based are some of the types of woven fabric patterns. With advanced textile technology and the usage of polymer, it is easy to get fast and automated production with higher mechanical properties. However, there might be various failures in composite surface interface due to cracking problems. The reduction in stiffness and strength of woven fabric composite can be caused due to delamination which is another major challenge. Delamination occurs due to applied load, such as compression, tension, bending and energy at different layers, which affect the polymeric structure. Another major challenge is crimp for the woven fabric composite which reduces the efficiency of the composites. It occurs due to local stress which can cause debonding of the fibre interface. The surface treatment of bio-fibers influences the mechanical properties of the fabric composite. The mechanical properties can also be increased by modifying the structure of fabric or its architecture. The progressive damage modeling (PDM) tool can be used to predict the distribution of the stress with the possible failure mechanism of the fabric composites.

The hybrid composite that is been manufactured can reduce the carbon footprint as well as increase the mechanical properties of the composite. Natural fibres have proven sustainable and economical having greater damping properties. The hybrid has more multifunctional properties than natural composites which are used in various applications. These composites are manufactured by melting the mixture with extrusion compression moulding, injection moulding, pultrusion, injection moulding or compression moulding. In both woven and non-woven fabrics, fibres are used to make composites. In Fig. 15.1 (a) shows woven fabric, (b) shows knitted fabric and (c) shows non-woven fabric.

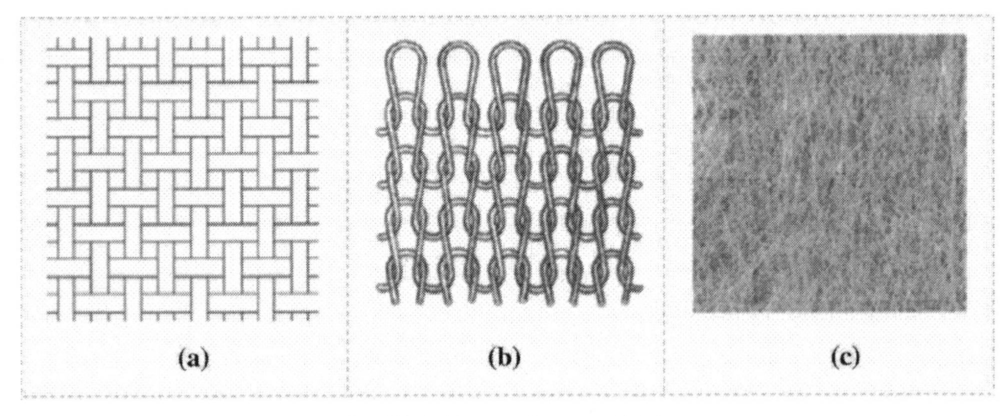

Fig. 15.1 Types of woven fabric

Source: Ref. [9]

The laminated composites made from a fabric base have proven applications in the automotive industry, defense, structure for the civil sector and transportation. They have advantages such as being light in weight, and low cost with greater strength which is of great interest for researchers. Fabrics that are used as laminates are of different types, that is, woven, knitted and non-woven. The woven fabrics are mostly used as a laminate. The properties of composite manufactured depend on the type of fabric, composition, the matrix used and the origin of the fabric. Natural and synthetic fabrics are used to manufacture the composite. Hemp, sisal, bamboo, flax, etc. are the natural fibres, whereas glass, nylon, carbon, polyethylene and polyester are the synthetic fibres. Nowadays, natural fibres are gaining interest, whereas for enhanced or specific properties the hybrid composite of natural and synthetic fabrics becoming popular. In the performance of composite made from fabric, the layers, fabric properties, areal density and structure play an important role. Cementitious material, thermosetting and thermoplastic are used to manufacture the composite. Each of the polymers has its properties and depending upon these they are selected.

Fabrics are excellent in drapability; hence they conform to the shape of the mould, tool or device. The methods which are used for manufacturing the fabric composites are compression moulding, machining, compression forming, roll forming and diaphragm forming. In the previous research, unidirectional fabric composites were developed. But now the researchers have begun to develop bi-directional fabrics and also knitted fabrics. Studies are been carried out to predict the potential for these composites with the simulation software.

Literature Review

Samal et al. presented work on bamboo-glass reinforced with polypropylene to enhance the properties, such as flexural and impact strength, and thermal and study dynamic mechanical analysis of the composite. Rao et al. reported that flexural and compression strength increases with an increase in glass percentage in BF-Glass hybrid composite. Ghossein et al. incorporated a new WL technique for manufacturing randomly oriented carbon fibre to mat for greater repeatability and with various combinations to manufacture the composite. T D Jagannatha et al. presented a vacuum bag method to manufacture carbon/glass-reinforced epoxy and its effect on the mechanical properties. Subhankar Biswas et al. studied the jute/bamboo/epoxy composite manufactured using the vacuum bag method for the mechanical, physical and thermal properties. P Tostra et al. demonstrated the electrical and mechanical properties of a composite made up of carbon reinforced with epoxy resin. Ramachandran M et al., Meenambika et al, G.B et al., Zhang, et al., Anigol et al. focused on bamboo-glass reinforced with polyester for structural application in an automobile. Y. S. Rao et al. investigated bamboo/carbon/epoxy composite for the properties, such as flexural and tensile strength of non-woven, and compared the result with the simulation result. Surbhi Kore et al. present the work on bamboo/carbon/PP composite of different compositions and studied the effect of fibre length and chemical treatment on the mechanical properties of the composite. Praveen Kumar J et al. studied the effect of SiC particulates on the mechanical properties of bamboo/carbon/epoxy composite. Guowei Zhou et al. studied the carbon/epoxy composite with plain and twill carbon fibre and its effect on the mechanical properties of the composite. Kannan Rassiah et al. investigated the layered laminate of bamboo/epoxy composites and their mechanical properties. Aidy Ali et al. studied E-glass/bamboo composite and the effect of stacking sequence on the mechanical properties. Ratim, Bonnia et al. studied the polyester composite reinforced with kenaf and the effect on the mechanical properties of different woven and non-woven fibre structures. W. Cantwell et al. investigated laminates having non-woven and mixed-woven layers of carbon fibre for post-impact fatigue performance. Meiling Chen et al. studied sustainable and innovative bamboo winding composite pipe products. William Nguegang Nkeuwa et al. presented a review on bamboo bonding fundamentals and the processes. Yan Maa, Masahito Uedab et al. presented the comparison between carbon fibre/epoxy and fibre/polyamide with mechanical properties and behaviour of failure in unidirectional composites. Hsuan-Hao Chiu et al. presented an investigation of bamboo fibre and bamboo mat with the behaviour of spring action and deformation under a simple bending test.

Methodology

Reinforcement

Bamboo fabric and UD carbon fibre were taken as the reinforcement and epoxy resin as the matrix. Both the reinforcement and matrix were 50%–50% by weight. The bamboo fabric of 140 GSM and carbon fibre of 400 GSM were used. The epoxy resin with a hardener of ratio 10:1 was used.

Composition

Eight compositions of reinforcement with varying from 0% to 10% of bamboo-carbon percentages were made finalised. The stacking order of the fibres was decided to gain the required size of the composite.

Manufacturing

Hand lay-up technique with compression moulding followed by post-curing was used to manufacture the composite. The mould of 400 × 400 mm was taken and the first silicon-coated film was put onto it. Epoxy resin with the bamboo fabric and carbon fibre was placed layer-by-layer. The mould was then compressed using 8 slabs of 10 kg for 14 hrs. The composite

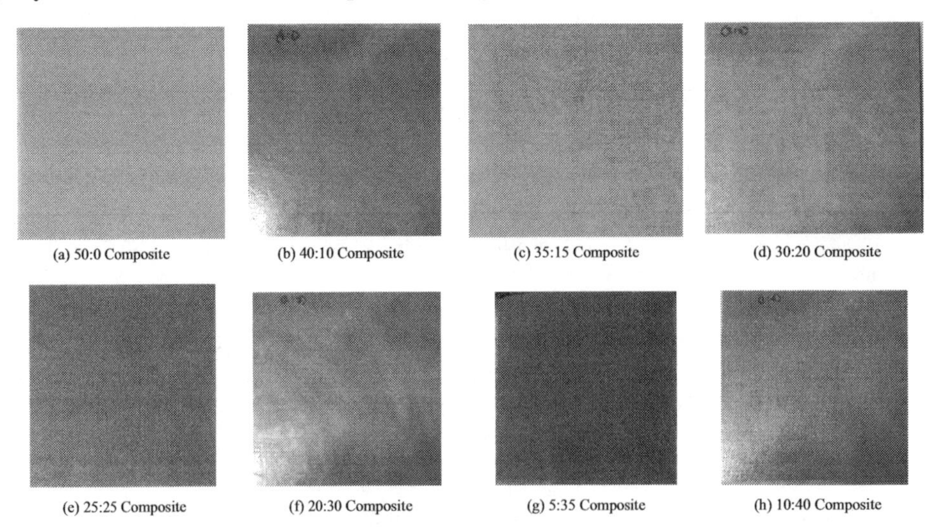

Fig. 15.2 Bamboo/carbon/epoxy composite

was then pulled out of the mould and then post-curing in the oven was done at a temperature of 110°C for 14 hrs. The Fig. 15.2 shows 8 composites manufactured.

Testing

The tensile and impact strength were tested using ASTM D3039 and ASTM D 256, respectively. The specimen of dimension 250 × 25 × 4 mm was used for tensile testing, and 63.5 × 12.7 × 4 mm was used for the Izod impact test. Three samples of each composition were tested and Fig. 15.3 and Fig. 15.4 show the specimen after the test.

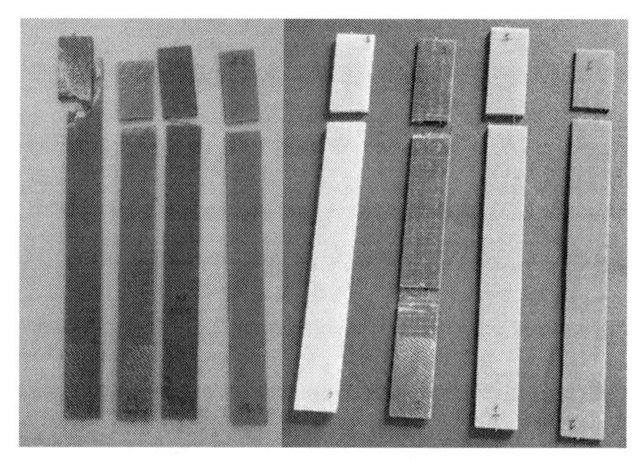

Fig. 15.3 Specimen after tensile test

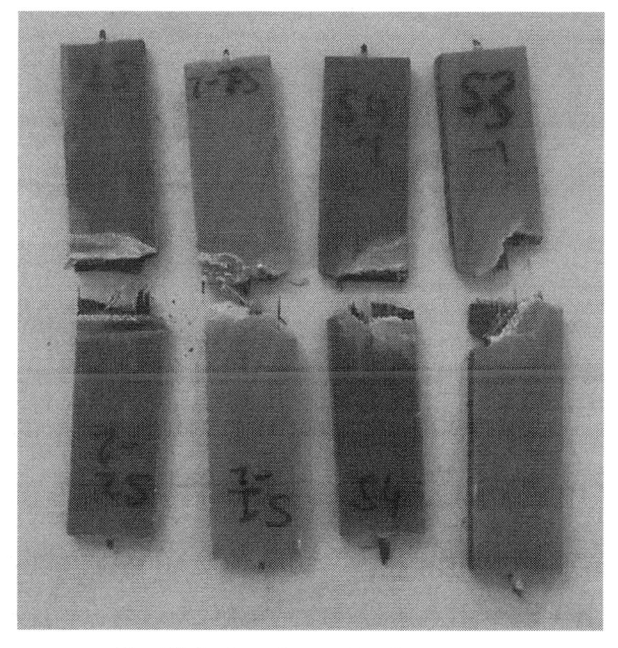

Fig. 15.4 Specimen after impact test

Results

Table 15.1 Result for tensile and impact strength

Specimen	Composition (Bamboo: Carbon) %wt.	Tensile strength (MPa)	Impact strength (J/m^2)
S1	50:0	49.88	3
S2	40:10	36.78	4
S3	35:15	37.67	4
S4	30:20	188.49	3.83
S5	25:25	12.61	4
S6	20:30	22.63	3.5
S7	15:35	313.21	3.5
S8	10:40	255.28	3.67

Conclusion

Eight different compositions of woven bamboo fabric and UD carbon fibre with epoxy (EITPL-505 and 10% PAM hardener) were manufactured and are shown in Table 1. ASTM 3039 and ASTM 256 standards were used for the tensile and impact test, respectively. The result showed that with number of carbon fibre (4:1) the tensile strength was 49.88 MPa, whereas with 35% of carbon fibre in the composite, the tensile strength was maximum, that is, 313.21 MPa. There was sudden fall in tensile strength at 25% of carbon fibre and slight increase at 30% of carbon fibre. An increase in tensile strength is shown at 35%, that is, 313.21 MPa, and slightly decrease with 40% of carbon. This variation is due to two reasons, firstly due to fibre agglomeration in the matrix caused due to poor adhesion in the matrix that ultimately results in the formation of cracks resulting uneven distribution of stress. Secondly, with a high percentage of weight of the fibre, there is an increase in fibre-to-fibre interaction that makes matrix to depress the fibre. There is not much significant variation in the impact strength for the four samples.

References

1. Aidy Ali, Kannan Rassiah, M.M.H Megat Ahmada. (2021). The effect of stacking sequence of woven bamboo on mechanical behaviour of fiber reinforced composites. Journal of Southwest 592 Jiaotong University / Vol.56 No.2.
2. Anigol,M.N.B., Anil, S.P. (2015). Study of the effect of various fillers on mechanical properties of carbon-epoxy composites. Int. Res. J. Eng. Technol. 02(03), 798–802.
3. Biswasa, S., Shahinura, S., Hasana,M., Ahsan, Q. (2015). Physical, mechanical and thermal properties of jute and bamboo fibre reinforced unidirectional epoxy composites. Procedia Eng. 105, 933–939
4. Guowei Zhou, Qingping Sun, Zhaoxu Meng, Dayong Li, Yinghong Peng, Danielle Zeng, Xuming Su.(2022).Experimental investigation on the effects of fabric architectures on mechanical and damage behaviors of carbon/epoxy woven composites, ISSN 2091-2730 ELSEVIER.

5. H. Ghossein, A. A. Hassen, V Paquit, L. J. Love, UK. Vaidya. (2018). Innovative method for enhancing carbon fibers dispersion in wet-laid nonwovens, Mater. Today Commun. 17 100–108.
6. H. Raghavendra Rao, A Varada Rajulu, G Ramachandra Reddy, K Hemachandra Reddy. (2010). Flexural and compressive properties of bamboo and glass fiber-reinforced epoxy hybrid composites. J. Reinf. Plast Compos. 29 1446–1450.
7. Hsuan-Hao Chiu and Wen-Bin Young. (2020). Characteristic study of bamboo fibers in preforming. Journal of Composite Materials.
8. Jagannatha, T.D., Harish, G. (2015). Mechanical properties of carbon/glass fibre reinforced epoxy hybrid polymer composites. Intern. J. Mech. Eng. Robot. Res. 4(2), 131–137.
9. K. M. Faridul Hasan, Peter Gyorgy Horvath, and Tibor Alpar. (2021). Potential fabric-reinforced composites: a comprehensive review, J Mater Sci 56:14381–14415
10. Kannan Rassiah, M. M. H. Megat Ahmad, Aidy Ali1, Abd Halid Abdullah, Sasitharan Nagapan. Mechanical Properties of Layered Laminated Woven Bamboo Gigantochloa Scortechinii/Epoxy Composites. J Polym Environ DOI 10.1007/s10924-017-1040-3 Springer.
11. Meenambika, G.B., Raghavendra, H.R. (2014). Mechanical and chemical properties of bamboo/ glass fibers reinforced polyester hybrid composites. Ind. Eng. Lett. 4(4), 39–42.
12. Meiling Chen, Yun Weng, Kate Semple, Shuxian Zhang, Yu'an Hu, Xiayun Jiang, Jianxin Ma, Benhua Fei, Chunping Dai. (2021). Sustainability and innovation of bamboo winding composite pipe products, 1 1364-0321/© Elsevier Ltd.
13. Praveenkumara J, Madhu P, Yashas Gowda T G, Pradeep S.(2018). Studies on Mechanical Properties of Bamboo/Carbon Fiber Reinforced Epoxy Hybrid Composites Filled with SiC Particulates. International Journal of Engineering Research and General Science Volume 6, Issue 5.
14. Ramachandran M., Bansal S., Fegade V., Raichurkar, P. (2015). Analysis of bamboo fibre composite with polyester and epoxy resin. Int. J. Text. Eng. Process. 1(4), 18–21.
15. Ratim, Bonnia, N.N., and Surip, S.N.(2012). The effect of woven and non-woven fiber structure on mechanical properties polyester composite reinforced kenaf. AIP conference.
16. S.K. Samal, S Mohanty, S. K. Nayak. (2009). Polypropylene-bamboo/glass fiber hybrid composites: fabrication and analysis of mechanical, morphological, thermal, and dynamic mechanical behavior. J. Reinf. Plast Compos. 28, 2729–2747.
17. Surbhi Kore, Ryan Spencer, Hicham Ghossein, Lee Slaven, David Knight, John Unser, Uday Vaidya. (2021).Performance of hybridized bamboo-carbon fiber reinforced polypropylene composites processed using wet laid technique. 20212666-6820/© 2021 Elsevier B.V.
18. Tsotra, P., Friedrich, K. (2003). Electrical and mechanical properties of functionally graded epoxyresin/carbon fibre composites. Compos. Part A: Appl. Sci. Manuf. 34(1), 75–82.
19. William Nguegang Nkeuwa, Jialin Zhang, Kate E. Semple, Meiling Chen, Yeling Xia, Chunping Dai. (2022). Bamboo-based composites: A review on fundamentals and processes of bamboo bonding. https://doi.org/10.1016/j.compositesb.2022.109776 ELSEVIER.
20. Y. S. Rao, B. Manikantesh, P. Sudheer Kumar and A. Yugandhar. (2020). Experimental Investigation on Mechanical Properties of Carbon/Bamboo/Epoxy Hybrid Laminated Composites. Springer Nature Singapore Pte Ltd. L. Li et al. (eds.).
21. Zhang, J. (2012). Glass/carbon fibre hybrid composite laminates for structural applications in automotive vehicles. Sustainable Automotive Technologies 2012, pp. 69–74, Springer Berlin Heidelberg.

CHAPTER 16

Management of Sludge Generated from Municipal Wastewater Treatment by Integrated Fixed-Film Activated Sludge Process

Sushovan. Sarkar*

Professor, Civil Engineering Department,
Budge Budge Institute of Technology, Budge Budge, Kolkata, India

Abstract

The sludge originated as an end product of municipal wastewater treatment by the hybrid bioreactor and mainly consists of organic matter and nutrients. The possibility of washing out the suspended biomass can be avoided by producing a less volume of sludge in the hybrid bioreactor. Treatment of sludge is aimed at reducing the volume of sludge, remove water and kill pathogenic organisms. The sequential treatment involves conditioning, thickening, dewatering and stabilisation. Conditioning is done either by adding chemical agents or by thermal conditioning for separating water from the sludge. Thickening is performed for removing free water from the sludge by the methods, such as sedimentation, flotation and centrifugation. The various dewatering processes are drying beds, centrifuging, filter belt and filter press. Stabilisation is done through the processes of anaerobic digestion, aerobic digestion, pasteurisation, composting and lime stabilisation for killing the pathogenic organisms in the sludge. By the process of anaerobic digestion, carbohydrates, fats and proteins are converted into methane gas and carbon dioxide. The produced biogas may be utilised in the electricity. In aerobic digestion, the sludge is decomposed in an aerated vessel by aerobic microorganisms generating heat in the vessel. Composting is the biological thermal oxidation of sludge by microorganisms. Thermal drying is a good alternative to composting because of the requirement of less land area. The usage of sludge as a fertiliser in agriculture is a process of recycling nutrients and thereby sustaining resources in the environment. The nutrients from sludge are found much cheaper than the artificial ones. Sludge can also be utilised as soil improver due to the presence of organic matter and nutrients in it. The six objectives for sludge waste management are for killing pathogens, dewatering the sludge, controlling the activity of metal, reducing the organic content and stabilisation, removing odour and reusing of sludge.

Keywords: sludge management, municipal wastewater treatment, integrated fixed-film, activated sludge process, aerobic/anaerobic stabilisation, reuse and recovery

*hod.ce@bbit.edu.in

Introduction

Sludge is the residue that accumulates in sewage treatment plants and is basically known as bio-solids. Treatment of municipal wastewater results in concentrating the impurities into a smaller volume of liquid called sludge [13]. Management of sludge from municipal wastewater treatment plants in an environmentally and economically acceptable manner is one of the challenging issues in modern society due to rapid increase in sludge production nowadays [6]. Integrated fixed-film activated sludge process (hybrid bioreactor) is thus practised in developed countries as an advance wastewater treatment for reducing the volume of sludge. Treatment and disposal of municipal sewage sludge are major factors for proper management of the sludge handling system, its reuse and recycling. The sludge generated from municipal wastewater have higher concentration of organic materials because of the source of their origin and thus can be used to increase the fertility in the agricultural land, crop production and recovery of degraded areas[10]. Two basic goals of the management of sludge before final disposal is to reduce its volume and to stabilise the organic materials. Stabilised sludge does not have an offensive odour and can be handled without causing a nuisance or health hazard. Smaller sludge volume reduces the costs of pumping and storage. Management of sludge disposal is done with the objective to kill pathogens, dewater the sludge, control the activity of metals, reduce organic content, stabilise the sludge, remove order and reuse of sludge for sustaining the development of the society as well as raw materials for construction. [1], [2], [3], [4], [5], [7]

Dumping sewage sludge was the most common practice in earlier days. However, this practice is limited due to non-availability of dumping sites, problems in handling, increasing costs and government regulations. The necessity of proper management of sludge disposal thus finds its relevance in this paper. Sludge management is an integral part of any modern municipal wastewater treatment plant: it is important not to lose the nutrients in the sludge, to make use of its material and energy and to make the disposal of the sludge very efficiently and sustainably. Treatment of sludge is often neglected in comparison with water-related parameters such as the outgoing load and the degree of removal of municipal wastewater. Sludge is a potential threat to the environment, for example, foaming sludge can be lost from the treatment process or sewage sludge maybe even deliberately disposed of into water courses. The practical and technical challenges of sludge handling are:

- Stabilising sludge is not inert and can have an unpleasant odour;
- Utilising the energy potential when economically possible;
- Reducing the amount of harmful microorganisms if people, animals or plants are in contact with the sludge; and
- Recovering nutrients, such as phosphorus, for agriculture.

Sludge management for municipal wastewater is more than only thickening, digestion, dewatering and disposal. It has consequences for the whole treatment plant:

- With sludge-originated biogas, it is possible to increase energy production (electrical and thermal) to over 100% of the power needed in the plant. Energy production and energy efficiency are thus very important issues. It is also possible to increase biogas production with certain pre-treatment methods.

- The retention time in primary sedimentation has a direct positive effect on biogas production. On the other hand, a higher retention time decreases the BOD load in the biological treatment; this decreases the denitrification capacity and may require an additional carbon source. Other possible effects are better dewaterability and lower disposal costs.
- In digestion, nitrogen is reduced to ammonia, which is in high concentration in the reject water that is separated from the sludge in dewatering. Better digestion causes a higher reject water load. If the nitrogen removal capacity of the wastewater treatment plant is too low, additional reject water treatment methods can be applied.
- Biological phosphorus removal reduces dewaterability up to 10% (Kopp, 2010). Some plants have problems operating a stable biological phosphorus removal or have other operational problems (e.g. bulking sludge). Chemical phosphorus removal, in turn, increases the amount of sludge.

The quality of sludge is of major importance for usage and thus it is necessary to investigate the sources of pollutants and sludge quality. In order to improve the quality of sludge, strategies for modern mechanism of pollution control are explored.

Origin of Municipal Sludge

There are two types of sources, point and diffused sources for originating sludge. One major point source is industry. Effluent water from the industry is monitored and controlled by some authority in order to decrease pollutants in the wastewater, hence the pollutants in the sludge. Mostly the metals in the influent wastewater are coming in the wastewater treatment plant from diffuse discharges in society which is difficult to trace and regulate. A significant source of metal is stormwater and the content of metals in the sludge could be increased if stormwater combines with municipal wastewater. Where copper pipes are used, in water lines, corrosion of pipes yield copper in the sludge. Corrosion of galvanised pipes yield zinc in the sludge. Sludge is generated at different stages in wastewater treatment. Sludge generated at primary sedimentation tank is called primary sludge. Sludge generated at biological treatment and post-precipitation stage is called secondary sludge and tertiary sludge, respectively. Sludge is of two types–liquid and solids. Liquid sludge consists of dissolved organic substances, such as fatty acids, carbohydrates and inorganic salts like ammonium. Solid sludge also contains organic and inorganic matter, such as metals and nutrients. Dry solids in the sludge remain as residue after evaporation and drying. Normally, sewage sludge is a mixture of primary sludge from primary clarifier and biological sludge from biological treatment units. Generally, it is found 5.6% solid content in the raw sewage, and 23% solid content in the dewatered sewage[10].

The general composition of sewage sludge from primary and secondary sludge in a developed country is shown in Table 16.1.

Sludge Treatment Facilities

The purpose of the treatment of sludge is to reduce the volume, stabilise the sludge, remove water and kill pathogenic organisms. The sequence of sludge treatments is conditioning, thickening, dewatering and stabilisation. Conditioning is a process of separating water

Table 16.1 Composition of primary and secondary sludge (Ref: European Commission 2001)

Contents	Units	Primary Sludge	Secondary Sludge
Dry solids (DS)	g/l	12	9
Volatile solids (VS)	% DS	65	67
pH		6	7
Carbon	% VS	51.5	52.5
Hydrogen	% VS	7	6
Oxygen	% VS	35.5	33
Nitrogen	% VS	4.5	7.5
Sulphur	% VS	1.5	1
Carbon/Nitrogen		11.4	7
Phosphorus	% DS	2	2
Chlorine	% DS	0.8	0.8
Potassium	% DS	0.3	0.3
Aluminium	% DS	0.2	0.2
Calcium	% DS	10	10
Iron	% DS	2	2
Magnesium	% DS	0.6	0.6
Fat	% DS	18	8
Proteins	% DS	24	36
Fibres	% DS	16	7
Calorific Value	kWh/t DS	4200	4100

from sludge. Conditioning is done either by adding chemical agents, such as lime, salts and polymers, or by thermal conditioning by heating the sludge at 150–200°C. Thickening is a process of removing free water from the sludge either by sedimentation or by centrifugation. In the process of dewatering, the volume of sludge is greatly reduced by removing water from the sludge. Dewatering is done by different methods, such as drying beds, centrifuging, centripress, filter belt and filter press. In centrifugal dewatering, centrigugal force is applied for separation of liquid from the solid in the sludge. In the centripress process, clarification by centrifuge and compaction by the filter press occur simultaneously. Filter belts are woven synthetic fibres used for dewatering the sludge continuously. The classification of dewaterability based on direct solid is shown in the following Table 16.2.

Table 16.2 Classification of dewaterability based on dry solid content (Ref. [13])

Dewaterability	Percentage of Dry Solids	Dewaterability	Percentage of Dry Solids
Good	26–30	Good	26–30
Sufficient to medium	22–26	Sufficient to medium	22–26
Bad	8–22	Bad	8–22

Stabilisation is a process where active biological sludge is transformed into inactive ones with reduced volume and odour through hygienisation. Sabilisation prohibits the re-growth of pathogens in the sludge. Stabilisation is done by different methods, such as anaerobic digestion, aerobic digestion, lime stabilisation and thermal sludge processing. In the process of anaerobic digestion, carbohydrates, fats and proteins are converted into methane gas and carbon dioxide in a digester (Nilsson, 2002). Biogas, by-product of anaerobic digestion is used as a source of electricity. The reduction of pathogens in anaerobic digestion depends on the temperature and retention time of digestion. Aerobic digestion is done in an aerated vessel in the presence of aerobic microorganisms generating heat. In the process of lime stabilisation, all active biological sludge is transformed into inactive one at pH value above 12. Lime stabilisation increases the content of dry solid, making it easy in handling the sludge. The addition of quicklime increases the temperature in sludge and thus stabilise the sludge. Thermal sludge processing is done on pressurised sludge under high temperature through different methods, such as pasteurisation, composting, thermal drying, pyrolysis and incineration (Lindquist, 2003). Composting is a process of biological thermal oxidation of sludge under the presence of microorganisms, such as mesophilic and thermophilic bacteria. Composting is done in two phases, such as active phase and curing phase. The rate of degradation of organic matter is faster in the active phase compared to the curing phase. The end product of sludge composting is a humus-like product that can be used as an amendment of soil, erosion control, etc. The process performance of composting depends on temperature, dry matters and volatile solids. Generally, 40%–60% dry solids and 60°C temperature is required for the satisfactory performance of composting. Thermal drying is a process of elimination of interstitial water and thereby reducing the volume of the sludge. In the process of pyrolysis, sludge is heated to 300–900°C producing the end product containing mineral matter, carbon and different gaseous compounds [1], [2], [3], [4], [5], [6], [7].

Different methods for pre-treatment of sludge before disposal have been illustrated in Fig. 16.1.

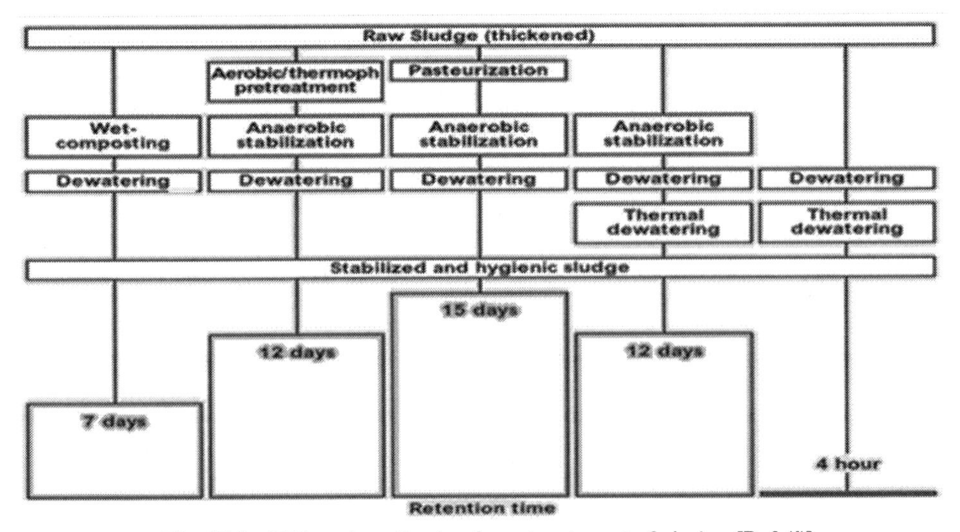

Fig. 16.1 Different methods of pre-treatment of sludge [Ref. (1)]

Reuse Potential and Method of Disposal of Sludge

Treated sludge can be used as fertiliser, soil improver and as construction materials. Sludge also may be used for landfill, gardening, traffic noise barriers, golf courses, reclamation sites and flower plantations after due treatment. Ash, the end product of the incinerated sludge is used in brick manufacturing. Sludge after dewatering in addition with lime can be directly used in brick manufacturing. The treatment and disposal method of sludge have some advantages and disadvantages depending on the local conditions. Different percentage of disposals of sludge in four (4) different countries is shown in Table 16.3.

Table 16.3 Disposal of sewage sludge in percentage in four (4) countries (Ref.[2])

Method of disposal	Sweden	EU	USA	China
Agricultural use and soil improvement	53	46	60	40
Landfilling	34	18	17	30
Incineration	0	12	20	-
Intermediate storage	8	-	-	-
Others	-	12	3	30
Not presented disposal	5	12	-	-

Composed or thermal-dried sludge being pelletised can be easier to handle and transport compared to the untreated sludge cake. After due treatment of composting and thermal drying, the sludge can be used as fertiliser in agriculture or as a soil improver, thereby creating recycle of nutrients back to the farmland. Sludge used as fertiliser in agriculture is comparatively inexpensive compared to artificial ones. Sludge being an organic fertiliser increases the water-holding capacity of the land due to increased organic matter in the soil. Sludge is also used in other green areas like golf courses, etc. where nutrients to some extent are made use of in an economical way. Poor-quality of sludge is used as construction materials, making it safe for the environment and public health due to the immobility of pollutants. Landfill normally produces biogas which can be utilised as energy. For preventing the waste from being transported away by the wind, waste sludge is mixed with soil to prevent such spreading and to improve the mineralisation of the waste. Sludge is kept at different wastewater treatment plants until sufficient amounts of sludge have been produced. Then trucks transport it away from disposal [1], [2], [3], [4], [5], [6], [7]. The sludge management scheme for the application of liquid, dewatered and dried municipal sludge in agricultural land is illustrated in Fig. 16.2.

In most European countries, landfilling is done 50%–75% and balance is disposed into agricultural land as a soil conditioner and fertiliser or other recycling outlets like parks, land restorations and landscaping Ref.[1]).

State-of-the-Art Techniques in Management of Sludge

Application of raw sludge in construction materials may avoid new facilities for incineration and thermal drying, thereby reducing the disposal cost. Use of sludge in construction

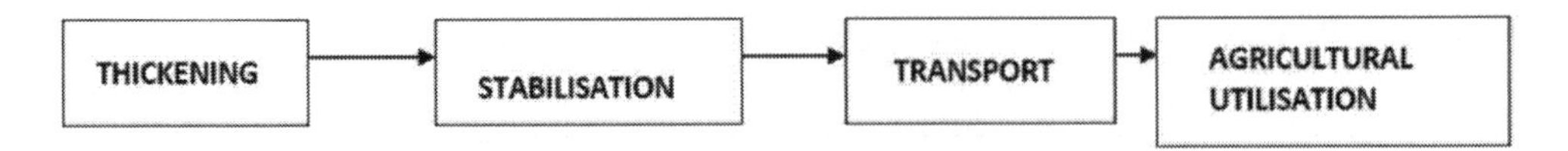

FLOWSHEET FOR APPLICATION OF LIQUID SLUDGE TO AGRICULTURAL LAND

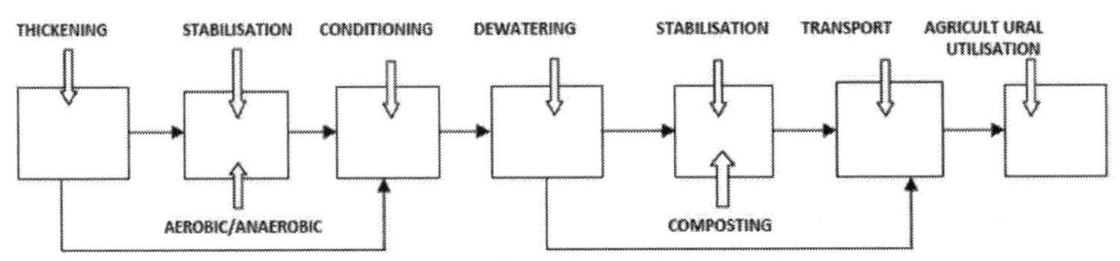

FLOWSHEET FOR APPLICATION OF DEWATERED SLUDGE TO AGRICULTURAL LAND

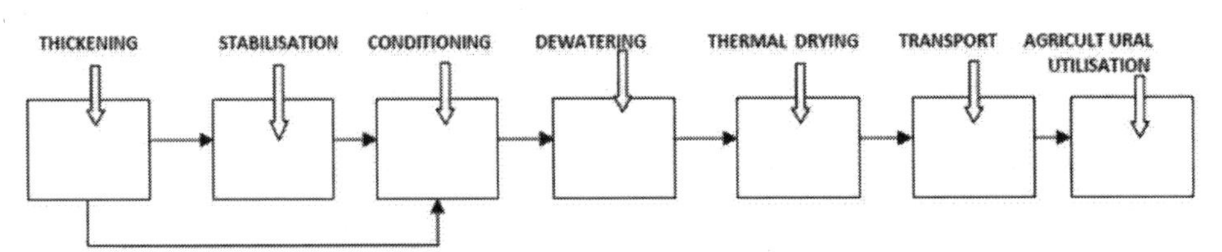

FLOWSHEET FOR APPLICATION OF DRIED SLUDGE TO AGRICULTURAL LAND

Fig. 16.2 Schematic diagram of sludge management for disposal to agricultural land (Ref. [8])

materials would decrease the withdrawal of raw materials, indicating a positive for the overall environment. In the latest technology for stabilisation of sludge, dual digestion process initiated by aerobic thermophilic digestion followed by mesophilic anaerobic digestion decreases levels of pathogens, overall volatile solids, organic matter and increases methane gas generation in anaerobic digester [8]. Compared to the single anaerobic digester, the reduction of volatile solids can be achieved one-third less in tankage in dual digestion stabilisation. In total 10 to 20% of the volatile solids are liquefied in the aerobic digester. The stabilised sludge, by the process of dual digestion, produces fewer orders compared to the single-stage anaerobic digester. Auto thermal thermophilic aerobic digestion is done with the heat generated from the oxidation of volatile solids through the process of exothermic microbial oxidation. The ATAD reactor ultimately decreases significantly the hydraulic retention time to achieve decreased suspended solids, reduces volatile solids by 30%–50%, and takes part in greater reduction of pathogenic viruses, bacteria and other parasites. In the latest modification of the aerobic digestion process, high-purity oxygen is used instead of air for digestion. This latest modification is mostly applicable in the cold region because of its relative insensitivity to changes in ambient air temperature. In high-purity oxygen aerobic digestion, closed tanks are used for maintaining high operating temperature through the exothermic process of digestion. The rate of destruction of volatile suspended solids is drastically increased by this modified digestion process.

Phosphorus recovery from sewage sludge or from sludge dewatering streams is receiving increased attention, particularly in developed countries, as phosphorus is a limited resource (a concept also known as 'peak phosphorus') and is needed as fertiliser to feed a growing world population. Phosphorus recovery methods from sludge can be categorised by the origin of the used matter (wastewater, sludge liquor, digested or non-digested sludge and ash) or by the type of recovery processes (precipitation, wet-chemical extraction, precipitation and thermal treatment). Research on phosphorus recovery methods from sewage sludge is still under development in research.

The Omni Processor is a process that is currently under development that treats sewage sludge and can generate a surplus of electrical energy if the input materials have the right level of dryness.

Thermal depolymerisation produces light hydrocarbons from sludge heated to 250°C and compressed to 40 MPa.

Approach to a Sustainable and Best Management Practices

To achieve a sustainable society, nutrients in the sludge can be made used in agricultural land and recycled. Sewage sludge can be superheated and converted into pelletised granules that are high in nitrogen and other organic materials. Sewage treatment plants should have dewatering facilities that use large centrifuges along with the addition of chemicals, such as polymer, to further remove liquid from the sludge. The product which is left is called 'cake' and that is picked up by companies that turn it into fertiliser pellets. This product is then sold to local farmers and turf farms as a soil amendment or fertiliser, reducing the amount of space required to dispose of sludge in landfills. The extent of dewatering from the sludge during its treatment significantly influences the cost and the selection of the process. Sometimes water in the sludge may be beneficial to crops and thus it is not necessary to remove water from the sludge before application to land. However, dewatering of the sludge is done for reducing the transportation cost of the same, that is, cost of dewatering is offset by the savings in transportation cost. Sludge drying although more expensive than the dewatering process, allows further lowers the transportation cost and makes it enable for storing and packaging (Ref. [8]).

Conclusion

Treatment methods should facilitate the reuse of sludge enabling the recycling of nutrients and maintaining a sustainable society. Thermal drying is a better alternative than composting because of the requirement for less area. Handling and transportation of sludge become easy when it is dry and pelletised. By expanding the composting treatment and the treatment with thermal drying recycling of sludge can be increased. Sludge can be used as fertiliser, soil improver and as construction materials. Sludge may be considered to be a valuable resource as a fertiliser and soil improver for future. The strategy of using composting and thermal drying for most of the sludge will enable larger recycling of important nutrients in the future.

The main objectives of the management of sludge disposals are as follows:

- Primary aim of the management of sludge from municipal wastewater treatment is removing solids from the wastewater.
- To produce a clean and stabilised sludge for optimising quality of sludge in agricultural use.
- Sludge volume should be minimised through innovative technical solution, such as treatment of wastewater by hybrid bioreactor.
- Recovery of products from sludge carrying intrinsic values, such as ammonium sulphate and biopolymers.
- Assessment of the effects of some micropollutants both organic and inorganic, in sludge-amended soil.

Each sanitary authority should carefully consider local conditions in evaluating the most suitable option for disposal of water treatment sludge from each plant under their control having regard to the volumes and characteristics of sludges and the availability of suitable outlets and the requirements of the Waste Management Regulations. Disposal of sludge without any treatment can suitably be used in case of landfilling. The treatment option for the sludge mostly depends on the ultimate disposal goal. However, the most desirable option for the management of sludge is the recycling of waste after due treatment for utilising it as a resource of nutrients in the agricultural field.

References

1. Bresters, A.R., Coulomb, I., Deak, B., Matter, B., et al, (1997),"Sludge treatment and disposal: Management approaches and experiences, European Environment Agency and International Solid waste Association, Denmark.
2. Dahlstrom, Hanna and Nilsson Chris, (2005),"Treatment and disposal methods of wastewater sludge in the area of Beijing, China", Minor field study, Master's Thesis Environment DG, 2000, Working document on Sludge, 3rd Draft Env.E.3/LM
3. European Commission, 2001, Disposal and recycling routes for sewage sludge scientific and technical sub-component report.
4. European Commission DG Environment, 2001, Evaluation of sludge treatments for Pathogen reduction.
5. Fuentes, A (2006), "Ecotoxicity, phytotoxicity and extractability of heavy metals from different stabilised sewage sludge, Environment Pollution 134, pp. 355–360.
6. Gudulas, K., Notara,P.M., Notaras,E., Papadimitriou,C., and Samaras, P(2007)," Management of sludge from municipal wastewater treatment plants in the regions of Thessaly and West Macedonia, Greece", Proceedings of the International Conference on Environmental Management, Engineering, Planning and Economics, pp. 1249–1254.
7. http://www.umweltbundesamt.de/publikationen/National-plan-fr-the-management-of-sewage-sludge, "Ntional plan for the management of sewage sludge from municipal wastewater treatment plants in Bulgaria.
8. http://www.nap.edu/catalog/5175.html, " Use of Reclaimed Water and Sludge in Food Crop Production", National Academic Press, Washington DC, 1996.
9. Metcalf and Eddy, "Wastewater Engineering Treatment and Reuse", A textbook reprint 2013.

10. National Food Processors Asociation(NEFA), 1993, Statement by the National Food Processors Association on the use of Municipal Sewage Sludge in the production of foods for human consumption, Dublin, Calif: NEFA.
11. Priscila, Lima., Raiza, Gianotto., Leonan,Arruda., and Fernando, M.F.,2015, " Alternatives to the disposal of sludge from water and waste water treatment plants", WASET, Vol 2, No. 7.
12. Spinosa, L; 2004,"EU Strategy and practice for sludge utilisation in agriculture, disposal and land filling, Journal of Residual Science and Technology, I(1), pp. 7–14.
13. Vigneswaran, S., and Kandasamy, J., Sludge treatment technologies

CHAPTER 17

Whey: An Emerging Solution for a Sustainable Environment

Thapar Parul[1] and Niharika G.[2]

Assistant Professor, Department of Food Science and Technology,
Gandhi Institute of Technology and Management (GITAM) Deemed University,
Hyderabad, India

Abstract

A continuous rise in air, soil and water pollution has created an alarming situation in the world. Despite of numerous steps and strategies implemented by the world's ministers of the environment during United Nations Environment Assembly, 2017 for working towards a pollution-free planet, it is estimated that air pollution is the leading environmental risk to health which has caused 95% of deaths in low and middle-income countries. Even an increase in COVID-19 cases has been found in areas associated with high levels of pollution. This chapter highlights a new strategy for overcoming this situation through the advancement of green technology using whey as a substrate. Whey is a yellow-green part of the milk (serum) that is obtained during the processing of cheese or paneer. Whey accounts for 85%–90% of the total volume of milk and contains 55% of the nutrients in the milk. Through an online survey, it is found that during cheese or paneer production, whey is directly disposed of in the drains and other water resources by the majority of Indian population. It itself is one of the causes of water pollution if not treated adequately. But whey has a characteristic property of being recycled or reused. It has been found that the components in the whey can be used as an alternative source of green technology. In this chapter, different approaches of utilising whey to generate green fuels and green plastics have been discussed, which can become an emerging solution for a sustainable environment.

Keywords: whey, sustainable environment, green technology, biogas, bioethanol, bioplastic

Introduction

The term 'environment' includes all the living and non-living elements present on the earth. The non-living components like air, water and land are the basic requirements for all living species, including humans. So, it is the prime responsibility of humans to contribute to conservation and sustainability of these resources. For the sake of development and success, humans have not only polluted these resources but also made the environment unsuitable for

[1]parul.thapar@yahoo.com; [2]ngemedar@gitam.edu

the life of some birds and animals, such as tigers, leopards and peacocks, which has put them on the verge of extinction.

Despite numerous steps and strategies implemented by the world's ministers of the environment during United Nations Environment Assembly, 2017 for working towards a pollution-free planet, it is estimated that air pollution is the leading environmental risk to health which has caused 95% of deaths in low and middle-income countries (World Bank Report, 2021). In a report by Andree and Johannes, 2020, it is found that there has been an increase in COVID-19 cases in areas associated with high levels of pollution.

The need of the hour is to develop certain environment-friendly and pollution-free products that- a. can be used for regular consumption, b. can control the situation of increasing pollution in the world and c. is sustainable. In view of the above-mentioned points, this chapter has indicated the advancement of green technology using whey as a substrate. Green technology is defined as the development of sustainable fuels from renewable resources. The whey (obtained from paneer or cheese) can be one of the solution towards a sustainable environment.

The rest of the paper is structured as follows: Section 2 reviews the extent literature. Section 3 describes the research methodology. Section 4 discusses the findings. Section 5 summarises the paper.

Literature Review

Whey is a yellow-green part of milk (serum) that is obtained during the processing of cheese or paneer (Zadona et al., 2021) (Fig. 17.1). Whey can be characterised as sweet and acid, depending upon the formation and coagulation (Skryplonek and Jasinska, 2017; Smithers, 2008; De Wit, 2001). Sweet whey is obtained by coagulating milk with an enzyme rennin during cheese production, having pH 6–7. Acid whey is obtained by coagulating milk with any acidic product, such as lemon or alum, during paneer production having pH <5 (Lievore et al., 2017). Whey accounts for 85%–90% of the total volume of milk and contains 55% of the nutrients in the milk (Zadona et al., 2021). The various nutrients present in whey (Papademas and Kotsaki, 2020) are represented in Table 17.1.

Fig. 17.1 Whey

Source: https://www.theprairiehomestead.com/2011/06/16-ways-to-use-your-whey.html

Table 17.1 Nutrients in the whey

Component	Sweet whey (g/L)	Acid whey (g/L)
Total protein	6.5–6.6	6.10- 6.20
Lactose (sugar)	46–52	44–47
Milk fat	0.2–0.5	0.3
Minerals	5–5.2	7.5–7.9
Lactic acid (organic acid)	2.0	6.4
Calcium	0.4–0.6	6–8
Free amino acid (organic acid)	0.133	0.45
Amino acid in protein	0.006	0.006

Besides the presence of these nutrients, as per global statistics, the cheese industry produces about 115 million tons of whey annually, adding it to be one of the causes of water pollution worldwide (Papademas and Kotsaki, 2020). Due to its high polluting capacity, it is recommended to the dairy industries for recycling or reuse of whey in a sustainable manner, which has become a great scientific challenge (Zadona et al., 2021).

Methodology

The study is partly empirical and partly descriptive. An attempt is made to evaluate the view of the Indian population in regard to whey and its management. An online survey was conducted in the month of May, 2022 in a form of a questionnaire which enquired about (1) awareness of respondents towards whey and (2) utilisation/discard of whey as a general routine.

A search on the literature related to data for the components in the whey as a source of green energy was done during the period of June and July 2022 using search engines, such as PubMed, Google Scholar, Science Direct and NCBI. The abstracts from different reviews and research papers were studied regarding nutrients in whey and their applications in different sectors of green technology. The search terms used were cheese or paneer whey, green energy and green technology.

Results and Discussion

Whey and its Management

An online survey conducted among 110 respondents in the Indian population revealed certain facts- their awareness towards the term 'whey' and their approach towards whey management during paneer/cheese making. Out of 110 respondents, 84% of them were aware about the term 'whey'. In a general routine, regarding whey management, 32% use it for kneading; 16% as a source of energy drink; 2% for cooking and 50% discard it (Fig. 17.2).

A study conducted by Papademas and Kotsaki, 2020, it has been reported that 47% of whey is being directly disposed in the drains and other water resources during paneer or cheese

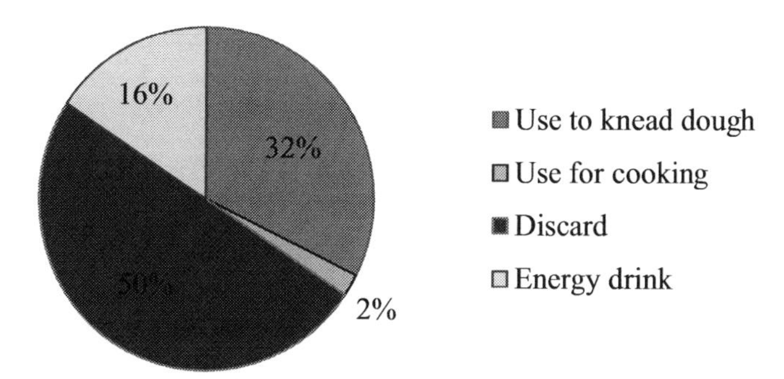

Fig. 17.2 Whey management

production. The BOD and COD content of whey can lead to a serious environmental hazard, adding it to be one of the causes of water pollution worldwide.

Not only this but also whey has a characteristic property of being recycled or reused. It has been found that the components in the whey can be used as an alternative source of green technology (Zadona et al., 2021).

Nutrients in the Whey as a Source for Green Technology

Biogas

Lactose- a component of whey has huge potential in the production of biohydrogen (Zadona et al., 2021). Biohydrogen is a by-product of the anaerobic conversion of organic wastes. The process is carried through the fermentation process of anaerobic and photosynthetic microorganisms, such as cyanobacteria, into organic acids. There are certain other organic acids like amino acids that are naturally present in the whey. These organic acids can be used as a substrate in the anaerobic digestion of whey, producing methane in the form of biogas (Hosseini and Wahid, 2016). Anaerobic digestion is carried out in anaerobic digestors (or fermenters), which involve the following steps (Caceres et al., 2012; Rojas et al., 2011) (Fig. 17.3):

1. *Hydrolysis:* This is the first step in the process of anaerobic digestion of whey. It involves the breakdown of complex substances like lactose into simpler forms of glucose and galactose by the addition of enzymes like lactase.
2. *Acidogenesis:* This is the second step which involves the use of fermentative microorganisms like cyanobacteria. Through the process of glycolysis, also called anaerobic respiration or fermentation, the bacteria use glucose (derived from metabolism of lactose in step (1) as a substrate within the cells and convert it into volatile acids, mainly pyruvic acid.
3. *Methanogenesis:* In this step, the pyruvic acid obtained is converted into methane and carbon dioxide through methanogens like *Methanobacterium formicicum.*
4. *Hydrolysis:* This is the first step in the process of anaerobic digestion of whey. It involves the breakdown of complex substances like lactose into simpler forms of glucose and galactose by the addition of enzymes like lactase.

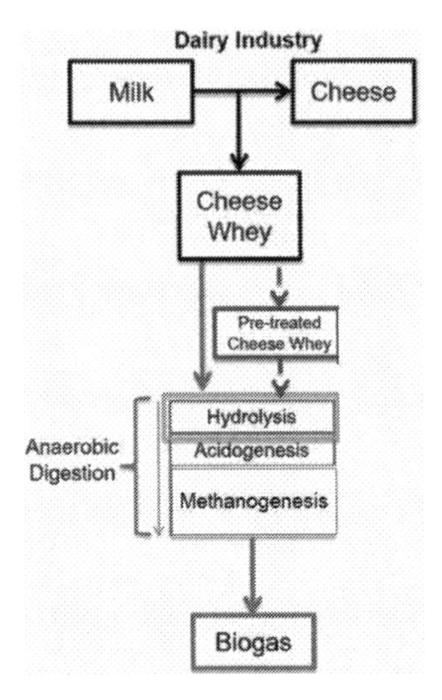

Fig. 17.3 Anaerobic digestion of whey for biogas production

5. *Acidogenesis:* This is the second step which involves the use of fermentative microorganisms like cyanobacteria. Through the process of glycolysis, also called anaerobic respiration or fermentation, the bacteria use glucose (derived from metabolism of lactose in step 1) as a substrate within the cells and convert it into volatile acids, mainly pyruvic acid.
6. *Methanogenesis:* In this step, the pyruvic acid obtained is converted into methane and carbon dioxide through methanogens like *Methanobacterium formicicum.*

But due to the low buffer capacity of these organic acids, there can be low production of biogas. Therefore, to increase the production of biogas, whey has to be mixed with wastes like kitchen waste or manure (Celik et al., 2016). Various studies were conducted for improving the production of biogas by utilisation of whey from different sources. Some of them are mentioned in Table 17.2.

Applications of Biogas

The biogas produced in the anaerobic digestors can be used for home heating, production of electricity and vehicle fuel.

Bioethanol

Bioethanol (green fuel) is an environmentally friendly fuel as it does not produce any toxic emissions on combustion. It has proved to be effective in decreasing air pollution and reducing global warming (Sharma et al., 2018). Bioethanol can be produced by fermenting lactose present in the whey into ethanol by the use of microbial species, mainly yeast. The microbial strain used is the species of *Kluevromyces marxianus* that has hydrolytic enzymes, such as β-galactosidase, pyruvate decarboxylase and galactose-1-phosphate uridyltransferase to metabolize lactose into bioethanol under low oxygen concentration (Sharma et al., 2018; Diniz et al., 2012) mainly using pentose phosphate pathway (PPP) (Vaidya et al., 2005) (Fig. 17.4).

Table 17.2 Studies conducted for improving the biogas production

S. No.	Studies	Results	Reference
1.	Use of digested cheese whey using swine wastewater as inoculum	At temperature of 32°C, reduction of volatile solids by 53.11% and biogas yield- 270 l with 63% generation of methane. At temperature of 26°C, reduction of volatile solids reduction- 45.76% and biogas yield- 171 l with 61% generation of methane	Antonelli et al., 2016
2.	Co-fermentation of cheese whey with glucose	Increased production of biohydrogen	Rosa et al., 2014
3.	Anaerobic sequence batch reactor with immobilised biomass on inert support of SBBR	COD – 540 mg/ L Hydrogen production- 660 ml/ day First reported that use of low temperature during digestion process produces high biohydrogen production	Lima et al., 2016
4.	Treatment of cheese whey using microbial electrolysis cells	Increased production of biohydrogen	Rivera et al., 2017
5.	Anaerobic structured bed reactor (ASTBR) – an alternative method for the production of biohydrogen by fermentative bed reactor	Biohydrogen yield- 2.4 moles hydrogen per lactose in 32-day period	Blanco et al., 2019

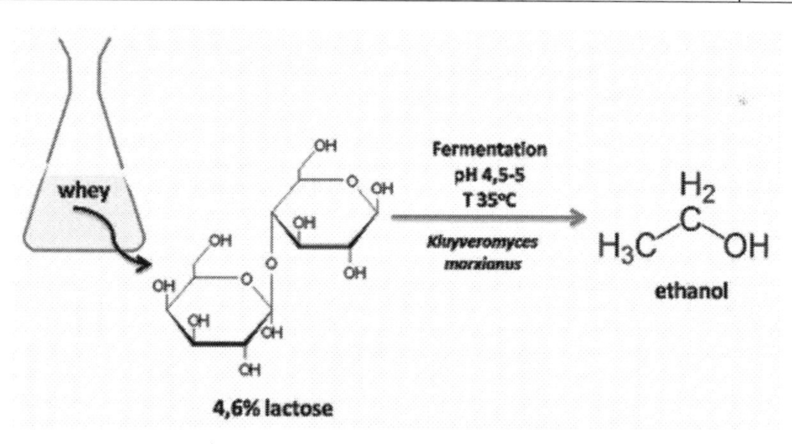

Fig. 17.4 Production of bioethanol

Source: https://ejournal.undip.ac.id/index.php/bcrec/article/view/4044

For using *Kluevromyces* spp., it should be noted that the whey should not contain more than 10% lactose (Wilharm and Sach, 1947) due to the limited physiological characteristics of these yeasts (Gawel and Kosikowski, 1978).

In one of the studies, it has been found that the production of bioethanol increases when used with different carbohydrate resources (Vaidya et al., 2005). Based on this, some of the studies were conducted as mentioned in Table 17.3.

Table 17.3 Studies conducted for improving the bioethanol production

S. No.	Substrate	Results	Reference
1.	Immobilisation of the cells of *Kluevromyces marxianus* with Ca alginate was done during a continuous fermentation process	Improved the yield of bioethanol production- 6.97 g/l	Sharma et al., 2018
2.	Using crude whey using a genetically engineered strain of *Saccharomyces cerevisiae* with lactose fermenting gene as a conventional strain of *Saccharomyces cerevisiae* lacks enzyme for hydrolyzing lactose	Improved yield of bioethanol	Beniwal et al., 2021
3.	Cheese whey permeate using a strain of *Kluevromyces lactis* CBS2359	Improved yield of bioethanol	Sampaio et al., 2020
4.	Cheese whey powder with a mixed culture of *Kluevromyces marxianus* and *Saccharomyces cerevisiae*	Improved yield of bioethanol	Farkas et al., 2019
5.	Cheese whey powder with *Escherichia coli*	Improved yield of bioethanol	Sar et al., 2019
6.	Whey permeates with a strain of *Lactococcus lactis*	Improved yield of bioethanol	Liu et al., 2016
7.	Mozzarella cheese whey and sugarcane molasses with mixed strain of yeast *Candida tropicalis* and bacterium *Bacillus capitatus*	Increased production of bioethanol	Balia et al., 2018

Applications of Bioethanol

Bioethanol obtained from whey can be applied in (Sharma et al., 2018):

- Food processing
- Manufacturing of drugs
- Chemical and cosmetic industries
- As an alternative to sustainable fuel

Bioplastics

The lactose present in the cheese whey can also be converted into polyhydroxyalkanoates (PHAs) and poly-lactic acid (PLA) (Ryan and Walsh, 2016). The PHAs and PLA are biodegradable plastics that can easily be digested by the microorganisms in the soil (Narayanan et al., 2017) and so these can be considered as environmental-friendly plastics and extremely valuable alternatives to commonly used petroleum-derived plastics (Dietrich et al., 2017; Mozejko-Ciesielska and Kiewisz, 2016; Li et al., 2016; Laycock et al., 2014; Akaraonye et al., 2010; Verlinden et al., 2007; Reddy et al., 2003; Sudesh et al., 2000; Madison and Huisman, 1999).

Polyhydroxyalkanoates

Polyhydroxyalkanoates include polyesters like polyhydroxybutyric acid (PHB), polyhydroxyvaleric acid (PHV), 3-hydroxy-2-methyl butyrate (3H2MB) and 3-hydroxy-2-methyl valerate (3H3MV) (Uckun et al., 2015). These can be synthesised from lactose in the whey (Fig. 17.5) by the three processes (Koller et al., 2017). These include:

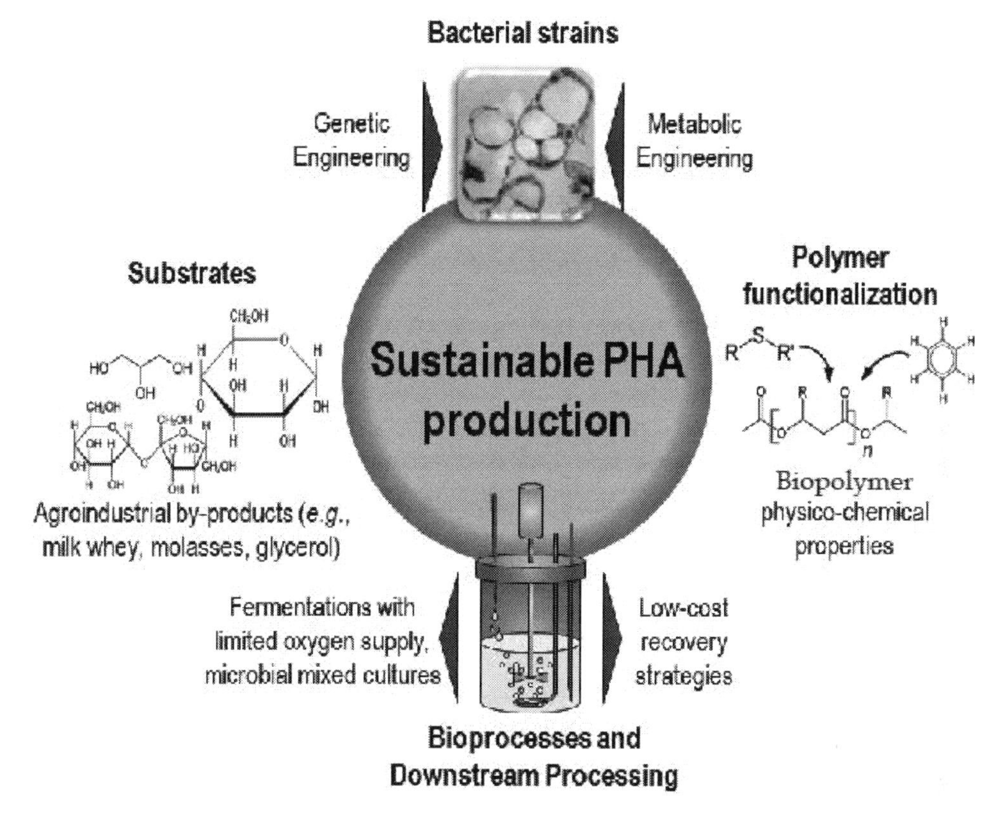

Fig. 17.5 Production of PHA using cheese whey and some additives
Source: https://bioplasticsnews.com/polyhydroxyalkanoates-or-pha/

(a) *Direct conversion of lactose to PHA:* The microorganisms involved in this process are the strains of *Hydrogenophaga pseudoflava* and recombinant *Escherichia coli,* as these microorganisms contain β-galactosidase enzyme ,which can directly convert lactose to polyhydroxyalkanoates.

(b) Breakdown of lactose into glucose and galactose and then into PHA: In this process, lactose can be hydrolysed into glucose and galactose through lactase enzyme. This can be converted further into PHA with the help of microorganisms, such as *Pseudomonas hydrogenovora* and *Haloferax mediterranei*.

(c) Fermentation of lactose to lactic acid and then conversion of lactic acid into PHA.

Polylactic Acid

Lactic acid is synthesised from lactose of whey by a fermentation process by *Sporolactobacillus laevolacticus*, *Lactobacillus plantarum*, *Sporolactobacillus ilulins* and *Lactobacillus bulgaricus* (Awasthi et al., 2018). The optimum pH required for lactic acid production is in the range from 5.5 to 6.5 (Chahal, 1990; Vickroy, 1985). When lactic acid monomers are polymerised by the process of condensation, polylactic acid formation occurs (Fig. 17.6). Polylactic acid (PLA)

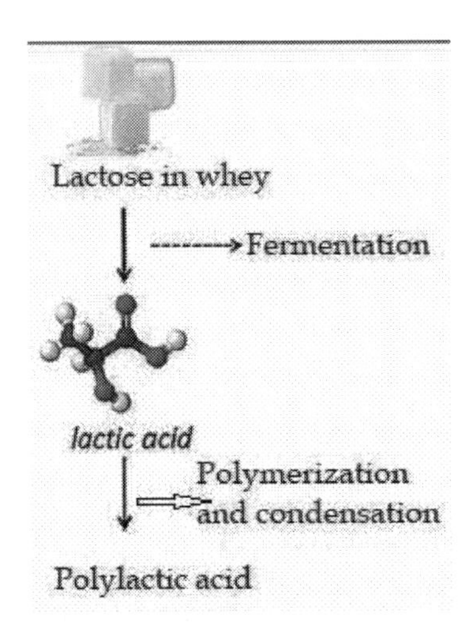

Fig. 17.6 Production of polylactic acid

has a generally recognised as safe (GRAS) status due to its low toxicity and so can also be used for food packaging (Zandona et al., 2021).

There are several other studies that are conducted using cheese whey for the production of biopolymers which are represented in Table 17.4.

Table 17.4 Application of cheese whey in development of bioploymers

S.No.	Substrate	Microorganisms	Biopolymers	Reference
1.	Cheese whey	Lactobacillus spp., *Rhodovulum sulphidofilum* DSM-1374	Poly 3-hydroxybutyrate	Carlozzi et al., 2021
2.	Ricotta cheese whey	Treatment by β-galactosidase enzyme using strain of *Haloferax mediterranei* DSM1411	Poly 3-hydroxybutyrate-co-3- hydroxyvalerate (PHBV)	Raho et al., 2020
3.	Fermented cheese whey	Mixed strain of bacteria and algae	PHA with hydroxyvalerate	Fradinho et al., 2020
4.	Sweet whey powder	Bacterial strains of Thaurea spp. and Lampropedia spp.	PHA	Oliveira et al., 2018

Applications of Bioplastics

These bioplastics may be applied in (Zandona et al., 2021):

- Packaging
- Spraying materials
- Device materials
- Electronic and agriculture products
- Chemical media and solvent industries

Conclusion

The planet earth is the mother of all living organisms on earth. Natural resources in the form of forests, glaciers, wildlife, etc. are blessings to humans, due to which they can survive on the basic needs of life. It should be a moral duty for all human beings that if they cannot contribute in sustainability, they should neither destroy it. During cheese or paneer production, the whey generated as a by-product is directly disposed of in the drains by the majority of the population, due to which it also acts as a major water pollutant. There are several ways by which some portion of natural resources can not only be utilised for the development but also for the protection of environment in a sustainable manner. The blessing in disguise is the 'whey' itself. The different nutrients present in it can be explored further for their utilisation as green technology in the form of biogas, biofuels and bioplastics. It is important to think about different approaches to the use of natural resources in a view towards a healthy environment for present and future generations. If managed in an efficient manner, swhey can be one of the sustainable solutions towards pollution control.

References

1. Akaraonye, E., Keshavarz, T., and Roy, I. (2010). Production of polyhydroxyalkanoates: the future green materials of choice. J. Chem. Technol. Biotechnol. 85:732–743.
2. Andree P. and Johannes, B. P. (2020). Incidence of COVID-19 and Connections with Air Pollution Exposure : Evidence from the Netherlands. Policy Research Working Paper; No. 9221. World Bank, Washington, DC. https://openknowledge.worldbank.org/handle/10986/33664
3. Antonelli, J., Lindino, C. A., Azevedo, J. C. R., Souza, S. N. M., Cremonez, P. A., and Rossi, E. (2016). Biogas production by the anaerobic digestion of whey. Rev. Cienc. Agric. 39(3):463–467.
4. Awasthi, D., Wang, L., Rhee, M. S., Wang, Q., Chauliac, D., Ingram, L. O., and Shanmugam, K. T. (2018). Metabolic engineering of *Bacillus subtilis* for production of d-lactic acid. Biotechnol. Bioeng. 115(2): 453–63.
5. Balia, R. L., Kurnani, T. B. A., and Utama, G. L. (2018). The combination of mozzarella cheese whey and sugarcane molasses in the production of bioethanol with the inoculation of indigenous yeasts. J. Japan Inst. Energ. 97(9): 266–9.
6. Beniwal, A., Saini, P., De, S., and Vij, S. (2021). Harnessing the nutritional potential of concentrated whey for enhanced galactose flux in fermentative yeast. Food Sci. Technol. 141: 110840.
7. Blanco, V. M. C., Oliveira, G. H. D., and Zaiat, M. (2019). Dark fermentative biohydrogen production from synthetic cheese whey in an anaerobic structured-bed reactor: Performance evaluation and kinetic modeling. Renew Energ. 139: 1310–1319.
8. Carlozzi, P., Giovannelli, A., Traversi, M. L., Touloupakis , E. (2021). Poly (3-hydroxybutyrate) bioproduction in a two-step sequential process using wastewater. J. Water Process Eng. 39: 101700.
9. Celik, K. and Onur, Z.Y. (2016). *Whey every aspect.* Istanbul, Turkey: Tudas Alapítvany
10. Chahal, S. P. (1990). Lactic acid, In Ullman's encyclopedia of industrial chemistry, ed. B. S. Elvers, J. K. Hawkins, and G. Schulz, 97–105. Weinheim Press.
11. De Wit, J. N. (2001). Lecturer's Handbook on whey and whey products. Eur. Whey Prod. Assoc. 91.
12. Dietrich, K., Durmont, M., Del Rio, F., and Orsat, V. (2017). Producing PHAs in the bioeconomy—Towards a sustainable bioplastic. Sustain. Prod. Consumpt. 9: 58–70.

13. Diniz, H. S. R., Silveira, B. W., Fietto, G. L., and Passos, M. L. F. (2012). The high fermentative metabolism of *Kluyveromyces marxianus* UFV-3 relies on the increased expression of key lactose metabolic enzymes. PubMed. 101(3): 541–50.
14. Farkas, C., Rezessy, S. M. J., Gupta, K.V., Bujna, E., Pham, M. T., and Pasztor, H. K. (2019). Batch and fed-batch ethanol fermentation of cheese-whey powder with mixed cultures of different yeasts. Energies. 12: 4495.
15. Fradinho, J. C., Oehmen, A., and Reis, M. A. M. (2019). Improving polyhydroxyalkanoates production in phototrophic mixed cultures by optimizing accumulator reactor operating conditions. Int. J. Biol. Macromol. 126: 1085–92.
16. Gadhe, A., Sonawane, S. S., and Varma, M. N. (2015). Enhanced biohydrogen production from dark fermentation of complex dairy wastewater by sonolysis. Int. J. Hydrogen Energ. 40(32): 9942–51.
17. Gawel, J. and Kosikowski, K. V. (1978). Improving alcohol fermentation in concentrated ultrafiltration permeates of cottage cheese whey. J. Food Sci. 43: 1717–1719
18. Hosseini, S. E. and Wahid, M. A. (2016). Hydrogen production from renewable and sustainable energy resources: promising green energy carrier for clean development. Renew. Sustain Energy Rev. 57: 850–66.
19. Koller, M., Marsalek, L., de Sousa, D. M. M., and Braunegg, G. (2017). Producing microbial polyhydroxyalkanoate (PHA) biopolyesters in a sustainable manner. New Biotechnol. 37: 24–28.
20. Laycock, B., Halley, P., Pratt, S., Werker, A., and Lant, P. (2014). The chemomechanical properties of microbial polyhydroxyalkanoates. Prog. Polym. Sci. 39: 397–442.
21. Li, Z., Yang, J., and Loh, X. J. (2016). Polyhydroxyalkanoates: opening doors for a sustainable future. NPG Asia Mater. 8: 265.
22. Lievore, P., Simoes, D. R. S., Silva, K. M., Drunkler, N. L., Barana, A. C., and Nogueira, A. (2015). Chemical characterisation and application of acid whey in fermented milk. J. Food Sci. Technol. 52(4): 2083–92.
23. Lima, D. M. F., Lazaro, C. Z., Rodrigues, J. A. D., Ratusznei, S. M., and Zaiat M. (2016). Optimization performance of an AnSBBR applied to biohydrogen production treating whey. J. Environ. Manage. 169: 191–201.
24. Liu, J., Dantoft, S. H., Wurtz, A., Jensen, P. R., and Solem, C. (2016). A novel cell factory for efficient production of ethanol from dairy waste. Biotechnol. Biofuels. 9(1): 33.
25. Madison, L. L., and Huisman, G. W. (1999). Metabolic engineering of poly (3- hydroxyalkanoates): from DNA to plastic. Microbiol. Mol. Biol. Rev. 63: 21–53.
26. Mozejko, C. J., and Kiewisz, R. (2016). Bacterial polyhydroxyalkanoates: still fabulous? Microbiol. Res. 192: 271–282
27. Narayanan, C. M., Das, S., and Pandey, A. (2017). Food waste utilization: Green technologies for manufacture of valuable products from food wastes and agricultural wastes. In *Food bioconversion*, ed. A. M. Grumezescu, and A. M. Holban, 1–54. London, UK: Academic Press
28. Oliveira, C. S., Silva, M. O., Silva, C. E., Carvalho, G., and Reis, M. A. (2018). Assessment of protein-rich cheese whey waste stream as a nutrients source for low-cost mixed microbial PHA production. Appl Sci. 8(10): 1817
29. Papademas, P. and Kotsaki, P. (2020). Technological Utilization of Whey towards Sustainable Exploitation. Adv. Dairy Res. 7(4): 1–10.
30. Raho, S., Carofiglio, V. E., Montemurro, M., Miceli, V., Centrone, D., and Stufano, P. (2020). Production of the polyhydroxyalkanoate PHBV from ricotta cheese exhausted whey by *Haloferax mediterranei* fermentation. Foods. 9(10): 1459.
31. Reddy, C. S. K., Ghai, R., Rashmi, C., and Kalia, V. C. (2003). Polyhydroxyalkanoates: an overview. Bioresour. Technol. 87, 137–146.

32. Rivera, I., Bakonyi, P., Cuautle, M. M. A., and Buitron, G. (2017). Evaluation of various cheese whey treatment scenarios in single-chamber microbial electrolysis cells for improved biohydrogen production. Chemos. 174: 253–9.
33. Rosa, P. R. F., Santos, S. C., and Silva, E. L. (2014). Different ratios of carbon sources in the fermentation of cheese whey and glucose as substrates for hydrogen and ethanol production in continuous reactors. Int. J. Hydrogen Energ. 39: 1288–96.
34. Ryan, M. P. and Walsh, G. (2016). The biotechnological potential of whey. Rev. Environ. Sci. Biotechnol. 15(3): 479–98.
35. Sampaio, F. C., de Faria, J. T., da Silva, M. F., de Souza, O. R. P., and Converti, A. (2020). Cheese whey permeate fermentation by *Kluyveromyces lactis*: A combined approach to wastewater treatment and bioethanol production. Environ Technol. 41: 3210–8.
36. Sar, T., Stark, B. C., and Akbas, M. Y. (2019). Bioethanol production from whey powder by immobilized *E. coli* expressing *Vitreoscilla* hemoglobin: Optimization of sugar concentration and inoculum inoculum size. Biofuels. 1–6.
37. Sharma, D., Manzoor, M., Yadav, P., Sohal, J. S., Aseri, G. K., and Khare, N. (2018). Bio-valorization of dairy whey for bioethanol by stress-tolerant yeast. In *Fungi and their role in sustainable development: Current perspectives*, ed. P. Gehlot, and J. Singh, 349-66. Singapore: Springer.
38. Skryplonek, K. and Jasinska, M. (2017). Whey-based beverages. Electron J Polish Agric. Univ. 20:4.
39. Smithers, G. W. (2008). Whey and whey proteins-from "gutter-to-gold." J. Int. Dairy. 18(7): 695–704.
40. Sudesh, K., Abe, H., and Doi, Y. (2000). Synthesis, structure and properties of polyhydroxyalkanoates: biological polyesters. Prog. Polym. Sci. 25: 1503–1555.
41. The world Bank report (2021). https://www.worldbank.org/en/topic/pollution#1
42. Uckun, K. E., Trzcinski, A. P., and Liu, Y. (2015). Platform chemical production from food wastes using a biorefinery concept. J. Chem. Technol. Biotechnol. 90(8): 1364–79
43. Vaidya, A., Pandey, A. R., Mudliar, S., Kumar, S. M., Chakrabarti, T., and Devotta, S. (2005). Production and Recovery of Lactic Acidfor Polylactide—An Overview. Critical Rev. Environ. Sci. Technol., 35: 429–467
44. Verlinden, R. A. J., Hill, D. J., Kenward, M. A., Williams, C. D., and Radecka, I. (2007). Bacterial synthesis of biodegradable polyhydroxyalkanoates. J. Appl. Microbiol. 102: 1437–1449.
45. Vickroy, T.B. (1985). Lactic acid, In Comprehensive biotechnology, In *Production and Recovery of Lactic Acid for Polylactide—An Overview*, ed. H. W. Blanch, S. Drew, and D.I.C. Wang, 761–778. Oxford: Pergamon Press
46. Wilharm G, Sach U (1947). The properties of several lactose fermented yeasts. Milchwissenschaft 10: 382
47. Zadona, E., Blazic, M., and Jambrak, R. A. (2021). Whey Utilisation: Sustainable Uses and Environmental Approach. Food Technol. Biotechnol. 59(2): 147–161.

CHAPTER 18

Working Environmental Impact of Personal Protective Equipment in Modern Industrial Sector

Ram Niwas[1]
Health Safety and Environmental Engineering Department,
Shri Rawatpura Sarkar University, Raipur, India

Omprakash Thakare[2]
Mechanical Engineering Department,
Shri Rawatpura Sarkar University, Raipur, India

Nitin Gudadhe[3]
Mechanical Engineering Department,
Shri Ramdeobaba College of Engineering and Management, Nagpur, India

Abstract

Personal protective equipment (PPE) is important for preventing workplaces' injuries. PPEs are playing a very important role in preventing injuries in all industries of India, including small, medium and large-scale industries as well. It is one of the best practice to eliminate any potential hazard at workplace. PPE is the last line of defence and means to avoid injury at the workplace. Generally, PPEs use of PPE implies working in at potentially hazardous and unsafe work environment. So, it is of the prime importance to ensure that the selection of equipment is both reliable and effective; it is being properly used and maintained and the user has undergone adequate training. The aim of this paper is to raise awareness related to occupational health and safety practices and the correct use of PPE by people from all walks of life. PPE provides good protection against the hazards of toxic exposure, splashing dust, water vapour and liquids, flying particles, hot substances, radiation and sharp edges. Many fatal accidents are caused by these reasons and the use of appropriate PPE can prevent and prevent many of them. The PPE provides a good defence against hazards of toxic exposure, dusting chemical splashes, steam water and liquids, flying particles, hot substances, radiation and sharp edges. Many fatal accidents are caused due to these reasons and the use of appropriate PPE can prevent and lessen many of them. The outcomes of this paper are to guarantee the best conceivable security for representatives in the work environment. The helpful endeavours of the representatives will help in laying out and keeping a protected and empowering and productive workplace. Finally maintained a healthful and fruitful work environment.

Keywords: personal protective equipment (PPE), risk assessment and control management, hazards assessment

[1]niwas2001@gmail.com; [2]omprakashthakare82@gmail.com; [3]gudadhenp@rknec.edu

Introduction

The world handles how to fight health-related issues. Each people have a different role in the survival conditions, but nowadays personal protective equipment (PPE) are the most important means of preventing work injuries. The best approach is to eliminate any potential hazard at workplace. PPE should only be relied upon as the last line of defence in places where it is not practicable to control the hazards at the source. Generally, PPEs use implies working in the potentially hazardous work environment and its use as a major means of injury prevention. So, it is most important to ensure that the selection of adequate and proper equipment is both reliable and effective; it is being properly used and maintained and the user has undergone adequate training. The aim of this term work is to increase awareness related to occupational safety and health best practices and the proper use of PPE by people from all walks of life. PPE provides good protection against the hazards of toxic exposure, splashing of dust, water vapour and liquids, flying particles, hot substances, radiation, sharp edges, welding, impact and stopping by objects, glare, falls of persons and injuries caused by falling bodies, noise, scrap cleaning, material handling, electrical shock, burns and firefighting. Many fatal accidents are caused by these reasons and the use of appropriate PPE can prevent and prevent many of them. The recommended methods are designed to be suitable to any number and type of PPE (e.g. not only helmet and safety vest but also gloves, goggles and steel-toed boots).

Literature Review

PPE acts as a basic barrier between workers and hazardous at the workplace. Depending on the areas protected by the body, standard classifications and examples of PPE include head, eye, face, hand, body, foot and hearing protection [7]. For example, workers who wear safety helmets can reduce the impact from falling objects to avoid injuries caused by accidental impacts with stationary objects. Gloves are essential to protect hands when handling rough or sharp materials or chemical or hot surface, etc. Likewise, the use of reflective safety vests could increase the visibility of workers in workplaces and thus reduce the likelihood of accidents, especially in low light or in the dark. Despite the high incidence of unsafe working conditions, compliance with the use of PPE in workplaces remains low. The Occupational Safety and Health Administration (OSHA) reported that the lack or improper use of PPE was one of the most violated OSHA standards during the fiscal year 2019 [8]. Bureau of Labor Statistics (BLS) statistics revealed that nearly 84% of workers who suffered head injuries due to the use of a different helmet, only 1% of nearly 770 workers who sustained facial injuries were wearing face protection correctly and the rate of use of protective footwear was 23% among workers who sustained leg injuries [9]. Companies and employers can also face substantial fines of up to $12,934 per violation for not complying with PPE [10]. Therefore, the detection of non-compliance with the requirements for the use of multi-class PPE is necessary for the workplace. There are two main techniques for PPE compliance verification on construction sites: vision-based and sensor-based [5]. Methods based on wearable sensors focus on the use of external position sensors and then analyse the recorded signals to monitor compliance with the principles of PPE use. Kelmetal. (2013) [11] introduced a mobile radio frequency identification (RFID) device to determine whether the use of PPE by workers complies with relevant safety regulations. Similarly, Li et al. (2017) [12] designed a helmet-less wear monitoring system by attaching silicone pressure sensors to helmet sweatbands (Kimetal 2018).

Problem Identifications

Even if engineering controls and safe systems of work have been applied, some hazards may still exist as shown in Fig. 18.1. These include injuries to:

- Control the lungs problem from breathing in contaminated air.
- The head and feet from the falling materials.
- Problems on eyes from flying particles or splashes of corrosive liquid.
- Problem on the skin from contact with corrosive materials.
- The body effect due to extremes of heat or cold.

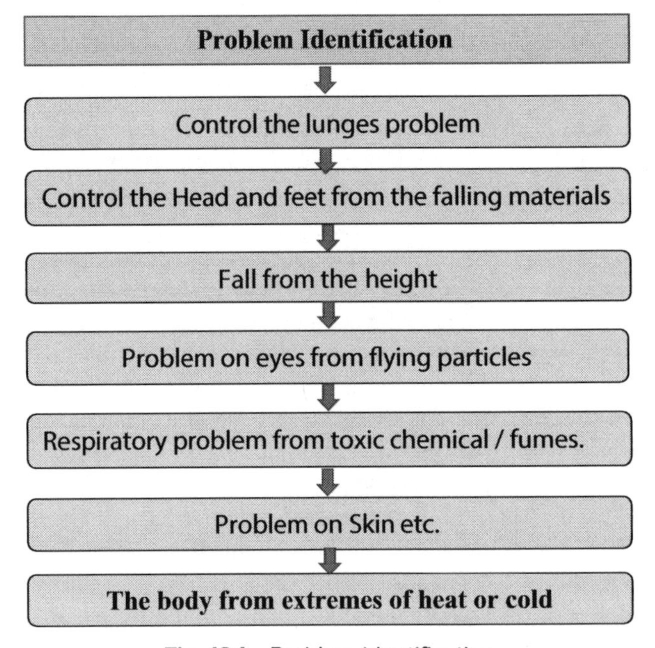

Fig. 18.1 Problem identification

Data and Variables

PPE

PPE should always be considered as a last resort, as it relies on people to use it. Only when there is no other alternative should PPE be provided. The characteristics of the PPE should be clearly defined to ensure that it is suitable to withstand the risks it is to encounter. Employees should be actively involved in the selection of PPE to ensure that it is easy to use and maintained in an effective condition. PPE includes goggles, respiratory protection, gloves, safety shoes and head protection.

Classification of PPE

PPE may be classified into two broad categories:

- Non-respiratory protective equipment
- Respiratory protective equipment

More importance is given to respiratory protection because even a small failure or violation in the use of respiratory protective equipment will affect the entire body system and will lead to unconsciousness or fatal injury. The non-respiratory protection equipment can be further classified according to the part of the body to be protected. General classification is given below:

Importance of PPE

Workplace safety involves providing instructions, procedures, training and supervision to encourage people to work safely and responsibly. Even if engineering controls and safe systems of work have been used, some hazards may still exist.

Safety Helmets

Safety helmets are hard hats or headgear of varying materials to design for protecting the workmen's head, not only from impact but from flying particles and electric shock or any combination of the three. These also protect the scalp, face and neck from spilled acids, other chemicals, hot liquid and also protect the hair from getting tangled in machines or exposed to irritating dust. Some of the hard hats are even provided with welding masks of face screens.

Safety Gloves

The use of hands is almost an indispensable part of any job, so there is always a risk of a hand injury. Therefore, you must choose suitable hand protection. Additionally, if your hands are likely to come into contact with harmful or caustic chemicals that result in rashes or skin inflammation, products such as protective ointments are to be used to provide additional protection. The regulations also defined that owners must provide an adequate supply of protective skin ointments for workers engaged in electrolytic chromium processes.

Eye Protection

Our eyes are highly susceptible to injury from external hazards that can lead to partial or total blindness. As per compliance, employers must provide suitable eye protection to employees undertaking certain processes and to others affected by those processes. Selection of eye and face protection equipment is the most suitable eye and face protection for employees should take into consideration the ability to protect against the specific workplace, it should fit properly and be reasonably comfortable to wear, it should be durable and cleanable and should allow unrestricted operation of any other required PPE comply with ANSI ZB7.1-19B9/ISI or be at least as effective as required by this standard.

Safety Shoes

Accident statistics from the Ministry of Labor show that many accidents occur each year due to stepping on objects and slipping. So how can we minimise the risk of employee foot injuries?

Wearing suitable and adequate safety shoes is an easy and effective way according to the nature of jobs.

Safety Belts, Harnesses, Lifelines and Ropes

The employee, unless protected by a perimeter guardrail or working on a portable ladder, must be protected by a safety harness attached to a lifeline or structure capable of supporting the load. However, there are conditions where to use harness and rope along with a handrail would be required, such as in a lifting platform or on scaffolding.

Fall Protection System

Fall protection is a backup system designed for a worker who might lose their balance at a height to control or eliminate the potential for injury. Fall protection systems can consist of devices that arrest a free fall or devices that hold the worker in a position to prevent a fall. A fall arrest system is used when a worker is at risk of falling from an elevated position.

Hearing Protection

Working with prolonged exposure to high noise levels can lead to hearing loss. Intermittent exposure to high noise levels can lead to irritability, reduced ability to concentrate and hearing damage and can even lead to accidents. If hearing is damaged, we cannot be restored it, so we must protect our hearing.

Protective Clothing

Protective clothing provides physical protection and can increase the level of comfort at work. The following are the protective clothing available to provide protection against various hazards:

(a) Universal protective clothing, including raincoats
(b) Heat-resistant work clothes/aprons.
(c) High-temperature work clothes
(d) Low-temperature work wear
(e) Anti-static work clothes
(f) Chemical impermeable work wear
(g) Lifejackets
(h) Reflective clothing
(i) Night visual safety belt

Respirators

The air quality of the working environment has a direct effect on the safety and health of employees. The ideal protective measures are to control any source of pollution and reduce the amount of pollutants entering the air supply. If unavoidable circumstances do not allow such measures to be taken immediately, the best strategy is to ensure the correct selection and use of appropriate respirators.

Common Types of Air Pollution

If you encounter the following hazardous situations, you may require protection with a suitable respirator:

(a) Insufficient supply of oxygen in air (less than 19% oxygen)
(b) Presence of toxic gases, including gases, such as hydrogen sulphate, and volatile substances, such as benzene.
(c) Harmful particles including harmful dust, such as marble and chalk., pneumoconiosis-causing dust particles, such as silica and asbestos, and toxic particles, such as lead dust and acid mist.
(d) Any combination of the foregoing.

Basically, three types of respirators to protect against the above-mentioned hazards:

1. Filtering facepiece respirator (FFR) disposable
2. Elastomeric half face piece respirator
3. Elastomeric full face piece respirator
4. Powered air-purifying respirator (PAPR)

Methodology

The following methodology is adopted to solve the above problems.

Step 1: Responsibilities

Head safety/HOD concerns are responsible for the development, implementation and management of the PPE procedure or policy and requirements. It includes:

Conduct risk assessment of the workplace to determine the presence of hazards that require the use of PPE.

- Selection and purchase of PPE.
- Checking, updating and conducting hazard risk assessment of PPE whenever:
- Work is changing.
- New gear released.
- An accident/incident has occurred.
- It is requested by a supervisor or an employee, or
- At least every year.
- Keeping hazard assessment records.
- Keeping records of PPE assignments and training.
- Providing training, guidance and assistance to managers and employees to use, care and cleaning of approved PPE.
- Regularly reassess the suitability of previously selected PPE.
- Review, update and evaluate the effectiveness of PPE use, training and policies.

Step 2: Team Leader has the primary responsibility to implement and enforce to use of PPE in work area. It includes:

- Providing suitable PPE and availability to employees.

- Ensuring employee training on proper use, care and cleaning of PPE.
- Ensuring that PPE training certification and evaluation forms are signed and given to the employee's supervisor.
- Ensure that employees use and maintain their PPE properly and follow the PPE procedure and rules.
- To ensure defective or damaged PPE to be disposed of immediately and replaced.

Step 3: Employees are responsible to follow the PPE procedure. It includes:

- Proper wearing of PPE as needed.
- Participation in required trainings.
- Proper care, cleaning, maintenance and inspection of PPE as required.
- Adherence to PPE procedure and rules.
- Informing the superior about the need for repair or replacement of PPE.
- Employees who repeatedly fail to respect and comply with PPE procedures and rules will face disciplinary action.

Step 4: Procedures

(a) Hazard Assessment for PPE: The safety manager, in cooperation with supervisors, will conduct an ongoing survey of each work area to identify sources of occupational hazards. Each survey will be documented and hazard assessment that identifies the work area surveyed, the person conducting the survey and the potential hazards identified.

(b) Selection of PPE: Once workplace hazards are identified, the safety officer determines whether first the hazard to be eliminated or reduced by methods other than PPE, that is, methods that do not rely on employee behaviour, such as engineering controls.

If such methods are not adequate or feasible in the workplace, the safety officer will determine the suitability of PPE currently available and, as necessary, select new or additional equipment that will provide a level of protection greater than the minimum required to protect our employees from hazards. Attention will be paid to recognising the possibility of multiple and simultaneous exposure to various hazards. Adequate protection against the highest level of hazard will be recommended upon purchase.

(c) Training of every worker should be trained to wear PPE and care of PPE before being allowed to perform work requiring the use of PPE. PPE users will be offered regular retraining as needed. Training will include, but may not be limited to, the following subjects:

- When is it necessary to wear PPE?
- What PPE is required
- How to properly fit, adjust and wear PPE.
- Limitation of PPE.
- Proper care, maintenance, service life and disposal of PPE.
- After training, employees demonstrate that they understand the correct use of PPE, or they will be retrained.

Step 5: Retraining

Retraining will be indicated when the employee's work habits or knowledge a lack of the necessary understanding, motivation and skills required to use PPE (i.e. he is using PPE incorrectly)

- A new device is installed.
- Changes in the workplace.
- Any changes in the types of PPE.

Step 6: Cleaning, Maintenance and Disposal of PPE

All PPEs to be properly cleaned and maintained is a very important step . Dirty or fogged lenses could impair vision, so eyes and face protection equipment cleaning is most important. Employees must inspect and clean their PPE before and after each use according to the manufacturer's instructions. The supervisor is responsible to ensure that every user maintains their PPE in good condition. He will ensure that PPE not be shared between employees until it has been properly cleaned and disinfected. PPE will be provided to individual use whenever possible. If employees provide their own PPE, ensure that it is appropriate for the risks at the workplace. It is kept in a clean and reliable condition. If any PPEs defective or damaged will not be used and immediately be discarded and replaced.

Steps 7: Security Disciplinary Measures

Without any disciplinary action, any procedure, especially PPE, procedure is not unenforceable. It is believe that in order to maintain a safe and healthy workplace, employees must be familiar with and to be aware about all health and safety regulations that apply to the specific job duties required. The following disciplinary action is in effect and will be applied to all health and safety violations.

The following steps will be taken against the PPE violation at workplace as per severity and instructions.

(a) The first violation will be discussed verbally between the safety officer, the supervisor/ HOD and the employee. This will be done as soon as possible.
(b) The second offense will be dealt with a written warning will be issued and a copy of this written documentation will be submitted to concern department head of the employee.
(c) For the third time, the offense will be dealt with in writing and a copy of this written documentation will be placed in the personal file of the employee through the concern departmental head.
(d) A fourth violation will result in time off or possible termination of the concern person depends on the violation's severity.

Note: All disciplinary measures will be taken according to the severity of the result and the causes.

Results and Discussions

During the study, it was found that the artificial respiratory kit was kept in packing condition, it should be ready to use in conditions in critical areas like MCC, battery and panels room, etc. Canister type's mask's cartridge was found expired. Its life is only 5 years after the manufacturing. High-voltage hand gloves were not tested, as per standards, these should be tested after 6 months after use. Some safety helmets were found expired, as per the manufacturer, these should be disposed of after 3 years of purchase. Some persons are not trained to use SCBA in critical areas. All persons should be trained and their frequently training/ mock drill should be organised. Minimum safe storage of PPE should be maintained. No provisions for fire helmets, fire proximity suits were made in emergency control center. Safety Policy is not displayed in the local language, that is, Hindi. The policy should be translated in Hindi and displayed. Associates are not well aware of appropriate PPE. Training sessions should be organised on the use and maintenance of PPE. The training module for casuals is not in the local language. It should be converted in Hindi for easy understanding. PPE visuals are missing at some workstations. PPE visuals should be displayed at each workstation. SOP is missing on some workstations. Weekly verification should be done by each team leader. Mention PPE in SOP with visuals. Low PPE compliance observed among contractor workers. Training sessions should also be planned for contractors as well. The involvement of associates is not evident during the selection of PPE. Associates must be involved during the selection of PPE. Safety shoes compliance is very impressive. Safety goggles and safety helmet compliance is very low and needs to be improved. A safety helmet with ventilation holes can be implemented. Video sessions should be organised to increase awareness about the use and importance of adequate PPE. Penalty provision should be made for repeated non-compliance of PPE. Area specific PPE signboard should be provided. People should be made at least once awareness section about the importance of PPE. Senior management should take the lead in PPE culture in the plant. PPE checklist to be prepared and properly maintained.

Conclusion and Scope of Further Work

To ensure the greatest possible protection for employees in the workplace, the cooperative efforts of both employers and employees will help in establishing and maintaining safe and healthful work environment. Employers are generally responsible for conducting a 'hazard assessment' in the workplace to identify and control physical and health hazards. Identification and provision of appropriate PPE for employees. Training of employees in the use and care of PPE. Maintenance of PPE, including replacement of worn or damaged PPE. Regularly review and update. Evaluation of the effectiveness of the program.

In general, employees should:

- Properly wear PPE.
- Attend PPE training sessions.
- Maintain, care for clean of PPE.
- If any PPE damaged and defective, inform his supervisor for repair or replace the same.

Notes on Contributors

Omprakash S.Thakar is Assistant professor of the Faculty of Engineering in Shri Rawatpura Sarkar University, Raipur, India. He has published 10 research articles in a reputed international journal and his area of interest is Manufacturing and Production Engineering.

References

1. Akbar, F., Khanzadeh (1998), "Factors contributing to discomfort or dissatisfaction as a result of wearing personal protective equipment, Journal of Human Ergology 27 (1998) 70–75".
2. Aybala,Y., A.¸ A.Y and Vildan, C(2021) "Determination of the effect of prophylactic dressing on the prevention of skin injuries associated with personal protective equipments in health care workers during COVID-19 pandemic" Journal of Tissue Viability 30 (2021) PP21–27.
3. Charles, J., Pfister, T., Everingham, M., and Zisserman, A (2014) "Automatic and efficient human pose estimation for sign language videos, Int. J. Comput. Vis." 110 (2014) 70–90, https://doi.org/10.1007/s11263-013-0672-6.
4. Chen, S., Demachi, K (2020) "A vision-based approach for ensuring proper use of personal protective equipment (PPE) in decommissioning of Fukushima Daiichi nuclear power station, Appl. Sci. 10 (2020) 5129," https://doi.org/10.3390/app10155129.
5. Clin Exp Dermatol 2021; 46(1): 142–4. https://doi.org/10.1111/ced.14397.
6. Coelho, M.M.F, Cavalcante, V.M.V and Morae, J.T (2020) "Pressure injury related to the use of personal protective equipment in COVID-19 pandemic". Rev Bras Enferm 2020; 73(Suppl 2):e20200670. https://doi.org/10.1590/0034-7167-2020-0670.
7. Du. S., Shehata. M and Badawy. W (2011) "Hard hat detection in video sequences based on face features, motion and color information" in: Proceedings of the International Conference on Computer Research and Development, 2011, pp. 25–29, https://doi. org/10.1109/ICCRD.2011.5763846.
8. Fang, H., Luo, X., Ding, L., Luo, H., and Rose, T.M (2018) "Detecting non-hardhat-use by a deep learning method from far-field surveillance videos, Autom. Constr. 85 (2018) 1–9", https://doi.org/10.1016/j.autcon.2017.09.018.
9. Fang, W., Ding, l., Luo, h., and Love, P (2018) "Falls from heights: a computer vision-based approach for safety harness detection, Autom. Constr. 91 (2018) 53–61", https:// doi.org/10.1016/j.autcon.2018.02.018.
10. Fischler, M. A., and Elschlager, R. A (1973) "The representation and matching of pictorial structures," in: IEEE Transactions on Computers, C–221, 1973, pp. 67–92, https:// doi.org/10.1109/T-C.1973.223602.
11. Hu, K., Fanm J., LiX (2020) "The adverse skin reactions of health care workers using personal protective equipment for COVID-19". Medicine 2020;99: e20603. https://doi.org/10.1097/MD.0000000000020603. 24.
12. Kiely, L., Moloney, E., and Sullivan, G (2021) "Irritant contact dermatitis in healthcare workers as a result of the COVID-19 pandemic: a cross-sectional study".
13. Long, H., Zhao, H. and Chen, A. (2020) "Protecting medical staff from skin injury/disease caused by personal protective equipment during epidemic period of COVID-19" experience from China. J Eur Acad Dermatol Venereol 2020, 19–21.
14. Mansour, A., Balkhyour , I.A and Mohammad, R(2019) "Assessment of personal protective equipment use and occupational exposures in small industries in Jeddah: Health implications for workers" Saudi Journal of Biological Sciences 26 (2019) PP653–659
15. Marraha, F., Faker, I. A, and Charif, F (2021) " Skin reactions to personal protective equipment among first-line covıd-19 healthcare workers" a survey in Northern Morocco. Ann Work Expo Health 2021:1–6. https://10.1093/annweh/wxab018.
16. OSHA (2020) "Personal protective equipment–Oregon OSHA online course 1241". https://ehs.oregonstate.edu/sites/ehs.oregonstate.edu/files/pdf/occsafety/or-osha_ppe_training.pdf. Accessed date: 3 April, 2020.
17. OSHA (2020) "penalties". https://www.osha.gov/laws-regs/oshact/section 17. Accessed date: 3 April, 2020.

18. OSHA's (2019) "top 10 most cited violations for 2019".
19. Oznur,G.K., Pakize and Ozyürek, (2022) " Skin-related problems associated with the use of personal protective equipment among health care workers during the COVID-19 pandemic: A online survey study" Journal of Tissue Viability 31 (2022) 112–118
20. Redmon, J., Divvala,S., Girshick, R., and Farhadi, A(2016) "You Only Look Once: Unified, real-time object detection," in: Proceedings of the IEEE Conference on Computer Vision and Pattern Recognition, 2016, pp. 779–788, https://doi.org/10.1109/ CVPR.2016.91.
21. Shrestha.K., Shrestha.P.P, Bajracharya. D, am Yfantis E.A(2015) " Hard-hat detection for construction safety visualization, Journal of Construction Engineering" (2015) 721380, https://doi.org/10.1155/2015/721380.
22. Tang, S., Roberts, D., and Golparvar-Fard, H (2020) " Human-object interaction recognition for automatic construction site safety inspection, Autom. Constr. 120 (2020) 103356", https://doi.org/10.1016/j.autcon.2020.103356.
23. Wu, J., Cai, N., Chen,W., Wang, H. and Wang, G (2019) " Automatic detection of hardhats worn by construction personnel: a deep learning approach and benchmark dataset, Autom. Constr. 106 (2019) 102894", https://doi.org/10.1016/j. autcon.2019.102894.
24. Yao, A., Gall, J., Fanelli, G., and Van G. L (2011) "Does human action recognition benefit from pose estimation? Proceedings of the British Machine Vision Conference, 2011", https://doi.org/10.5244/C.25.67, 67.1–67.11.
25. Yao, J. G., Fanelli, L. and Van. G (2011) "Does human action recognition benefit from pose estimation? Proceedings of the British Machine Vision Conference, 2011" https://doi.org/10.5244/C.25.67, 67.1–67.11.
26. Yeung,S., Rinaldo, J., Jopling, J., Liu, B., Mehra, R N.L. and Downing, A. M(2019) " A computer vision system for deep learning-based detection of patient mobilization activities in the ICU, NPJ Digital Medicine 2 (2019) 11", https://doi.org/10.1038/ s41746-019-0087-z.
27. Yuen, K., and Trivedi, M. M (2019) "Looking at hands in autonomous vehicles: a convnet approach using part affinity fields, IEEE Transactions on Intelligent Vehicles" 5 (3) (2019) 361–371, https://doi.org/10.1109/TIV.2019.2955369.
28. Yuen, K., and Trivedi, M.M (2019) " Looking at hands in autonomous vehicles: a convnet approach using part affinity fields, IEEE Transactions on Intelligent Vehicles 5 (3) (2019) 361–371", https://doi.org/10.1109/TIV.2019.2955369.

CHAPTER 19

Assessment of Urban Drainage Water Quality: A Stretch from Aligarh to Agra via Hathras, Uttar Pradesh, India

Shruthi T. S. and Puneet Kumar Singh
CSIR-National Environmental Engineering Research Institute, Zonal Centre, Delhi and National Institute of Technology, Warangal (NITW), Telangana, India

Papiya Mandal*
CSIR-National Environmental Engineering Research Institute, Zonal Centre, Delhi

P Sridhar
National Institute of Technology, Warangal (NITW), Telangana, India

Abstract

Rapid urbanisation, industrialisation and commercialisation in association with economic growth in India, play a vital role in deteriorating the urban drainage water quality and its transportation. In most of the Indian cities, the construction of unplanned drainage system, encroachment of existing natural drainage corridors, water ways, dumping of solid waste, discharging of raw or partially treated sewage or industrial effluent exacerbates the problem of urban drainage system. As a result, drainage system is becoming less efficient and several intricacies are appearing due to water logging, urban flooding, unprecedented degradation of the drainage infrastructure and enormous socio-environmental hazards. The river is by and large is contaminated due to the contamination of the connecting drains. A comprehensive approach is being in need to know the facts involved in appropriate functioning of the urban drainage system. In view of above, a pilot study was carried out to compare the drain water quality with Yamuna river water quality in Uttar Pradesh, India. Drain water samples were collected from the main drain at the U/S, D/S point of stretch Aligarh, Hathras and Agra. Yamuna river water samples were collected from U/S and D/S point of confluence point of drain and Yamuna river at Agra during September and October month of 2021. The samples were analysed for pH, dissolved oxygen (DO), biochemical oxygen demand (BOD), chemical oxygen demand (COD), total dissolved solids (TDS) and total suspended solids (TSS). The study revealed that drain water quality exceeded the discharge norms of the Central Pollution Control Board (CPCB). Yamuna river water quality is not suitable for drinking purposes as per the water quality index. The outcome of the study will help the researchers/academicians/ policymakers to formulate the policy to rejuvenate the Yamuna river.

Keywords: drain, Yamuna river, confluence, sustainability

**Corresponding Author:* p_mandal@neeri.res.in/ papiya.mandal1942@gmail.com

Introduction

Rapid urbanisation and industrialisation originate river water pollution across the world. The foremost source of water is surface water available in rivers, reservoirs and lakes. The rivers are an indispensable part of both ecology and environment and are closely associated with a circular economy. In India, due to the excessive pollution load of industrial and domestic sewage discharged into rivers, the Indian rivers are extremely polluted (Hassan et al., 2017). Both natural and anthropological activities have a great impact on river water quality (Carpenter et al., 2016). Anthropogenic activities, such as urban expansion and development, contaminated effluents from industries, refining and mining activities, drainage from agricultural sectors and domestic discharges. The deterioration of river water quality presages not only worsens potential conflicts and water shortages over inadequate supplies, but shoots up ecological damage (Mulk et al., 2015). According to WHO report, approximately 1.8 million infants are dying annually due to consumption of contaminated drinking water (Parween et al., 2017). The inefficient existing sewerage system and inappropriate treatment plants, generation of wastewater from both domestic and industrial activities are discharged directly or indirectly into rivers which is the prime factors to increase the pollution load and cause adverse environmental and health effects. In Aligarh city, near about 25% of the areas has covered with sewer lines, while in the remaining areas, the generated sewage is directly discharged into open drains. In the city, due to inadequate sewerage infrastructure, sewage is transported to a sewage farm located on Mathura road through pumping stations. Kaalidesh area is located on Agra road and Naveen colony located on Kanpur road faces flooding of sewage issues from time to time, especially during the monsoon season. In the Master Plan 2021, it has been proposed to establish the sewage farm and sewage treatment plant both at Mathura road to treat the sewage. The waste water pollution is not only causing health impacts on public but also responsible for the depletion of already existing water sources. Hence, it is desirable to reuse the wastewater with proper treatment which may reduce the utilisation of fresh water. To reduce the river water pollution, a sustainable approach should be prepared by carrying out a comprehensive study to know the spatial variation of wastewater quality of drain and river surface water quality at the confluence point of river and drain. Keeping the above criteria in mind, an attempt was made to assess the wastewater quality of the main drain passing from Aligarh to Agra through Hathras and also the water quality of river Yamuna at Agra at the confluence point of drain and river.

The paper will meet the following objectives:

- To assess the pollution load of the main drain from Aligarh to Agra via Hathras.
- To predict the impact on the water quality of Yamuna river due to contamination of the Aligarh major drain wastewater at Agra.
- To compare the obtained results with the standards.

Materials and Methods

Weekly basis wastewater samples were collected from the identified most polluting locations of drains in the month of September and October, 2021. Yamuna river water samples were collected at the U/S point before meeting the drain and also at the D/S point after the confluence of the drain with Yamuna. Sampling points are identified where pollution levels were predicted to be high. The major industries located in this particular stretch are slaughterhouse, textile and dye, etc. Grab sampling was followed to collect the samples. Samples are collected in jerry cans and kept in an ice box where ice cubes were used to maintain the temperature. For DO measurement, DO was fixed at the site itself. To evaluate the quality of drain and river water, collected samples of drain and Yamuna river water was analysed for physico-chemical parameters, such as pH, TSS, TDS, BOD, DO and COD. The selected parameters were analysed as per standard methods of the American Public Health Association (APHA, 1998). The water quality index (WQI) was estimated by using the weighted arithmetic method to estimate the appropriateness of Yamuna river water quality for suitability for drinking purposes in the study area. Each of the parameters was given a weight (W_i), based on its relative importance in the overall quality of drinking water. The parameters and their unit and ideal weights are shown in Table 19.1.

Table 19.1 Parameters and their unit and ideal weights

Parameters	S_n	W_i	Ideal value (V_o)
pH	8.5	0.13507	7
Conductivity (Micro Siemen)	250	0.00459	0
Total Dissolved Solids (mg/L)	500	0.0023	0
Total Hardness (mg/L)	200	0.00574	0
Alkalinity (mg/L)	200	0.00574	0
Chloride (mg/L)	250	0.00459	0
Turbidity (NTU)	5	0.22963	0
DO (mg/L)	4	0.22963	14.6
BOD (mg/L)	3	0.38271	0

Water Quality Index

WQI is defined as 'a rating that reveals the composite influence of a number of water quality parameters on the overall water quality' for human consumption (Shankar & Sanjeev, 2008). WQI is a risk communication tool that is normally used to explain the state of water by converting a huge amount of disparate data into a single value. The single value indicates overall water quality in a specific area based on numerous water quality criteria. It is one of the most effective tools for communicating information on the overall quality state of water to the concerned users, policymakers and key stakeholders of the water sector. WQI is considered as a critical element in decision-making and planning for surface water assessment and management. The steps involved in WQI calculation are explained below. Water classes according to WQI are mentioned in Table 2.

Step 1: Parameter selection; in this work, four physico-chemical variables (pH, TDS, BOD and COD) were chosen from the list of possible parameters due to the lack of a specified permissible drinking water limit by WHO (2011) and BIS (2012).

Table 19.2 Water classes according to WQI

WQI Value	Rating of Water Quality	Grading
0–25	Excellent	A
26–50	Good	B
51–75	Poor	C
76–100	Very Poor	D
Above 100	Unsuitable for drinking purposes	E

Source: (Calmuc et al 2018)

Step 2: The second step is to calculate the sub-index or quality rating (q_i), which is done using the following equation:

$$qi = \frac{Va - Vi}{Vs - Vi}$$

where

q_i = ith parameter's sub-index

V_a = Actual value of the ith parameter at a particular sample station

V_i = Ideal value for the ith parameter

V_s = Standard value for the ith parameters

Step 3: The third step is to calculate the ith parameter's unit weight (W_i), which is inversely proportional to the standard value of that variable.

$$Wi = \frac{K}{Si}$$

where

S_i = parameter's standard value

K = proportionality constant, which can be computed using the following formula:

$$K = \frac{1}{\sum \frac{1}{Si}}$$

Step 4: Computation of WQI using the below formula:

$$WQI = \frac{Wiqi}{Wi}$$

where

q_i = ith parameter's sub-index

W_i = unit weight of selected parameters

Site Description

Aligarh city is located at 27.8974 degrees north and 78.0880 degrees east in terms of latitude and longitude in Uttar Pradesh state. It has an elevation of 178 m and the city covers an area of 3,788 km^2 (approx.). It is well-known for the production of homes and commercial locks as well as hardware items in terms of non-polluting industries. As such, Aligarh city is the hub of polluting industries and major polluting industries are electroplating, slaughterhouses, dye, textiles, etc. The sanitary condition of Aligarh city is not proper because the town infrastructure is such a way that it is not able to take extra pollution load. Aligarh-Agra drain is known as Mathura Bypass Drain in Aligarh. This drain passes through the outskirts area of Aligarh at Mathura bypass road, where most of the large and medium-scale of slaughterhouse industries are located. The untreated or partially treated effluent of slaughter houses is directly discharged into the existing sewer lines (Muriuki et al., 2020). The city sewage, which accounts for about half of the total sewage, is also dumped into the drain. Hathras is situated at a distance of approx. 40 km in the south direction of Aligarh city. This drain passes from the outskirts of Hathras and Sadabad and finally falls into the Yamuna river. The sampling point of the drain with latitude and longitude of drains of Yamuna river is shown in Table 19.3 and Fig. 19.1 and 19.2.

Table 19.3 Sampling stations of drain and Yamuna river

Sampling points	Name of Sampling Site	Latitude	Longitude	Source
S1	U/S Aligarh Drain, Panjipur, Aligarh	78.04691667	27.93028889	Drain
S2	D/S Aligarh Drain, Mathura Road, Aligarh	78.05055000	27.84109722	Drain
S3	U/S Hathras, Aligarh Drain, Ruheri, Hathras	78.06858611	27.63794444	Drain
S4	Aligarh-Hathras Drain, Madhapithu, Hathras	78.07462500	27.43871111	Drain
S5	Jaitai, Sadabad	78.06485556	27.40836667	Drain
S6	Jharna Nala before meeting Yamuna, Agra	78.07925556	27.20163611	Drain
S7	U/S Yamuna, Morewali Dargah, Agra	78.07916111	27.20088889	Yamuna river
S8	D/S Yamuna after confluence of Jharna Nala and Yamuna	78.08218611	27.20183333	Yamuna river

Results and Discussions

Monthly and spatial variations of drain water quality with respect to pH, TSS, TDS, BOD and COD are shown in Fig. 19.3 to Fig. 19.7. Monthly and spatial variations of Yamuna river water quality with respect to pH, TSS, TDS, BOD, DO and COD are shown in Fig. 19.8 to Fig. 19.13.

Drain Water Quality

pH is one of the significant parameter, which is generally used to assessed wastewater and it has a direct influence on wastewater treatment irrespective of the type of treatment. From Fig. 19.3, it can be noticed that the pH value for drain samples varied between 7.43 and 7.83

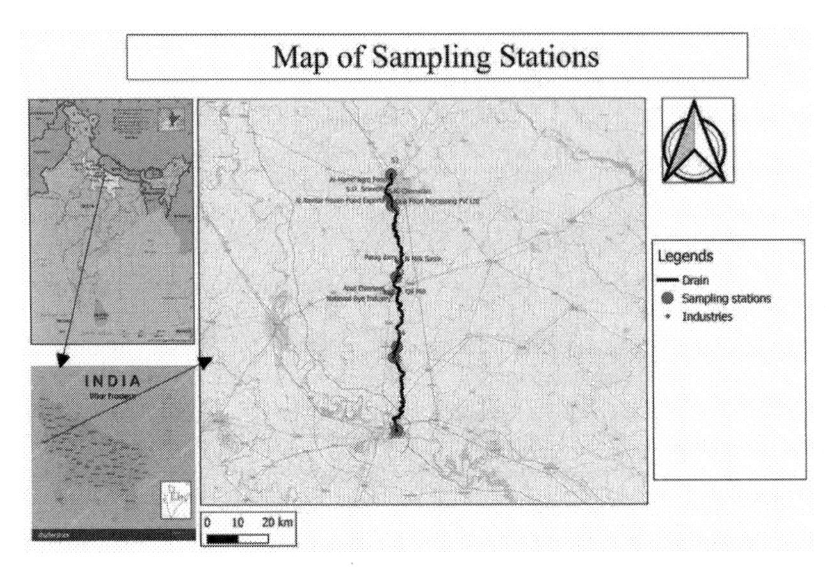

Fig. 19.1 Map of study area of drain

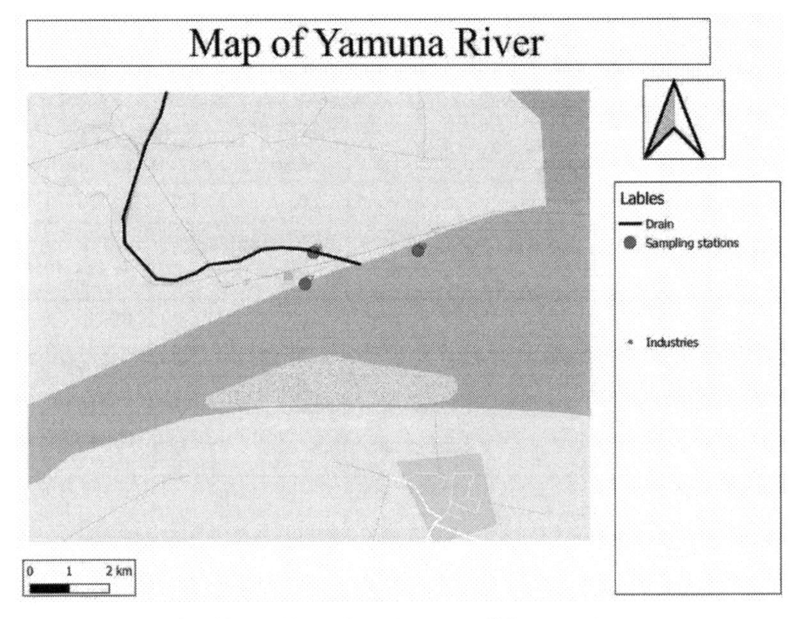

Fig. 19.2 Map of study area of Yamuna river

in the month of September, 2021 and 7.6 and 7.9 in the month of October, 2021 for all the six drain sampling points. It was noticed that pH value of the drain increased as the post-monsoon period approaches but there is no much spatial variation. The maximum value of pH was observed in the month of October, 2021. The value of pH was observed within the permissible limits (5.5 to 9) for both September and October month of 2021 as standard specified by the

Central Pollution Control Board (CPCB, 2006). The monthly and spatial variation of drain water quality w.r.t pH, TSS, TDS, BOD and COD is presented in Figs 19.3 to 19.7.

From Fig. 19.4, it was observed that the values of TSS ranged from 285 to 490 mg/L in the month of September, 2021 and 291 mg/L to 492 mg/L in the month of October, 2021. The maximum value of 492 mg/L is found in S4, that is,a Aligarh-Hathras Drain, Madhapithu and Hathras location. The maximum value of TSS at S4 might be due to the discharge of Hathras city sewage within the drain.

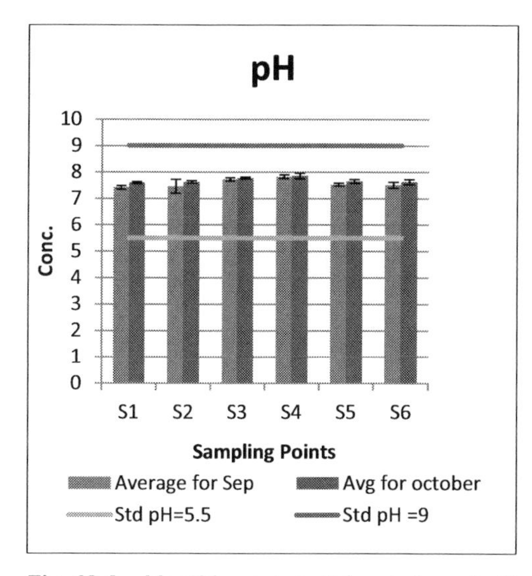

Fig. 19.3 Monthly and spatial variation of pH

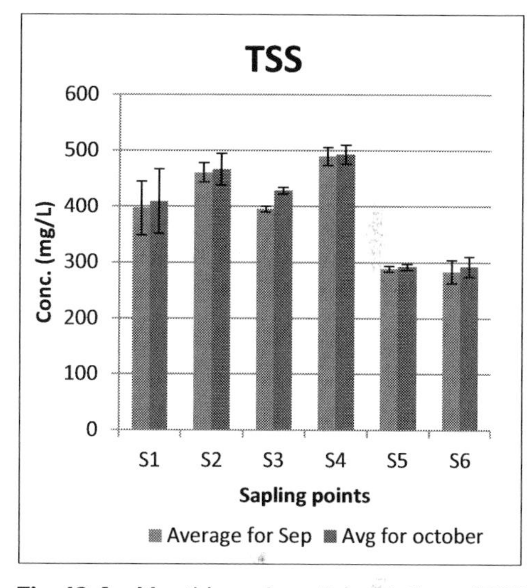

Fig. 19.4 Monthly and spatial variation of TSS

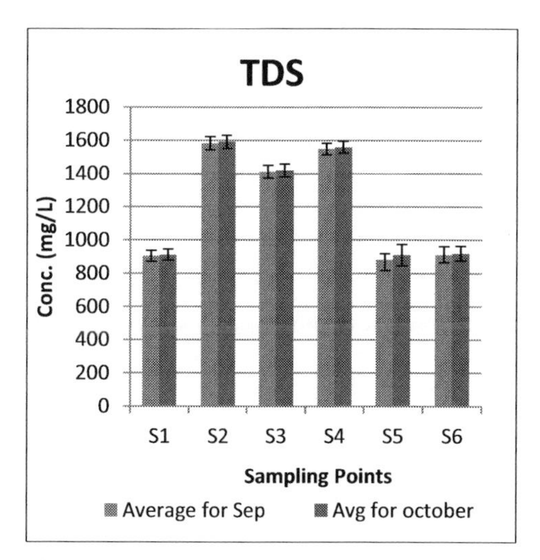

Fig. 19.5 Monthly and spatial variation of TDS

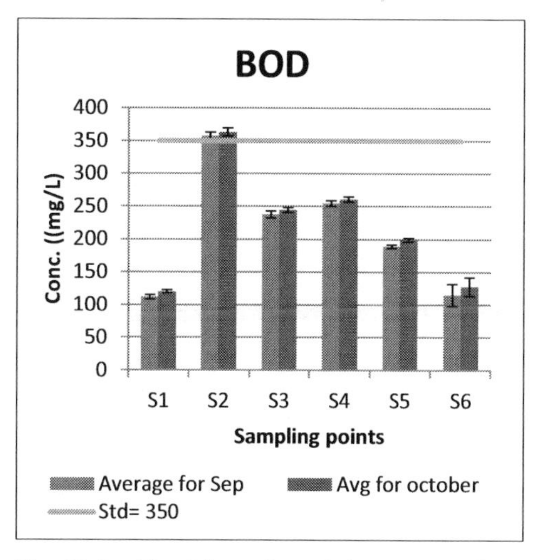

Fig. 19.6 Monthly and spatial variation of BOD

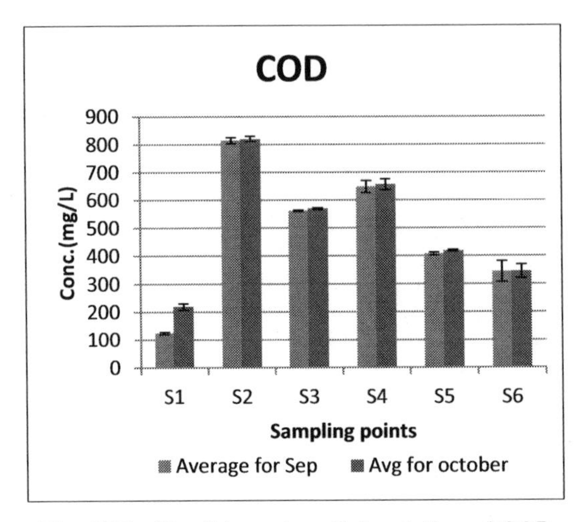

Fig. 19.7 Monthly and spatial variation of COD

From Fig. 19.5, it was noticed that the significant value of TDS was 1583 mg/L in the month of September, 2021 and observed in location S2, that is, D/S Aligarh Drain, Mathura Road, Aligarh. The reason for higher concentration at S2 might be due to the presence of pollution load of industries and city sewage from Aligarh. The minimum value of TDS of about 883 mg/L was observed at the sampling location of S6, that is, Jharna Nala before meeting Yamuna, Agra at Jaitai.

BOD is an important water quality parameter and represents an indicator of organic pollution load in wastewater. From Fig. 19.6, it was noticed that the maximum BOD concentration was 364 mg/L in the October month of 2021 at the sampling location of S2, that is, D/S Aligarh Drain, Mathura Road, Aligarh and it was exceeding the permissible limits (350 mg/L) as specified by CPCB. The minimum value of BOD was observed in September month of 2021 at the location of S1,that is, U/S Aligarh Drain, Panjipur, Aligarh (112 mg/L). Discharge of effluents into the drain from industries, especially from slaughterhouse industries, was the major contributor to the increase BOD load at the location of S2. In the drain, DO concentration was negligible, so it is not reported. D/S Aligarh drain, Mathura Road, Aligarh was significantly contaminated as compared to U/S Aligarh drain, Panjipur, Aligarh, Uttar Pradesh.

COD is a measure of water quality that is not only used to determine the quantity of biologically active substances in water, such as bacteria, but also the quantity of biologically inactive organic matter. It was observed from Fig. 19.7 that COD concentration increased in October month of 2021 and a maximum value of 820 mg/L was found in the sampling location of S2, that is, D/S of Aligarh drain, Mathura Road, Aligarh and minimum value of 124 mg/L was found in the month of September, 2021 at the sampling location of S1, that is, U/S Aligarh Drain, Panjipur, Aligarh. The significant amount of COD was observed at the sampling location of S2 might be due to discharging of raw or partially treated effluents from slaughter houses and meat processing industries.

Yamuna River Water Quality

The monthly and spatial variation of Yamuna river water quality w.r.t pH, TSS, TDS, DO, BOD and COD is presented in Figures 19.8 to 19.13 is a measure of the amount of hydrogen (H^+) and hydroxyl (OH^-) ions present in water and also indicates the alkaline and acidic nature of water. pH value ranged from a minimum of 7.3 at sampling point U/S Yamuna to a maximum of 7.7 at sampling point D/S Yamuna. From Fig. 19.8, it is noticeable that pH value increased for almost all samples as October, 2021 approaches. The maximum value of pH for Yamuna river samples was obtained in the month of October, 2021. pH values of all the samples for both September and October months were found to be within the permissible limits (6.5-8.5) as per BIS standard (2012).

From Fig. 19.9, it was observed that TSS concentration varied from 341 to 378 mg/L in the month of September, 2021 and 460 to 526 mg/L in the month of October, 2021. The minimum value of TSS was found at sampling point U/S Yamuna river, that is 341 mg/L, whereas maximum value of 526 mg/L was reported at D/S of Yamuna river. The increase in TSS value from U/S to D/S might be due to the discharge of wastewater from the drain to D/S of Yamuna river.

From Fig. 19.10, it was noticed that a significant concentration of TDS was 732 mg/L in the month of October, 2021 in the D/S of Yamuna. The minimum value of TDS was 479 mg/L observed at U/S of Yamuna. The value of TDS was moderately increased for Yamuna river water samples in the month of October, 2021. It was perceived that TDS concentration in almost all the Yamuna river water samples exceeding the permissible limits (500 mg/L) as recommended by BIS standard (2012).

DO plays a vital role in the survival of aquatic species. If the oxygen level falls below 4 ppm, fish go extinct, and the river water becomes anaerobic, resulting in bad odour and developing of septic conditions. From Fig. 19.11, it was evident that DO values started decreasing at the D/S side of Yamuna after the confluence of the drain with Yamuna. A maximum DO value of 6 mg/L was reported in the month of September, 2021 at U/S of Yamuna. A minimum value of DO was 5.25 mg/L was recorded at D/S of the Yamuna in the month of October, 2021.

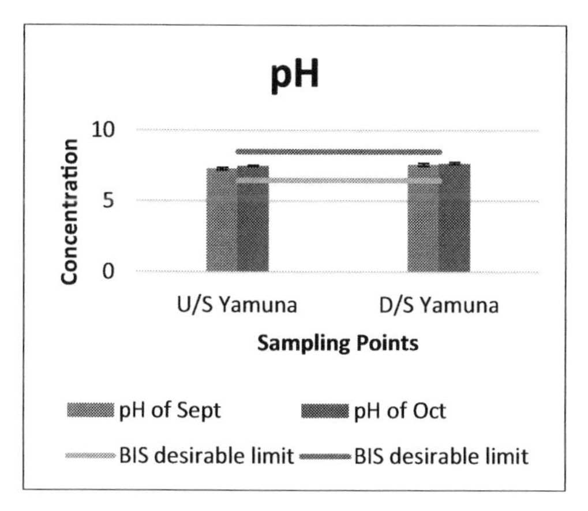

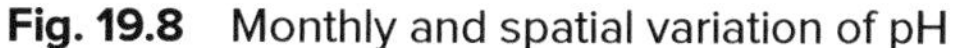

Fig. 19.8 Monthly and spatial variation of pH

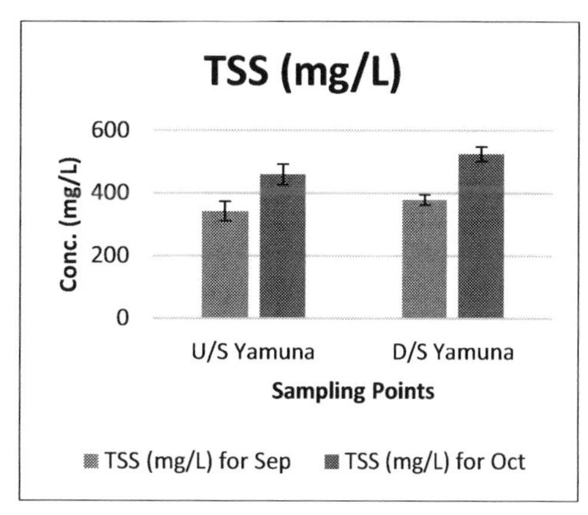

Fig. 19.9 Monthly and spatial variation of TSS

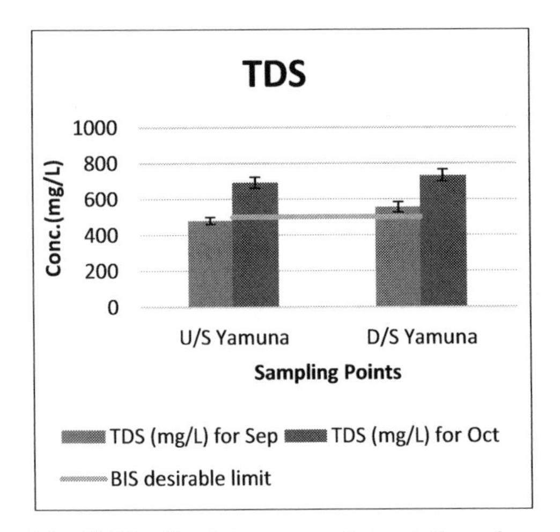

Fig. 19.10 Monthly and spatial variation of TDS

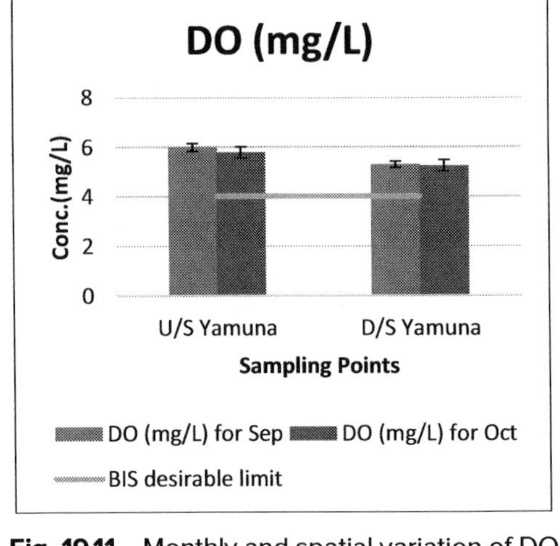

Fig. 19.11 Monthly and spatial variation of DO

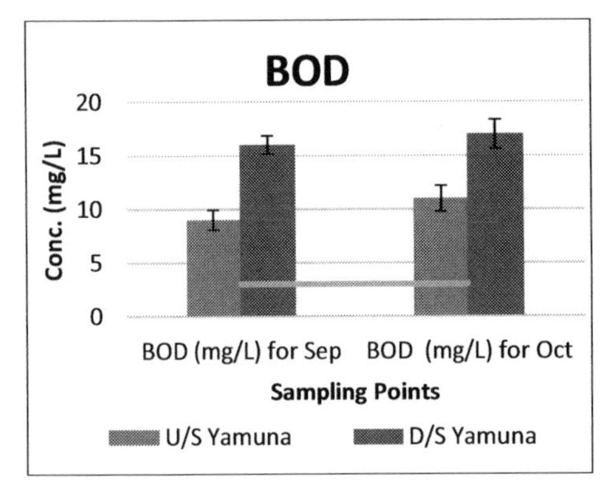

Fig. 19.12 Monthly and spatial variation of BOD

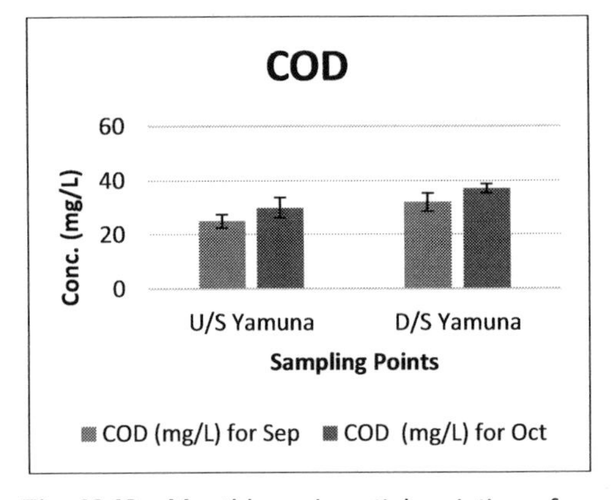

Fig. 19.13 Monthly and spatial variation of COD

BOD is one of the methods to assess the quality of water by calculating the oxygen requirement for the decomposition of its organic matter. BOD is one of the significant parameters used to evaluate the quality of water A considerable increase in a water body's BOD has a negative impact on aquatic life. According to the BIS standard for drinking water, the permissible range of BOD is 3 mg/L. In Figures 19.12 and 19.13, it was observed that the BOD and COD concentrations recorded maximum in the D/S of Yamuna. Maximum BOD and COD concentrations were 17 mg/L and 37 mg/L, respectively. For both U/S and D/S Yamuna samples, BOD crossed the permissible limit (3 mg/L) specified by BIS standards. The reason behind is to increase in the concentrations of BOD and COD in Yamuna river water significantly might be due to the discharge of untreated sewage into the drain, which contains partially treated or untreated effluents. It was observed that river water is mostly contaminated due to excessive organic and inorganic loading, that is, BOD and COD (Srivivas et al., 2018).

From Table 19.4, it was noticed that WQI increased from U/S of Yamuna to D/S of Yamuna. The maximum WQI of 685 was recorded at D/S of Yamuna in the month of October, 2021. From the WQI value, it may be predicted that the water quality at the particular selected stretch of both U/S and D/S location of Yamuna river was not suitable for drinking purposes. The reason for the deterioration of river water quality might be due to the discharge of partially or untreated effluent from industries, domestic sewage, agricultural runoff, encroachment of existing natural drainage corridors, waterways and dumping of solid waste, etc. to deteriorate the river water quality and increase the WQI value. The precipitation of natural rainfall in the month of September, 2021 might have diluted the pollutant concentration of the Yamuna river and in parallel decreased the WQI value as compared to WQI in October month of 2021. However, U/S and D/S location of Yamuna river, the variation of WQI was marginally variable.

Table 19.4 WQI value of Yamuna river water for September and October months, 2021

Sampling points	September, 2021	October, 2021
U/S of Yamuna river	526	616
D/S of Yamuna river	545	685

Source: Călmuc et al., 2018

Conclusions

The short-term assessment of the water quality of the drain and river Yamuna, and a stretch from Aligarh to Agra via Hathras, Uttar Pradesh, India for two months (September to October, 2021) revealed that drain and river water quality varied from months to months. The study revealed that parameters, such as BOD and TDS, were abnormally exceeded the prescribed standards of CPCB and BIS. It could be only possible when industries or STPs are not discharging the effluent into the drain as the prescribed standard. The abatement of Yamuna river water quality is mandatory. Therefore, at regular intervals, an assessment of Yamuna river water quality and its connected drains water quality is necessary to improve or purify the river water quality. In general, industries and STPs are treating the effluents partially and discharging into the drain. WQI value of U/S and D/S of the river suggested that the water is not suitable for drinking purposes. The agricultural runoff also increases the pollution load of the river. Natural precipitation of rain reduces the pollution load as well as lowers the WQI value. The improvement of treatment of industrial effluent and domestic sewage, and prevention of agricultural runoff, will improve Yamuna river water quality.

References

1. APHA (1998). American public health association, Standard Methods for Examination of waters and wastewaters, 20th Edition, Washington, DC, USABIS (2012). Drinking Water Specification (IS10500, 2012) Second Revision
2. Călmuc, V. A., Călmuc, M., Ţopa, C. M., Timofti, M., Iticescu, C., and Georgescu, L. P. (2018). Various methods for calculating the water quality index. Annals of the "Dunarea de Jos" University of Galati. Fascicle II, Mathematics.

3. Carpenter, S. R., Caraco, N. F., Correll, D. L., Howarth, R. W., Sharpley, A. N., & Smith, V. H. (2016). Nonpoint pollution of surface waters with phosphorus and nitrogen. *Ecological applications*, *8*(3), 559–568.
4. CPCB. (2006). *Water quality status of Yamuna River.* Government of India: Central Pollution Control Board.
5. Hussain, M., Mumtaz, M. W., Hussain, S. M., Abbas, M. N., Mehmood, S., & Imran, M. (2017). Comparative physico-chemical characterization and spatial distribution of pollutants in rural and urban drains water. *Soil & Environment*, *34*(1).
6. Muriuki, C. W., Home, P. G., Raude, J. M., Ngumba, E. K., Munala, G. K., Kairigo, P. K., ... &Tuhkanen, T. A. (2020). Occurrence, distribution, and risk assessment of pharmerciuticals in wastewater and open surface drains of peri-urban areas: Case study of Juja town, Kenya. *Environmental Pollution*, *267*, 115503.
7. Parween, M., Ramanathan, A. L., & Raju, N. J. (2017). Waste water management and water quality of river Yamuna in the megacity of Delhi. *International Journal of Environmental Science and Technology*, *14*(10), 2109–2124. M.N. Murthy and Surendar Kumr. (2011). *Water pollution in India: An Economic Appraisal.* Environmental science.
8. Mulk,S. (2015). Impact of marble industry effluents on water and sediment quality of Barandu River in Buner District, Pakistan. *Environmental Monitoring and Assessment* , p. 8.
9. Srinivas, R., Singh, A. P., Gupta, A. A., & Kumar, P. (2018). Holistic approach for quantification and identification of pollutant sources of a river basin by analyzing the open drains using an advanced multivariate clustering. *Environmental Monitoring and Assessment*, *190*(12), 1–24.WHO (2011) Guidelines for Drinking-water Quality (Fourth Edition)

CHAPTER 20

Assessment and Characterisation of the Service Quality of Existing E-bus System Based on the Road Users' Perception Encouraging the Sustainable Transport in Kolkata City

Rupam Sam
PhD Scholar, Department of Civil Engineering,
IIEST, Shibpur, India

Sudip Kumar Roy
Professor, Department of Civil Engineering,
IIEST, Shibpur, India

Abstract

Most cities across the world do not have the potential to increase road infrastructure to confront the growing urbanisation resulting the road traffic congestion, which causes further consequences covering economic, environmental and social aspects, such as travel delays, air pollution, public health concerns, greenhouse gas emissions and noise pollution. Public transportation, with a focus on electric buses, offers the opportunity for green mobility for a greater number of passengers cheaply. Unlike their ICE (Internal combustion engine) counterparts, electric buses have zero tailpipe emissions and lower noise pollution, providing a better commuter experience and encouraging sustainable green mobility of urban commuters. As a result, these buses have been receiving a large push from policymakers at both national and sub-national levels. It is targeted towards contributing significantly in terms of sustainable mobility by shifting a large number of commuters from their current transport mode of choice in most cases private vehicles. This research paper is based on an in-depth survey of the responses of E-bus commuters to the questionnaire designed for assessing the service quality of E-buses based on 31 important attributes of bus service along with bus stop facilities selecting the important corridors of E-bus routes in Kolkata. Exploratory factor analysis (EFA) is used to analyse the data and identify the effect of customers' perception about the quality of performance of various factors on customer satisfaction with existing E-bus service. The results of EFA were further evaluated by confirmatory factor analysis (CFA) in order to improve the weakness of EFA, which leads to reasonable results. To improve the infrastructures, facilities, services as well as demand for public buses, transit agencies require an understanding of passengers'

[1]rupam.sam1993@gmail.com; [2]sudip@civil.iiests.ac.in

expectations by considering the service quality parameters that have actually needed to be fulfilled. This study will help the transport providers and policymakers to make a significant effort by improving the determining principle components (derived from the 31 parameters) to attract more existing and potential commuters (by focusing on passengers' overall satisfaction) towards the E-bus (along with bus stop) regarding service quality. This research work focuses on the commuters' demand management strategy for the achievement of sustainable mobility.

Keywords: sustainable transport, E-bus, users' perception, service quality, exploratory factor analysis, confirmatory factor analysis

Introduction

Background: India is the second most populated nation in the world, having more than one billion people residing in it. In India, around 32% of the population is residing in urban areas and this is likely to increase to 40% by 2030 (NIPFP, 2007). According to the census 2011, there was an increment in million-plus cities from 35 in 2001 to 55, consisting of 107.9 million urban (39%) population (Census of India, 2011). Urbanisation in Indian cities is putting enormous pressure on transportation infrastructures to respond to an increasing travel demand with greater strength and efficiency of the public transport system (Madhav and Haide, 2007). Transportation plays a very crucial role in terms of socio-economic and sustainable growth in any nation. In the 21st century, most countries are facing critical issues and challenges related to urbanisation. The transportation system is considered the backbone of the cities as it connects the people as well as goods (Onokala, 2001). This rapid growth in population in the cities has resulted in an increased demand for transportation infrastructure, which has caused increased use of vehicles across the city resulting in congestion. In India, the number of motor vehicles has doubled every 4 years for the last three decades (MORTH, 2004) to meet the increasing demand for transportation. The vital problem is not the increasing number of vehicles in the country but the maximum concentrations of motor vehicles in the cities (Singh, 2005). The modal shifts (towards private modes) of people have been witnessed mainly in developing countries, such as India, which has ultimately resulted in major urban transportation problems, such as traffic congestion, environmental pollution, increase in fuel consumption, inadequate parking spaces and increase in the number of accidents. In response to such problems, the public transport system (PTS) can be considered an effective solution (Transportation Research Board, 1994). This increasing number of motorised vehicles in many cities caused some detrimental effects on the environment. Transport sectors in India emit nearly 261 Tg of CO_2, of which 94.5% was emitted by road transport (Sharma et al., 2011). Worldwide, emissions from transportation play an outsized role in urban air quality (Molina and Molina 2004). Vehicular traffic is frequently the single largest contributor to particulate matter emissions in major cities, and it accounts for 37% of Indian cities (Karagulian et al. 2015). Most Indian cities are struggling to keep the air breathable for inhabitants (Nature 2018). Transportation systems impact significantly on the environment accounting for between 20% and 25% of world energy consumption and carbon dioxide emissions, respectively. Road transport almost contributes almost 4.6 tons of particulate matter (PM) per day. In general, large cities in developing countries are highly dependent on road transport. India is a developing nation and with this ongoing development of the country is the rising needs and wants of the customer. The increasing shift in motorised modes of transport in megacities of developing countries is contributing to increased levels of air pollutants, attributed to vehicular sources (Jain & Khare, 2008; Kumar et al., 2013). Vehicular traffic is frequently the single largest contributor to particulate matter emissions in major cities, and it accounts for 37% of Indian cities (Karagulian et al. 2015).In order to prevent more problems caused by the increase in motorisation, it is highly recommended by many researchers as well as public decision-makers to provide an attractive public transport service as an alternative transport mode in many cities. Kolkata has recorded the highest amount of PM and nitrogen oxide (NO_x) emissions per 0.1 million of the vehicular population despite having fewer on-road vehicles than other metropolitan cities in the country which is primarily due to the operation of fleets of older vehicles.

Public Transport as Sustainable Mode

Sustainable transport refers to the broad subject of transport that is sustainable in terms of social, environmental and climate impacts. A sustainable transportation system makes a positive contribution to the environmental, social and economic sustainability of the commuters they serve. A sustainable transport system is one that is accessible, safe, environmentally friendly and affordable (ECMT, 2004). The sustainability of future transport should rely on public transportation as one of the key components. A study estimated that the increased use of public bus transport, bicycle transportation and walking for commuters within city limits could contribute to reducing emissions by a staggering 40% by the year 2050 (Replogle and Fulton). For many years, PT has been viewed as an inevitable need in several parts of the world (Cervero & Golub, 2007). Transport modes available for public use; a transport system (of buses, trains, etc.) that runs on fixed routes at set times and may be used by anyone with a valid ticket or pass, termed as PT (Oxford English Dictionary, 2022). The PT available to the inhabitants of a town, city or even a region can vary in its diversity. Public transportation plays a significant role as it improves the quality of our day-to-day life by expediting traffic, saves money, creates new jobs and helps environmentally due to these benefits it is the main reason for focusing on the present research. However, PT infrastructure in Indian cities has failed to keep pace with population growth in these urban spaces. Current PT usage is still much lower than automobile usage in many regions around the world, and thus, Private Cars (PC) are becoming the most common way to travel daily. This is one of the effects of poor service quality of PT. Though various transport modes are available, road-based transport is the most popular since its peculiar advantages such as flexibility, capital requirements, capacity, infrastructure, accessibility and adaptability, route, direction, time and speed (Friman et al., 2001). A bus is a very much popular and widely used public transportation mode in India because of its low cost and has made a significant contribution to the national economy. With the prevailing conditions of city buses in India, bus transport may not suit the needs of most passengers, especially car users. A NITI Aayog study (2018), estimated that India has only 1.3 buses for every 1,000 people, much lower than other developing countries, such as Brazil (4.74 per 1,000) and South

Africa (6.38 per 1,000). The lack of buses along with poor service quality has led to commuters increasingly moving away from bus transport. The severe shortage of buses has increasingly pushed commuters towards personal transport. The overall increased demand for travel has triggered the increase in usage of PC, leading to rapid motorisation worldwide. Increased private motorisation has resulted in increased traffic congestion, leading to longer travel times. Indeed, India is the world's largest market for two-wheelers, which are significant contributors to increased traffic congestion, greenhouse gas emissions and the worsening of air pollution. In India, transportation is treated as a part of public service and the primary focus is a provision of an affordable, safe and reliable bus service to the people (Nagadevara and Ramanyya, 2009). As it is serving a large number of passengers, the quality of service is the main concern issue, and therefore understanding customer expectations is of immense importance. Service quality assessment has received increased attention amongst transport planners in recent years due to the increased importance to improve user satisfaction and demand for public transport services. The development of techniques for customer satisfaction analysis is necessary as they allow the critical aspects of the supplied services to be identified and customer satisfaction to be increased (Cuomo, 2000). As performance measurement is an essential tool to evaluate the existing condition of the public transport system of the city and therefore, it has become a subject of major concern in the last decade for urban and transport planners (Eboli & Mazzulla, 2011). Service quality is of increasing importance to all businesses, including public transport organisations, as it influences customer satisfaction, commuter choices, passenger demand, investment decisions and revenue (Anderson et al., 2013). PTS ensures the movement of a large number of people, especially in the densely populated urban area, and also provides mobility to people who cannot access private vehicles. However, in order to keep and attract more passengers, PT must have high service quality to satisfy and fulfil more wide range of different customers' needs (Oliver 1980; Anable 2005). Recommendations to improve the quality of buses (the bus infrastructure) were given by researchers to increase ridership, Desai and Joshi (2016). In India, the urban road transport sector is currently facing the issues of increased travel demand, increased use of private vehicles, traffic congestion as well as alarming levels of air pollution. The high consumption of fossil fuels resulting in recent rising oil prices, greenhouse gas emissions and global warming, has encouraged metropolises to use modern technologies in the urban transport sector resulting in a need of different strategies to decarbonise the mobility systems. Therefore, PT needs reliable and efficient methods by identifying the determinants of service quality from the customers' perspective. In order to prevent more problems caused by this increase in motorisation it is highly recommended by many researchers as well as public decision makers to provide an attractive eco-friendly public transport service as a prior road transport mode in many cities. Public transport should become a part of a solution for achieving sustainable transport.

Electric Mobility in Public Transport (E-bus)

Electrical vehicles will control both noise and sound pollution in the city these vehicles are undoubtedly an eco-friendly mode of traveling. Many cities in the world are considering electric buses for PT as one of the potential solutions to address these issues. Unlike their ICE counterparts, electric buses have zero tailpipe emissions, lower noise pollution and are touted to provide a better commuter experience. As a result, these buses have been receiving a large push from policymakers at both national and sub-national levels. The government of India is committed to reducing carbon emissions by 45% by the year 2030, and bringing them down to 0% by 2070. Many experts believe these targets are ambitious for a country like India, which is mostly dependent on fossil fuels for energy needs. However, the electric vehicle policy is a step in the right direction. In many ways, electric buses have brought back attention to the historically neglected public bus service, with many state governments announcing electric bus procurement programs. In March 2019, the Indian government formally launched the second phase of the Faster Adoption and Manufacturing of (Hybrid &) Electric Vehicles (FAME) scheme, announcing plans to procure 5,585 electric buses, to help clean up the environment. BYD-Electra is a leading player in India's electric bus market with more than 200 buses plying across the country. PMI electro mobility solutions has a technical tie-up with Chinese CV maker Beiqi Foton Motors, which is the second most popular choice of STUs for E-bus. It has closed to 800 E-bus orders till date. Presently, under the battery segment, the lithium-iron-phosphate (LFP) category dominates the Indian electric bus market due to it's safest and fastest charging feature, which is why they remain popular among bus manufacturers and buyers. The deployment of electric buses is seen as an opportunity to decouple the negative externalities of bus travel. The market leaders of conventional buses, Tata and Ashok Leyland have developed hybrid as well as pure electric buses but have very few electric buses plying on Indian roads at present. The increased use of private modes is also influenced due to low service quality of existing mostly conventional ICE buses.

Purpose of this Study

As E-buses has a huge potential to be an alternative to replace polluting vehicles in an age when controlling vehicular emissions is critical. Several attractive and convenient features (along with the low fares) as compared to other buses would encourage more people to opt for E-buses for their commute. Novel strategies need to be developed to promote the use of this sustainable mode of road transport. One of the major benefits of developing the existing system is to reduce commuters' overwhelming dependency on private vehicles in order to make the urban transportation system sustainable. Considering satisfaction is one of the key factors, which influence customer loyalty, this is assumed that the satisfied customer will be willing to use the same service provider again. Capturing the demand side quality of services is very important based on the commuters' perceptions regarding the quality of service of existing E-Bus systems in different demographic cohorts in different Indian cities. No research work, to the best of our knowledge, has attempted to study this assessment of E-bus service quality based on the detailed parameters regarding the bus service quality assessment in present scenario in the Indian road context. This study addresses one such research gap to assess the present service quality of the E-buses on a medium-scale, in the tier I Indian city, Kolkata. A structural equation model (SEM) approach is used to reveal the unobserved latent aspects describing the service and the relationships between these aspects with the overall satisfaction level of the commuters. This could be done using a multivariate technique, EFA in particular principal component analysis which could help in this regard by way of data reduction.

Objective of this Study

With this motivation, the present paper attempts to assess the influencing factors regarding the PT system in view of both demand and supply measures together. Therefore, the major objectives of the present research are as follows:

- To analyse and determine the principle components regarding the service quality of E-buses in Kolkata, based on the public transport users' satisfaction with several parameters influencing the bus service quality.
- To develop a model by establishing the relationship between the principle components (latent) and the perceived service quality of that type of public bus considering the satisfaction level of the observed variable.

For this purpose, the study was undertaken to carry out a segmental analysis of latent factors through SEM. EFA was used in this study to group variables into the dimension of bus service and interpret in-depth based on the questionnaire items. More importantly, the results of EFA were further evaluated by CFA for the purpose of improving the weakness of EFA which leads to reasonable results.

After the introduction, where the main research question and objective are determined this paper is structured into five sections. Section 2 presents the literature review, Section 3 discusses the study area and data collection and Section 4 presents the analysis and result. The conclusions of the main findings, together with ideas for further research with the limitation, are covered in Section 5.

Literature Review

Past literature has shown that many studies are being conducted to assess the public transport service quality in both developed countries or developing countries. Methods for increasing the use of PT have been researched in diverse contexts around the globe, with no standardised solution being achieved. Increasing travel demand and preferences in using private vehicle is causing rapid motorisation in many counties around the world. Most people are now highly dependent on private motorise travel (Ellaway et al., 2003). This phenomenon was caused because of attractiveness of car and people love to drive (Beirão & Sarsfield Cabral 2007). An increased private motorisation has resulted in an increased traffic congestion which in turn results in longer travel times for many people (Beirão & Sarsfield Cabral 2007; Asri & Hidayat 2005). In addition to congestion, private motorisation is also affecting the safety of vulnerable road users (Kodukula 2009), high consumption of non-renewable resources (Abmann & Sieber 2005), and causes a serious threat to the quality of human environments (Goodwin 1996; Greene & Wegener 1997). Several attempts have been made to develop the performance measure of the transit services based on different aspects. There are many ways to definite 'service quality', for example, Oliver (1997) defined satisfaction as the customer's fulfilment. Many experts over the world, in a way, agree with Parasuraman et al., that service quality relies on directly users' expectations as well as real experience. Basically, performance is measured with reference to passengers, operating agency and community. The perception of users regarding transit service, business perspective of operator and role of transit service to achieve the broad community goal are the major concern of performance measure. (Transportation Research Board, 2003b). Karlaftis et al., (2001) focused on the cleanliness of the interior and seats of the bus. Lawa and Mazzulla (2007) studied reliable driving in terms of bus service assessment. Karlaftis et al., (2001) considered safety against crimes in determining the bus service quality. Hensher et al. (2003) studied crowd management of bus and bus stop distance in service quality research. Fellesson and Friman, (2008) carried out an extensive study on customers' satisfaction in the domain of public transportation covering eight European cities, which have recognised four general categories of satisfaction: systemic (concerning satisfaction with traffic supply, reliability and information); design (a category that includes comfort and overall travel experience); staff (deals with skill, knowledge and attitude toward customers) and safety (on the bus, at the bus stop, as well as safety from traffic accidents). Colob, (2003) focused on the physical condition of bus stops in bus service quality. Rahaman and Rahaman, (2009) focused on the railway transportation sector to develop a model defining the relationship between overall satisfaction and 20 service-quality attributes and eight principle components found after performing the PCA. Ching-Chiao et al. (2009) have used PCA to identify crucial resources and logistics service capabilities in container shipping services. A complaint provides an opportunity for service recovery which, in turn, has the potential to educate the customer, strengthen loyalty and induce positive word-of-mouth comments (Edvardsson & Roos 2003; Friman & Edvardsson 2003). Although attracting new customers is vital, successful service companies recognise that retaining current customers and building loyalty is even more important for profitability; as such, successful service companies actually encourage the dissatisfied customer to complain (Tax et al. 1998). Singh (1991) argues that providers recognise the extent of customer dissatisfaction in the marketplace and the handling of service recovery as key indicators of customer loyalty and welfare. Belonging to the LOS concept, the potential values for a performance measure are divided into six ranges denoted by the capital letters from A (the best level) to F (the worst level) (Das & Pandit, 2012). Zaimah Ubaidillah et al., (2022) investigated the role of public bus service quality (tangibility, reliability, assurance, empathy and responsiveness) by utilising the SERVQUAL model for users' satisfaction using Partial Least Square-Structural Equation Model. Public transportation has taken the hardest blow in the COVID-19 situation (Molloy et al. 2020; Astroza et al. 2020). This virus can accumulate and remain infectious in indoor air for hours (Prather, Wang, and Schooley 2020), which is the greatest challenge for public transportation. The existence of multiple surfaces, such as seats, handrails, doors and ticket machines, easily transfer germs (UITP 2020). The success and the market share growth of public passenger transport operators are determined by the quality of supplied services and, moreover, by the passenger's perception of the provided quality (Dragu et al., 2013). Chen and Lai (2011, p.308) note that travellers who perceive the good quality of public transit service are more likely to have a higher level of perceived value and satisfaction, and so continue to use this service.

Study Area and Data Collection

The user perception survey conducted as part of this research was limited to the city of Kolkata, West Bengal. As per Census 2011, Kolkata is the largest megapolis in eastern India covering an area of 1480 sq. km. with a population of 44,96,694. The busy route in Kolkata city has been focused on in this study, considering the most used land use demand of the daily commuters during peak hours. The identification of input parameters has been considered through brainstorming, experience and thorough knowledge are vital for evolving an exhaustive and sufficient list of questions or parameters (Chidambaranathan et al., 2009). The user is asked to evaluate each statement (each parameter), in terms of their own perception and expectation of performance (satisfaction level) regarding several parameters of the service being analysed.

When a customer survey is to be conducted a questionnaire has to be developed. In a pilot survey, the list of questions evolved was initially administered to a group of commuters to ascertain if all of the parameters are necessary. Mainly, the data were collected through a survey using a questionnaire regarding the satisfaction level (in a 6-point scale) of 31 qualitative observed variables influencing the service quality of public transport (E-bus). As there is no record of secondary perceptive data regarding E-buses in Kolkata, an in-depth field survey was conducted to collect the primary data for this study. The size of any sample can be determined from the following relation, assuming that Central Limit Theorem is followed:

$$N = k\frac{s^2}{e^2}$$

(where N = sample size, k = constant depending upon the 95% confidence level, s = standard deviation of the population, e = tolerance level)

In this study, a total of 320 responses were obtained by conducting an in-depth survey regarding their socio-economic characteristics and perceived bus service quality. A statistical descriptive analysis of the sample was carried out. The socio-economic categorisation of the sample is listed in Table 20.1.

Table 20.1 Socio-economic characteristics parameters

Sl No.	Parameters	Description
1	Gender	Male (M), Female (F)
2	Age	<25 years, 25–40 years, 40–55 years, > 55 years
3	Level of education	10th or below, 12th, Bachelors, Masters, Doctorate or higher degree
4	Purpose of travel	Home-based work (HBW), Home-based education (HBE), Home-based others (HBO), Non home-based (NHB)
5	Frequency of travel	1 day/week, 2–3 days/week, 4–5 days/week, 6–7 days/week

The sample was spread over 49% male and 51% female respondents. Of the total collected sample, 25% was below 25 years old, 37% between 25 and 40 years old, 22% between 40 and 55 years old and 16% was above 55 years old. 72% commuters' who belonged to the education level Bachelor and Master degree (completed or pursuing) were considered for this study, relying on mostly the educated commuters' perception to enhance the response reliability of the survey. The sample considered 41% of commuters travel 6–7 days/week, 55% commuters travel 4–5 days/week –and rest 23 days/week. Commuters who use E-bus service less than once a week are not considered as input data for this study, assuming them irregular E-bus users.

The aim of this study is to measure overall customer satisfaction and investigate the related service quality attributes that influence the most. The questionnaire is the most common tool to investigate such research and the usual measure of getting this is by conducting a survey with a set of statements using a Likert or ordinal scale. In this step, a perception survey was conducted (in a 6-point scale) on 31 service quality parameters on board on E-bus commuters in Kolkata, during peak periods for all working days, as shown in Table 20.2.

Table 20.2 Observed parameters

Sl No.	Parameters	Description	Rating	N	
				Valid	Missing
1	Overcrowding in bus during peak hour	How much overcrowding during peak hour	1 = Always to 6 = Never	314	6
2	Seat comfort in bus	Rating of the seats comfort for travel	1 = Worst to 6 = Best	316	4
3	Seat availability in bus	Availability of seat in bus during peak hour	1 = Never to 6 = Always	319	1
4	Seating arrangements in bus for special cases	Whether elder people, handicapped or physically challenged people provide separate seating	1 = Worst to 6 = Best	315	5
5	Cleanliness in bus	Upkeep of the buses and bus stop, such as dusting, cleaning	1 = Worst to 6 = Best	320	0
6	Safety level in bus	Safety level inside bus against theft, molestation, etc.	1 = No safety to 6 = Full safety	311	9
7	Safety level in bus stop	Rating on safety felt at bus stop mainly in late-hours	1 = Worst to 6 = Best	318	2
8	Window glass and bus lighting condition	Window glass condition of the bus and illumination inside the bus	1 = Worst to 6 = Best	319	1
9	Breakdown issue	Frequency of breakdown of buses	1 = Always to 6 = Never	320	0
10	Waiting time at signal	Waiting time in the traffic signal	1 = Very long to 6 = Very short	312	8
11	The punctuality of bus	Rating on punctuality in arrival and departure at each stoppage	1 = Worst to 6 = Best	317	3
12	Usage of mask and sanitizer	Tendency of using mask, sanitizer by co-passengers and staffs	1 = Worst to 6 = Best	311	9
13	Bus service information to passengers	Rating on proper information for passengers on smartphone, web, etc.	1 = Worst to 6 = Best	320	0

Sl No.	Parameters	Description	Rating	N	
				Valid	Missing
14	Waiting time in bus stop	Waiting time for bus	1 = Very long to 6 = Very short	311	9
15	Exterior digital display in bus	Exterior display facility of the bus	1 = Worst to 6 = Best	318	2
16	Late night availability	Availability of bus service late night	1 = Worst to 6 = Best	319	1
17	Convenience in payment of fare	Convenient way of ticket issuing, cashless transaction ticketing facility	1 = Worst to 6 = Best	320	0
18	Easy to transfer	Rating on the required time and convenience in terms of switching modes by passengers	1 = Worst to 6 = Best	316	4
19	Inside amenities in bus	Such as luggage loft, digital display, audio, charging, Wi-Fi and entertainment is provided or not	1 = Worst to 6 = Best	320	0
20	Seat comfort in bus stop	Seat comfort in the bus	1 = Worst to 6 = Best	312	8
21	Shelter of the bus stop	Rating on shelter capacity at bus stop Proper way of complaining about any issue	1 = Worst to 6 = Best	315	5
22	Bus stop location proximity	Distance of bus stop	1 = Worst to 6 = Best	311	9
23	Ascending and descending convenience	Level of convenience while getting onto or off to the bus in several aspects	1 = Worst to 6 = Best	319	1
24	Behaviour from staff and co-passengers	Overall behaviour from staff co-passengers	1 = Worst to 6 = Best	316	4
25	Bus stop cleanliness	Cleanliness at the bus stop	1 = Worst to 6 = Best	320	0
26	Facility available at bus stop	Such as drinking water, bus time-table, etc.	1 = Worst to 6 = Best	320	0
27	Comfort while standing in the bus	Rating in terms of comfortable grab handles and poles legroom and foot space while travelling	1 = Worst to 6 = Best	320	0
28	Ventilation in bus	Rating on proper comfortable ventilation inside the vehicle	1 = Worst to 6 = Best	311	9
29	Extra facility in bus	Whether medical, emergency button, fire extinguisher, emergency exit, etc.	1 = Worst to 6 = Best	319	1

Sl No.	Parameters	Description	Rating	N	
				Valid	Missing
30	Dealing with complain	How good is the way of complaining and handling that issue	1 = Worst to 6 = Best	315	5
31	Drive safely and smoothly	Rating on comfortable and safe driving by the driver	1 = Worst to 6 = Best	320	0

Analysis and Results

The users' perceived quality of satisfaction data was initially analysed using EFA. In EFA, factors are extracted using PCA. The EFA for service quality attributes was conducted on the bus commuters' perceived data. SEM/Path analysis has been a multivariate technique used to test the hypotheses framed for interaction between variables. The method combines factorial analysis with regression to interpret complex multiple relationships between the observed variables (indicators) and unobserved variables (latent construct). SEM can evaluate the indirect effects on the latent variables, which cannot be done in regression analysis. The framework comprises (1) a measurement model for the unobserved/dependent variable, (2) a measurement model for the observed/ independent variable and (3) a structural model that is simultaneously estimated. SPSS AMOS 4.0 package was employed for the scale measurements. The data that were collected will be analysed using statistical methods (Statistical tool and SPSS). To summarise and rearrange the data several interrelated procedure are performed during the data analysis stage (Muijs 2004). To determine the number of factors, only the eigenvalues greater than or equal to one were considered (Cuttman, 1954; Kaiser, 1960). In addition, the KMO test (Kaiser, Mayer and Olkin) and the Barlett sphericity test were considered (Fabbris 1997) for testing the validity of data for EFA. KMO values closer to 1.0 are considered as ideal while values less than 0.5 are unacceptable. An orthogonal rotated solution (Quartimax) was adopted (Carroll 1953). The scree plot is used to determine the number of factors to retain in EFA or principal components to keep in a principal component analysis (Cattel 1966).

From the output result, the obtained KMO value of 0.814 indicates that it is plausible to conduct factor analysis. The finding of statistically significant factors or components of this study, using a scree plot (known as a scree test) is shown in Fig. 20.1. The total variance table also indicates the total variance and eigenvalues (greater than 1) are shown in Table 20.3. The seven components together explained 81.945% of the cumulative variance of PCA.

Component 1 has an eigenvalue of 9.565 and accounts for 30.856% of the total variance. As component 1 has high positive loadings on the mentioned factors under this category. Component 1 points to the fact that comfort during travel has a significant influence on passengers' satisfaction and therefore can be called 'comfort in buses and bus stops'. Component 2 has an eigenvalue of 5.383 and accounts for 17.364% of the total variance, pointing to the fact that convenience during travel has a significant influence on passengers' satisfaction. This component can be called as 'Convenience in buses and bus stops'. Component 3 has an eigenvalue of 3.373 and accounts for 10.881% of the total variance, pointing to the fact that

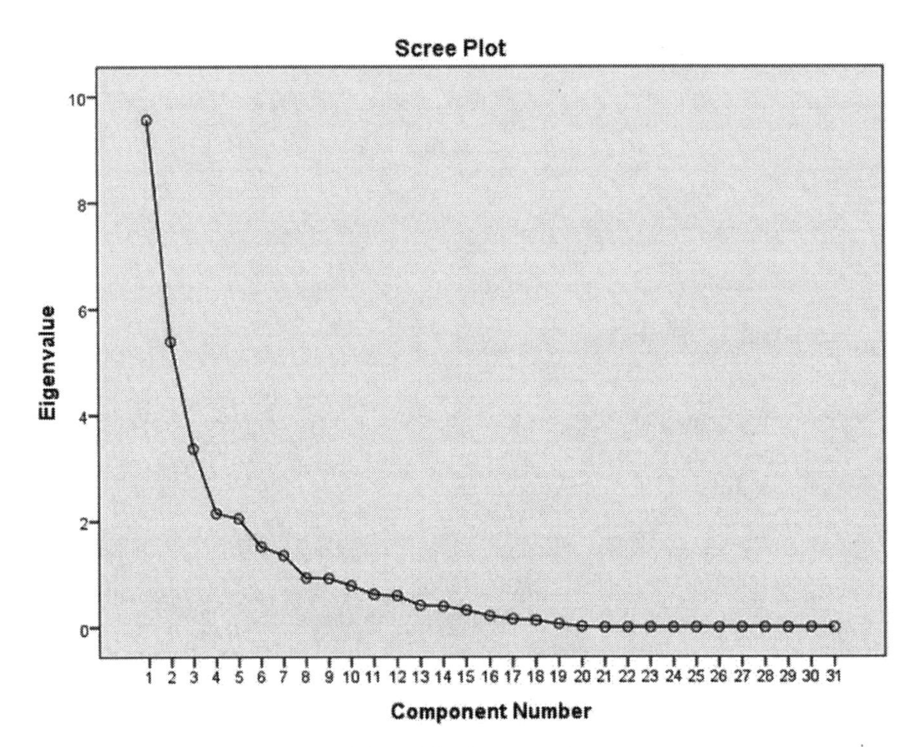

Fig. 20.1 Scree test

Table 20.3 Total variance table

Total Variance Explained						
Component	Initial Eigenvalues			Extraction Sums of Squared Loadings		
	Total	% of Variance	Cumulative %	Total	% of Variance	Cumulative %
1	9.565	30.856	30.856	9.565	30.856	30.856
2	5.383	17.364	48.220	5.383	17.364	48.220
3	3.373	10.881	59.101	3.373	10.881	59.101
4	2.149	6.934	66.035	2.149	6.934	66.035
5	2.050	6.614	72.649	2.050	6.614	72.649
6	1.523	4.912	77.562	1.523	4.912	77.562
7	1.359	4.383	81.945	1.359	4.383	81.945

comfort and safety during travel have a significant influence on passengers' satisfaction. This component can be called as 'Safety and amenities in buses'. Component 4 has an eigenvalue of 2.149 and accounts for 16.675% of the total variance, pointing to the fact that comfort and safety during travel have a significant influence on passengers' satisfaction. This component can be called as 'Availability of buses'. Component 5 has an eigenvalue of 2.050 and accounts for 6.614% of the total variance, pointing to the fact that accessibility has a significant influence on passengers' satisfaction. This component can be called as 'Accessibility in buses'. Component 6 has an eigenvalue of 1.523 and accounts for 4.912% of the total variance, pointing to the fact that management has significant influence on passengers' satisfaction. This component can be called as 'Management of bus'. Component 7 has an eigenvalue of 1.359 and accounts for 4.383% of the total variance, pointing somewhat to the fact that maintenance of the bus has significant influence on passengers' satisfaction. This component can be called as 'Maintenance in buses'.

The rotated component matrix (sometimes referred to as the loadings) contains estimates of the correlations between each of the variables and the estimated components, as shown in Table 20.4. The standardised regression coefficients further showed that comfort-convenience has the greatest impact on overall satisfaction, followed by availability and accessibility. The service quality variables were narrated by the EFA in the form of principal component analysis. The latent variables are the unobserved service quality aspects that can be explained by the observed variables. The latent variables were defined by means of an EFA implemented in the form of principal component analysis.

Table 20.4 Rotated component matrix

	Principle Components						
	1	2	3	4	5	6	7
Overcrowding	.912						
Bus seat comfort	.587						
Bus seat availability	.716						
Seating for special cases		.643					
Cleanliness in bus	.915						
Safety level in bus			.677				
Safety level in bus stop			.452				
Window glass and bus lighting		.458					
Breakdown issue							.562
Waiting at traffic signal		.531					
Punctuality				.437			
Mask and sanitizer usage						.635	
Bus service information					.706		
Waiting time				.867			
Exterior digital display		.795					
Late night availability				.519			
Convenience in fare payment		.608					
Easiness to transfer					.497		

	Principle Components						
	1	2	3	4	5	6	7
Inside amenities	.625						
Seat comfort at bus stop			.559				
Shelter of bus stop		.502					
Bus stop proximity	.795						
Ascending and descending					.611		
Overall behaviour	.704						
Bus stop cleanliness						.526	
Facility at bus stop	.862						
Comfort while standing			.631				
Ventilation in bus	.573						
Extra safety facility in bus				.604			
Dealing with complain						.702	
Drive safely and smoothly							.429

Principal component analysis (PCA) is a very popular technique in this regard for analysing such large datasets containing a high number of dimensions per observation, increasing the interpretability of data while preserving the maximum amount of information. Using PCA, seven underlying factors were extracted that influence passenger satisfaction with public bus transport services in the city Kolkata, as shown in Table 20.5.

Table 20.5 Principle component analysis

Principle Components	Mean	Standard Deviation	Parameters
Comfort (C_1)	4.206	0.7933	Overcrowding in bus (C_{11}) Availability of seat (C_{12}) Comfort while seating (C_{13}) Comfort standing in bus (C_{14}) Bus lighting, window glass condition (C_{15}) Cleanliness at bus (C_{16}) Cleanliness at bus stop (C_{17}) Availability seating at bus stop (C_{18}) Ventilation in bus (C_{19})
Convenience (C_2)	3.7816	0.7554	Exterior digital display (C_{21}) Convenience in fare payment (C_{22}) Bus stop shelter (C_{23}) Seating special cases (C_{24}) Waiting time at traffic signal (C_{25}) Ascending descending (C_{26})

Principle Components	Mean	Standard Deviation	Parameters
Safety (S_1) **Amenities (A_1)**	2.6791	0.5891	Safety at bus (S_{11}) Safety at bus stops (S_{12}) Extra safety facility in bus (A_{12}) Facility available bus stop (A_{12}) Inside amenities in bus (A_{13})
Availability (A_2)	2.6333	0.5686	Availability on late night (A_{21}) Waiting time (A_{22}) Bus punctuality (A_{23})
Accessibility (A_3)	2.7166	0.6714	Easiness to transfer (A_{31}) Proximity of bus stop (A_{32}) Bus service information (A_{33})
Management (M_1)	3.5867	0.8358	Behaviour by staff & passengers (M_{11}) Dealing with complain (M_{12}) Mask and sanitizer usage (M_{13})
Maintenance (M_2)	4.7771	0.3634	Breakdown of bus (M_{21}) Drive safely and smoothly (M_{22})

The level of perception on the various constructs are ranging from moderate to satisfy since their respective mean scores are from 2.6333 to 4.771. The correlation between all pair of constructs is significant at a five percent level ensuring the discriminant validity.

The tests on the goodness of fit are quite satisfactory. The goodness-of-fit index (GFI) is 0.9574, the adjusted goodness-of-fit index (AGFI) is 0.9234, and the comparative fit index (CFI) is 0.9022. The indices are bounded above 1, which indicates a perfect fit. Therefore, the indices obtained from the model are very good. The root mean square residual (RMR) index has a value of 0.1317, and the root mean square error of approximation (RMSEA) has a value of 0.0431; the value of these indices is low and therefore are quite good (Bollen, 1989). The minimum value of the discrepancy function is 316.763, which is statistically significant according to the chi-square test.

The model has seven latent exogenous variables namely comfort, convenience, safety and amenities, availability, accessibility, management and maintenance of the bus service. All the observed parameters assume a value statistically different from zero at a good level of significance ($P < 0.05$)) as shown in Table 20.6. Most of the standard regression weights are reasonably high, having standard regression weights of more than 0.5. It is to be noted that the attributes ventilation in the bus (C_{19}), bus stop shelter (C_{23}) , mask and sanitizer usage (M_{13}) and proximity of bus stop (A_{32}) was initially taken into consideration because of their importance and overall influence in user's choice, both PCA and CFA results showed low factor loading and significant cross-loadings. As a result, these four attributes (C_{19}, C_{23}, M_{13}, A_{32}) have been removed during analysis of the final model. The results of this study also highlight the irrespective of variation in socio-economic and trip characteristics of individuals. Bus service attributes are categorised primarily to these mentioned seven exogenous variables which establish the relationship with the endogenous construct variable indicating the

Table 20.6 Path coefficient of the significant paths in SEM

Sl No.	Significant Path	Unstanderdised Weight	Standard Error	Critical Ratio	p-Value	Standardised Weight
1	Comfort (C_1) ⟶ Satisfaction	0.5345	0.1329	4.0128	0.0016	0.6604
2	Convenience (C_2) ⟶ Satisfaction	0.4041	0.1502	2.6904	0.0258	0.5349
3	Safety (S_1) Amenities (A_1) ⟶ Satisfaction	0.2364	0.0916	2.5808	0.0054	0.2987
4	Availability (A_2) ⟶ Satisfaction	0.2104	0.0721	2.9182	0	0.2736
5	Accessibility (A_3) ⟶ Satisfaction	0.4367	0.1652	2.6434	0	0.3569
6	Management (M_1) ⟶ Satisfaction	0.6105	0.2012	3.0343	0	0.4742
7	Maintenance (M_2) ⟶ Satisfaction	1.1046	0.3248	34,009	0.0013	0.4504
8	C_{11} ⟵ Comfort (C_1)	0.3152	0.2103	1.4988	0	0.5196
9	C_{12} ⟵ Comfort (C_1)	0.4293	0.1321	3.2498	0.0258	0.2952
10	C_{13} ⟵ Comfort (C_1)	0.6249	0.2405	2.5983	0.0034	0.4657
11	C_{14} ⟵ Comfort (C_1)	0.5013	0.0967	5.1841	0	0.5329
12	C_{15} ⟵ Comfort (C_1)	0.5834	0.2106	2.7702	0	0.5579
13	C_{16} ⟵ Comfort (C_1)	0.8146	0.1521	5.3557	0.0052	0.6359
14	C_{17} ⟵ Comfort (C_1)	0.1392	0.0315	4.419	0	0.3427
15	C_{18} ⟵ Comfort (C_1)	0.1193	0.0721	1.6546	0.0038	0.2478
16	C_{19} ⟵ Comfort (C_1)	0.2246	0.1652	1.3596	0	0.1097
17	C_{21} ⟵ Convenience (C_2)	0.3724	0.1204	3.093	0.0065	0.3592
18	C_{22} ⟵ Convenience (C_2)	0.7345	0.2038	3.604	0.0031	0.2205
19	C_{23} ⟵ Convenience (C_2)	0.5042	0.2103	2.3975	0.0418	0.1161
20	C_{24} ⟵ Convenience (C_2)	0.6345	0.1265	5.0158	0.0227	0.3648
21	C_{25} ⟵ Convenience (C_2)	0.1147	0.0405	2.832	0	0.2135
22	C_{26} ⟵ Convenience (C_2)	0.6305	0.0967	6.5202	0.0016	0.5642
23	S_{11} ⟵ Safety (S_1)	0.5301	0.1652	3.2088	0.0073	0.2039
24	S_{21} ⟵ Safety (S_1)	0.3345	0.1012	3.3053	0	0.3048
25	A_{11} ⟵ Amenities (A_1)	0.6434	0.2443	2.6336	0.0018	0.5874
26	A_{12} ⟵ Amenities (A_1)	0.5014	0.1193	4.2028	0.0006	0.4461
27	A_{13} ⟵ Amenities (A_1)	0.4783	0.1325	3.6098	0	0.3124
28	A_{21} Availability (A_2)	0.5344	0.1405	3.8036	0.0191	0.4952
29	A_{22} ⟵ Availability (A_2)	0.2347	0.0967	2.4271	0	0.3657
30	A_{23} ⟵ Availability (A_2)	0.3345	0.2106	1.5883	0.0013	0.3329

Sl No.	Significant Path	Unstan-derdised Weight	Standard Error	Critical Ratio	p-Value	Standardised Weight
31	$A_{31} \longleftarrow$ Accessibility (A_3)	0.4352	0.1521	2.8613	0.0224	0.5076
32	$A_{32} \longleftarrow$ Accessibility (A_3)	0.6045	0.1252	4.8283	0	0.2094
33	$A_{33} \longleftarrow$ Accessibility (A_3)	0.1173	0.0721	1.6269	0	0.6427
34	$M_{11} \longleftarrow$ Management (M_1)	0.6345	0.1652	3.8408	0.0028	0.2478
35	$M_{12} \longleftarrow$ Management (M_1)	1.0079	0.2002	5.0344	0	0.4266
36	$M_{13} \longleftarrow$ Management (M_1)	1.0345	0.3248	3.185	0.0511	0.1302
37	$M_{21} \longleftarrow$ Maintenance (M_2)	0.1742	0.0821	2.1218	0.0006	0.7245
38	$M_{22} \longleftarrow$ Maintenance (M_2)	0.4237	0.1479	2.8648	0	0.5802

passenger overall satisfaction. However, there exists a difference in prioritisation of service attributes among individual user groups. The EFA results for all socio-economic groups have been validated through CFA and the standardised regression weights (SRW) for the best-fit output model. By effecting some preparatory calibrations, the present study has proposed the final model shown in Fig. 20.2.

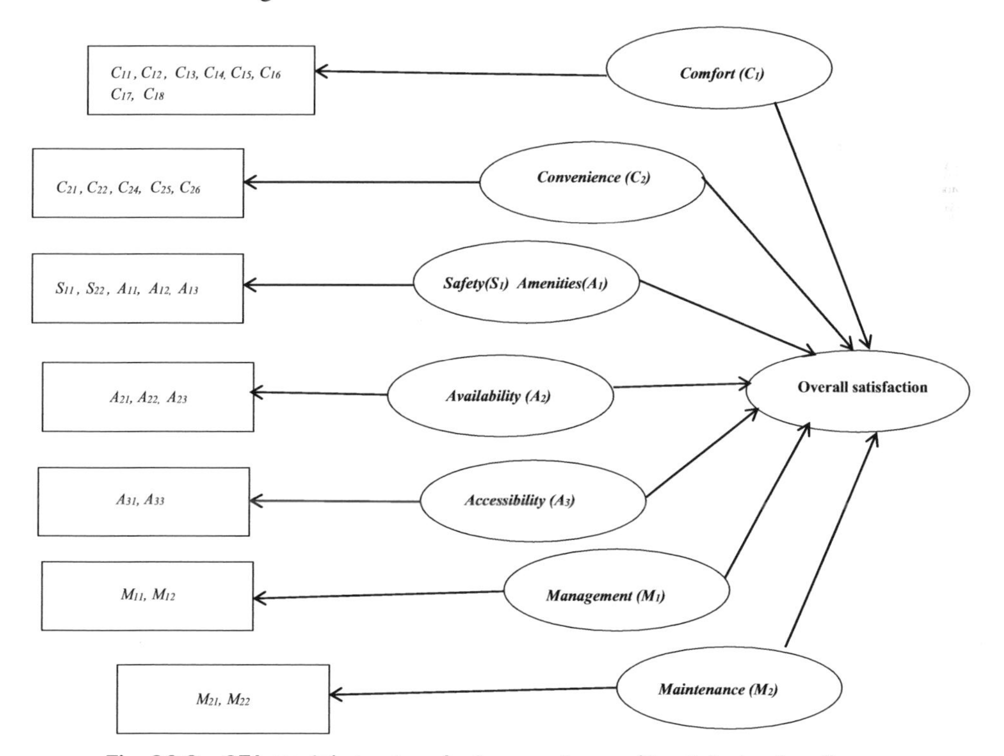

Fig. 20.2 CFA model structure for bus service quality attributes for all users

Conclusions and Limitations

The present study used the structural equation model to evaluate the passengers perception on service quality in bus services and then overall satisfaction towards bus services. The proposed model identifies service quality factors to improve, with the aim of offering bus services characterised by high level of quality. The present study used the SEM to evaluate the passengers' perception of service quality in bus services and then overall satisfaction towards bus services. The proposed model identifies service quality factors to improve, with the aim of offering bus services characterised by high level of quality. This study proposed various contributions towards the improvement of conceptual framework and particular knowledge regarding the service quality of bus services. The findings of this research can provide operating companies and transport managers valuable information for designing appropriate transport policies to attract new passengers and retain the current ones considering the users' overall satisfaction on the basis of these findings. Thus, based on this result, necessary recommendations and steps could be considered in order to improve public bus transport services in the city of Kolkata, resulting in sustainable mobility for commuters.

The major limitation of this study is the scope of the study is limited to Kolkata city, West Bengal. A more accurate analysis of service quality in PT should be based on a survey addressing all the cities at the national level. However, the present study can be considered as a reference for future research on the service quality of public and private transportation. As the results may vary across various geographic locations, a detailed study on the heterogeneity of perception needs to be conducted across geographies, mostly considering the potential users' perspective for future analysis.

References

1. Anderson, H. C., and Gerling, D. W., (1988), "Structural equation modeling in Practice: A Review and recommended two step approach", Psychological Bulletin, Vol.103, pp. 411–423.
2. Anumita Roychowdhury and Sayan Roy (2021), "Electric bus towards zero-emission commuting"
3. Bonsall, P., Beale, J., Paulley, N., Pedler, A. (2005), "The differing perspectives of road users and service providers." Transport Policy, 12(4), 334–344.
4. Bromley, R. D., Matthews, D. L., Thomas, C. J. (2007), "City centre accessibility for wheelchair users: The consumer perspective and the planning implicatons." Cities, 24(3), 229–241.
5. Cirilli, C., Eboli, L., Mazzulla, G. (2011), "On the asymmtric user perception of transit service quality." International Journal of Sustainable Transportation, 5(4), 216–232.
6. Das, S., Pandit, D. (2013) "Importance of user perception in evaluating level of service for bus transit for a developing country like India: a review." Transport Reviews, 33(4), 402–420.
7. De, J., de Ona, R. (2015), "Quality of service in public transport based on customer satisfaction surveys: a review and assessment of methodological approaches." Transportation Science, 49(3), 605–622.
8. De, J., de Ona, R., Eboli, L., Mazzulla, G. (2013), "Perceived service quality in bus transit service: A structural equation approach." Transport Policy, 29, 219–226.
9. Dell'.O, L., Ibeas, A., Cecin, P. (2010), "Modelling user perception of bus transit quality." Transport Policy, 17(6), 388–397.

10. Domencich, T. and Mc Fadden, D. (1975), "Urban travel demand, a behavioural analysis", North Holland Publishing, Oxford.
11. Eboli, L., Mazzulla, G. (2007), "Service quality attributes affecting customer satisfaction for bus transit." Journal of Public Transportation, 10(3), 21–34.
12. Eboli, L., Mazzulla, G. (2011), "A methodology for evaluating transit service quality based on subjective and objective measures from passenger's point of view." Transport Policy, 18(1), 172–181.
13. Garrido, A., Ortuzar, J. (1994), "Deriving public transport level of service weights from a multiple comparison of latent and observable variable." Journal of the Operational Research Society, 45(10), 1099–1107.
14. Golob, T., Canty, E., Gustafson, R., Vitt, J. (1972) "An analysis of consumer preferences for a public transportation system." Transportation Research,6(1), 81–102.
15. Guirao, B., Garcia-Pastor, A., Lopez-Lambas, (2016) "The importance of service quality attributes in public transportation: Narrowing the gap between scientific research and practitioner's needs." Transport Policy, 49, 68–77.
16. Guo, Z., Wilson, N. H. (2011), "Assessing the cost of transfer inconvenience in public transport systems: A case study of the London underground." Transportation Research Part A; Policy and Practice, 45(2), 91–104.
17. Guo, J. Y., Sen, S., Weston, L. (2005), "Measuring access to public transportation services: Review of customer-oriented transit performance measures and methods of transit submarket identification." Austin, Texas: Centre for Transportation Research, The University of Texas.
18. Hooper, D., Coughlan, J. Mullen, M. R., (2008), "Structural Equation Modelling: Guidelines for Determining Model Fit." The Electronic Journal of Business Research Methods Volume, 6 (1)
19. Kinsella and Caulfield, (2011), "An examination of the quality and ease of use of public transport in Dublin from a newcomer's perspective."
20. Norhisham, S., Bakar, M.F.A., Huri, A.H., Mohammad, A.M., Khalid, N.H.N., Basri, H.; Mardi, N.H., Zaini, N.A.(2019) "Evaluting thetravel demand using E-bike on Campus."
21. Pira, M.L., Attrad, M., Ison, S.G. (2021), "Urban transport planning and policy in a changing world: Briding the gap between theory and practice."
22. Saikat Deb, Mokaddes Ali Ahmed,Debasish Das (2022) "Service quality estimation and improvement plan of bus Service: A perception and expectation based analysis"
23. Sakamoto, K., Belka, S., (2010), "Financing Sustainable Urban Transport, Transport Policy Advisory Services."
24. Sánchez-Atondo, A. et al. (2020), "Understanding public transport ridership in developing countries to promote sustainable urban mobility: A case study of Mexicali, Mexico".
25. Santos,G., Nikolaev,N (2021), "Mobility as a service and public transport: a rapid literature review and the case of Moovit" Sustainability, 13 (7) (2021), p. 3666
26. Shahid, S., Minhans, A., et. al. (2014), "Assessment of greenhouse gas emission reduction measures in transportation sector of malaysia", Jurnal Teknologi, 70(4).
27. Sharifi and Khavarian-Garmsir, (2020), "The COVID-19 pandemic: Impacts on cities and major lessons for urban planning, design, and management."
28. Shivalashetti, A.S. and Hugar, S.S., (2008), "Passengers' satisfaction towards Karnataka State Road Transport Corporation in Godag District: An Empirical Study", The ICFAI Journal of Services Marketing, 6 (3), pp. 29–37.
29. Sinha, S., Shivanand Swamy, H.M., and Modi, K (2020), "User Perceptions of Public Transport Service Quality."

30. Stradling, S., Carreno, M., Rye, T., & Noble, A., (2007), "Passenger perceptions and the ideal urban bus journey experience." Transport Policy
31. Timokhina, G., Ivashkova, N., Skorobogatykh, I., Murtuzalieva, T., Musatova, Z (2020), "Management of competitiveness of metropolis public transport in the COVID-19 pandemic based on core consumers' values"
32. Too, L., and Earl, G., (2010), "Rathic transport service quality and sustaibale development: a community stakeholder perspective", Sustainable development, 18(1), pp. 51–61.
33. Wang, H.; Peng, Z.; Lu, Q.; Sun, D. J.; Bai, C. (2017) "Assessing Effects of Bus Services Quality on Passenger' Taxi-Hiring Behavior."
34. Wilbur Smith Associates ((2008), "Study on Traffic and Transportation Policies and Strategies in urban areas in India" Ministry of Urban Development.

CHAPTER 21

Ecological Aspects of Fear and Hunting Cooperation in Prey-Predator Dynamics: A Case Study Analysis

Shilpa Samaddar[1] and Paritosh Bhattacharya[2]
Department of Mathematics, National Institute of Technology Agartala, Jirania, West Tripura, India

Abstract

Recent studies on terrestrial vertebrates indicate that the fear of predators can induce anti-predator behaviour in the prey population, which drastically reduces their reproduction. Moreover, investigation into the cooperative nature of predators in prey hunting is attracting researchers remarkably. Numerous works have been carried out considering these two effects separately. There are very few studies that reveal the combined effects, but any real data comparison has not yet been performed. We have developed a mathematical model to describe a prey-predator system incorporating the cost of hunting cooperation with fear and presented a real case study by fitting the actual species data with our proposed model. We have discussed the robustness of each parameter that affects the system, which endows the visualisation of dynamical consequences of major parameters. Additionally, our analyses show that high fear and low cooperation can bring sustainability to the system. Also, the increment of cooperation, along with fear, can transform the subcritical limit cycle (bi-stability) to a supercritical one.

Keywords: fear effect, predator cooperation, prey refuge, Hopf bifurcation, bi-stability

Introduction

We will discuss the prey-predator dynamics by formulating a mathematical model concentrating on their natural phenomena. Over the time, prey-predator interaction, specifically the predation process, achieves the centre of attraction in the domain of mathematical biology. To understand this mechanism, many mathematicians developed various differential equation models. Prey always has some predation threat. Additionally, limited resources influence the prey population for migration and induce hiding behaviour that may provide a certain amount of protection. This behaviour is considered as a survival strategy from getting extinct (Dill & Fraser, 1997; González-Olivares, & Ramos-Jiliberto, 2003; Ma et al. 2009; Ripple, & Beschta, 2004; Schneider, & Harrington; 1981; Sih et al., 2010). While considering survival

[1]shilpasmddr@gmail.com; [2]pari76@rediffmail.com

hacks, predators also achieved some hacks to survive. One of the most popular strategies is cooperation during prey hunting. Two or more predator individuals increase their fitness and attack rate by cooperating among themselves. Cooperative behaviours with increased adults effectively influence the hunting rate and decrease chasing distances (S. Creel, & Creel, 1995). Additionally, it helps to capture large prey (Bednarz, 1988), to reduce the time of prey finding (Pitcher, Magurran, & Winfield, 1982), to prevent the stealing of carcasses from outsiders (Brockmann & Barnard, 1979; Vucetich, Peterson, & Waite, 2004) and in many other aspects.

Holling type II functional response is then vigorously used to authenticate the existence of a unique stable limit cycle, to understand the need to abandon the unreasonable assumption of unsaturated attack rates and many more things (Berryman, 1992; Kooij & Zegeling, 1997; Kuang & Freedman, 1998). Some other functional responses like Michaelis–Menten type (Huang, Ruan, & Song, 2014; May, 1972), Gause-type (Kuang, & Freedman, 1998) functional response are only prey dependent, whereas both species-dependent functional response is Beddington–DeAngelis response (Beddington, 1975; DeAngelis, Goldstein, & O'Neill 1975). Meanwhile, all these responses assume that prey is affected by predators through a direct attack. Regardless of revolutionary models formed to improve the prey-predator dynamics, these fail to justify the later studies which have confirmed that a mere sense of predator presence can affect the prey behaviour drastically. It affects the prey population physiologically and psychologically to such an extent that the behavioural changes are more recognisable than direct killing (Creel & Christianson, 2008; Lima, 2009). Zanette et al. (Zanette, White, Allen, & Clinchy, 2011) highlighted the fear effect solely by examining the behavioural changes in song sparrows during the season of their entire breeding while keeping no direct attack. The authors considered two sets of birds and broadcasted predator sound to one set while the other remained with no predator sound. From this study, it came out that the first set of birds loose reproduction by 35%. Concerning these results, Wang et al. (Wang, Zanette, & Zou, 2016) firsts made a breakthrough by demonstrating the reproduction loss mathematically. Recently, many authors have worked on this phenomenon in different systems (K. D. Prasad & Prasad, 2019; Samaddar, Dhar, & P. Bhattacharya, 2020; J. Wang, Cai, Fu, & Wang, 2019).

Fear effect offsets the prey reproduction that consequences predator density loss. On the other hand, predator density improves through cooperative hunting. These results suggest that to analyse any system, both effects should be considered. Few studies describe both the effects as listed in refs. (S. Pal, Pal, Samanta, & Chattopadhyay, 2019a,b) but none of them considered these two effects with Holling type II functional response. Moreover, any real case study has not also been taken care of previously. This inspires us to propose a model considering all the phenomena with type II response and test the accuracy by data fit a real case study in Bow valley.

We have assembled the paper as follows: We mathematically formulate the system in Section 2. Then, we analyse the biologically relevant equilibrium points through their stability conditions in Section 3. Section 4 demonstrates a detailed explanation of Hopf bifurcation and its direction. Then, all the analytical results are numerically verified and the simultaneous effects of different parameters are discussed in Section 5. Most importantly, a real case study of Bow valley is analysed in Section 6 by data fitting. Finally, we conclude our finding through discussion in Section 7.

Mathematical Model Formulation

Experimental studies suggest that hunting cooperation advantages the prey-capturing success of the predators. Moreover, the presence of predators alone induces fear in the prey colony. The fear affects the prey so drastically that forces them to reduce the growth rate. Here, we develop a two-dimensional prey-predator model by ordinary differential equation (ODE) to analyse the effect of hunting cooperation under the threat of predation fear. The variables we introduce to explain the model are prey density (x) and predator density (y). The prey naturally increases with a growth rate α while the predation fear effect (K) reduces the growth rate to $\alpha/(1 + Ky)$. The loss rate due to intra-specific competition among the prey population is b. Most of the prey populations seek refuge to avoid predator contact. $m \in (0, 1)$ denotes the prey refuge rate. Hence, $(1 - m)x$ preys are exposed to predators. β is the individual attack rate of the predators where the cooperation effect (σ) increases the attack rate to ($\beta + \sigma y$). Consideration of the saturation response and prey refuge transform the prey loss due to predator attack as $\frac{(\beta+\sigma y)(1-m)xy}{1+a(1-m)x}$, where a represents half-saturation constant. The prey biomass converses to predator biomass with a rate c. From the above assumptions we propose the following dynamical system:

$$\frac{dx}{dt} = \frac{\alpha x}{1+Ky} - bx^2 - \frac{(\beta+\sigma y)(1-m)xy}{1+a(1-m)x} \equiv F(x,y), \tag{2.1}$$

$$\frac{dy}{dt} = \frac{c(\beta+\sigma y)(1-m)xy}{1+a(1-m)x} - \gamma y \equiv G(x,y); \tag{2.2}$$

$$x(0) > 0,\ y(0) > 0;$$

The study (Samaddar, Dhar, & Bhattacharya, 2020) confirms that the prey growth decreasing function $f(K, y) = \frac{1}{1+Ky}$ and the predator attack rate increasing function $g(\sigma, y) = \sigma y$ are biologically meaningful. Concerning the ecological perspective, all the model parameters are taken positive.

Mathematical Analysis

In this section, we analyse some basic results of the system (2) that confirm its well behaviour.

Positivity: An interacting system satisfies positivity; the statement biologically means that if the populations interact with positive initial densities then both of the populations will never extinct.

We consider the non-negative quadrant of $\mathbb{R}_+^2 = [0, \infty)$. The right-hand side of the system $F(x, y)$ and $G(x, y)$ are continuously differentiable and locally Lipschitz in $\mathbb{R}_+^2$. Hence, there exists a sufficiently large M such that if the population starts with positive initial conditions then satisfies non-negativity conditions and exists uniquely in $[0, M)$ ((Thieme, 2018) Theorem A.4).

Boundedness: A system of prey-predator interacting populations is bounded means none of the species can grow abruptly for a long time. This happens due to limited resources and predator attacks.

Theorem: All the solutions of the system (2) with positive initial conditions remain uniformly bounded.

Proof: Let us consider the function

$$w(t) = x + \frac{1}{c}y$$

Then

$$\begin{aligned}\frac{dw}{dt} &= \frac{dx}{dt} + \frac{1}{c}\frac{dy}{dt} \\ &= \frac{\alpha x}{1+Ky} - \frac{(\beta+\sigma y)(1-m)xy}{1+a(1-m)x} - bx^2 - \frac{\gamma}{c}y + \frac{(\beta+\sigma y)(1-m)xy}{1+a(1-m)x} \\ &= \frac{\alpha x}{1+Ky} - \frac{\gamma}{c}y - bx^2\end{aligned}$$

Let τ be any positive constant such that $\tau < \gamma$, then

$$\begin{aligned}\frac{dw}{dt} + \tau w &= \frac{\alpha x}{1+Ky} - \frac{\gamma}{c}y - bx^2 + \tau x + \frac{\tau}{c}y \\ &\le (\alpha + \tau - bx)x + \frac{1}{c}(\tau - \gamma)y \\ &\le (\alpha + \tau - bx)x\end{aligned}$$

Let $P = (\alpha + \tau - bx)x$, then

$$\frac{dP}{dx} = \alpha + \tau - 2bx$$

$$\frac{dP}{dx} = 0 x = \frac{\alpha+\tau}{2b}; \frac{d^2P}{dx^2} = -2b < 0$$

$$Max(P) = (\alpha + \tau - \frac{\alpha+\tau}{2} \cdot \frac{(\alpha+\tau)}{2b}) = \frac{(\alpha+\tau)^2}{4b}, \text{ and}$$

$$\frac{dw}{dt} + \tau w < M = \frac{(\alpha+\tau)^2}{2b}$$

Applying Gronwall's inequality (Birkhoff & Rota, 1989) we have,

$$0 < w(t) < \frac{M}{\tau}(1 - e^{-\tau t}) + w(0)e^{-\tau t}$$

This implies $0 < w(t) < \frac{M}{\tau}$ for $t \to \infty$.

Hence, all the solutions, initiating from $\mathbb{R}_+^2 \setminus \{0,0\}$, confines in the region

$$B = \left\{ (x, y) \in \mathbb{R}_+^2 : x + \frac{1}{c} y \leq \frac{M}{\tau} + \varepsilon \right\}.$$

This confirms the system's uniform boundedness.

Dissipativeness:

Theorem: The system (2) is dissipative if $\alpha > b$.

Proof: The first equation of system (2) shows

$$\frac{dx}{dt} = \frac{\alpha x}{1 + Ky} - bx^2 - \frac{(\beta + \sigma y)(1 - m)xy}{1 + a(1 - m)x}$$
$$\leq \alpha x - bx^2$$

We assume $\alpha > b$, this provides $\limsup_{t \to \infty} x(t) \leq \frac{\alpha}{b}$. Thus, for any arbitrary small $\varepsilon_1 > 0$, we get a positive real no $T_1 > 0$ such that $x(t) \leq \frac{\alpha}{b} + \varepsilon_1$, for all $t \geq T_1$. Now, in the region $t \geq T_1$, we have

$$\frac{d}{dt}\left(x + \frac{1}{c} y \right) = \frac{dx}{dt} + \frac{1}{c}\frac{dy}{dt}$$
$$= \frac{\alpha x}{1 + Ky} - bx^2 - \frac{\gamma y}{c}$$
$$\leq \alpha x - \frac{\gamma y}{c}$$
$$\leq G - \gamma\left(x + \frac{1}{c} y \right), \text{ where } G = (\alpha + \gamma)\left(\frac{\alpha}{b} + \varepsilon_1 \right)$$

Thus, we have, $\limsup_{t \to \infty}\left(x + \frac{1}{c} y \right) \leq \frac{G}{\gamma}$.

This ensures the existence of a positive real number M_1 such that

$$\limsup_{t \to \infty} y(t) \leq M_1.$$

So, for any arbitrary small $\varepsilon_2 > 0$, there exists a positive real number $T_2 > T_1$ such that

$$y(t) \leq M_1 + \varepsilon_2, \forall t \geq T_2.$$

Hence, this confirms the dissipativeness of the system (2) for $\alpha > b$.

Uniform Persistence: Geometrically, uniform persistence implies the existence of such a region with a non-zero distance from the boundary in which all species coexist for a long time. Hence, uniform persistence means the permanence of species survival.

We call a system uniformly persistent if every solution initiating from $\mathbb{R}_+^2$, satisfies the conditions below:

1. $x(t), y(t) \geq 0 \ \forall \ t \geq 0$,
2. There exists $\varepsilon > 0$ such that $\liminf_{t\to\infty} x(t), y(t) \geq \varepsilon$.

Theorem: *The system (2) is uniformly persistent for $\alpha > \alpha^*$ and*

$$c\beta(1-m)(p_1-\varepsilon_3) > \gamma\left\{1+a(1-m)\left(\frac{\alpha}{b}+\varepsilon_3\right)\right\}.$$

Proof: *From the first equation of system (2), for $t > T_2$, we have*

$$\frac{dx}{dt} = \frac{\alpha x}{1+Ky} - bx^2 - \frac{(\beta+\sigma y)(1-m)xy}{1+a(1-m)x}$$

$$\geq x\left[\frac{\alpha}{1+K(M_1+\varepsilon_2)} - bx - (1-m)\{\beta+\sigma(M_1+\varepsilon_2)\}(M_1+\varepsilon_2)\right]$$

$$= x\left[\left\{\frac{\alpha}{1+K(M_1+\varepsilon_2)} - (1-m)\{\beta+\sigma(M_1+\varepsilon_2)\}(M_1+\varepsilon_2)\right\} - bx\right]$$

Let us assume that $\alpha > (1-m)\{1+K(M_1+\varepsilon_2)\}\{\beta+\sigma(M_1+\varepsilon_2)\}(M_1+\varepsilon_2) = \alpha^*$ (say).

Then, we have

$$\liminf_{t\to\infty} x(t) \geq \frac{1}{b}\left[\frac{\alpha}{1+K(M_1+\varepsilon_2)} - (1-m)\{\beta+\sigma(M_1+\varepsilon_2)\}(M_1+\varepsilon_2)\right] = p \text{ (say)}.$$

Thus, for any arbitrary small $\varepsilon_3 > 0$, there exists $T_3 > T_2$ such that $x(t) \geq p - \varepsilon_3$, $\forall t \geq T_3$. Again, for this ε_3 there exists a positive real number T_4 such that $x(t) \leq \frac{\alpha}{b} + \varepsilon_3, \forall t \geq T_4$. Now, we choose T_5 such that $T_5 = \max\{T_3, T_4\}$. So, for $t \geq T_5$ from the second equation of system (2), we have

$$\frac{dy}{dt} = \frac{c(1-m)(\beta+\sigma y)xy}{1+a(1-m)x} - \gamma y$$

$$\geq y\left[\frac{c(1-m)(\beta+\sigma y)(p-\varepsilon_3)}{1+a(1-m)\left(\frac{\alpha}{b}+\varepsilon_3\right)} - \gamma\right]$$

$$\geq y\left[\left\{\frac{c\beta(1-m)(p-\varepsilon_3)}{1+a(1-m)\left(\frac{\alpha}{b}+\varepsilon_3\right)} - \gamma\right\} + \frac{c\sigma(1-m)(p-\varepsilon_3)y}{1+a(1-m)\left(\frac{\alpha}{b}+\varepsilon_3\right)}\right]$$

$\frac{dy}{dt} \geq 0$ for $c\beta(1-m)(p-\varepsilon_3) > \gamma\left\{1+a(1-m)\left(\frac{\alpha}{b}+\varepsilon_3\right)\right\}$. Hence, $y(t)$ is a monotonically increasing function of t for all $t \geq T_5$. Therefore, $y(t) > y(T_5)$ for all $t \geq T_5$. Now, we choose $\varepsilon = \min\{p, y(T_5)\}$. Thus for $\varepsilon > 0$, $\liminf_{t\to\infty} x(t), y(t) \geq \varepsilon$. This completes the proof.

Existence of Equilibria

The zero growth rate isoclines $F(x, y) = G(x, y) = 0$ provide three ecologically meaningful equilibria. The no-species equilibrium $E_0(0, 0)$, always exists and unstable saddle. The predator-free equilibrium $E_1\left(\frac{\alpha}{b}, 0\right)$, which corresponds to the no predator state, always exists and is locally stable if $\alpha(1 - m)(c\beta - a\gamma) - b\gamma < 0$, else unstable. The coexistence equilibrium $E_2(x^*, y^*)$, is the positive intersection of the nullclines. Here, $x^* = \dfrac{\gamma}{(1-m)\{(c\beta - a\gamma) + c\sigma y^*\}}$, and y^* is a positive solution of

$$A_0 y^5 + A_1 y^4 + A_2 y^3 + A_3 y^2 + A_4 y + A_5 = 0, \tag{3.1}$$

where

$A_0 = Kc^2 \sigma^3 (1 - m)^3 > 0,$

$A_1 = c^2\sigma^3 (1 - m)^3 + Kc^2 \beta\sigma^2 (1 - m)^3 + 2Kc\sigma^2 (1 - m)^3 (c\beta - a\gamma),$

$A_2 = K\sigma(1 - m)^3 (c\beta - a\gamma)^2 + 2Kc\beta\sigma(1 - m)^3 (c\beta - a\gamma) + 2c\sigma^2 (1 - m)^3 (c\beta - a\gamma) + 2c^2 \beta\sigma^2 (1 - m)^3$

$A_3 = K\beta(1 - m)^3 (c\beta - a\gamma)^2 + \sigma(1 - m)^3 (c\beta - a\gamma)^2 + 2c\beta\sigma(1 - m)^3 (c\beta - a\gamma) + \sigma(1 - m)\{Kb\gamma - c\alpha\sigma(1 - m)\},$

$A_4 = \beta(1 - m)^3 (c\beta - a\gamma)^2 - c\sigma(1 - m)\{\alpha(1 - m)(c\beta - a\gamma) - b\gamma\} + c\beta(1 - m)\{Kb\gamma - c\alpha\sigma(1 - m)\},$

$A_5 = -c\beta(1 - m)\{\alpha(1 - m)(c\beta - a\gamma) - b\gamma\}.$

This equilibrium is locally stable if

$bc^2x^{*2} (\beta + \sigma y^*) > (a\gamma + c^2\sigma x^*)\gamma y^*$ and

$$\frac{K\alpha x^*(\beta + \sigma y^*)}{(1 + Ky^*)^2} + \gamma\{\beta + (2 + ax^*)y^*\} - bc\sigma x^{*2}\{1 + a(1 - m)x^*\} > 0.$$

Remark 1:

(a) For $(c\beta - a\gamma) < 0$, x^* exists if $y^* > \dfrac{a\gamma - c\beta}{c\sigma}$. Additionally, in this condition $A_0 > 0$ as well as $A_5 > 0$. This indicates that Eq. 3.1 has no solution or more than one solution.

(b) If $(c\beta - a\gamma) > 0$, x^* exists. Additionally, in this condition $A_0, A_1, A_2 > 0$.

Case 1: If $\alpha(1 - m)(c\beta - a\gamma) - b\gamma > 0$, $A_5 < 0$. Here, at least one solution exists.

Case 2: If **Case1** holds along with $Kb\gamma - c\alpha\sigma(1 - m) > 0$, then $A_3 > 0$. Hence, exactly one solution exists.

Hopf Bifurcation Analysis

Hopf bifurcation describes the value of a parameter that ensures the stability and unstability of any system steady state. In a two-dimensional system, Hopf bifurcation occurs at some particular point as it switches from unstable to stable (or, vice-versa) and a periodic solution disappears or, appears. Here, we analyse all the conditions to explore the Hopf bifurcation occurrence possibility and to identify its direction.

Theorem 1: When σ crosses the threshold value $\sigma^{[h]}$, the system (2) experiences Hopf bifurcation under the necessary and sufficient conditions stated as:

1. $\dfrac{K\alpha x^{*}(\beta+\sigma y^{*})}{(1+Ky^{*})^{2}}+\gamma\{\beta+(2+ax^{*})y^{*}\}-bc\sigma x^{*2}\{1+a(1-m)x^{*}\}>0$

2. $\dfrac{d}{d\sigma}\text{Trace}(J_2)|_{\sigma=\sigma^{[h]}}\neq 0$

Here, J_2 is the Jacobian matrix at E_2.

Proof: *We consider the Jacobian matrix J_2 at the equilibrium point E_2.*

The Jacobian matrix of the system corresponds to E_2 is

$$J_2=\begin{pmatrix} -bx^{*}+\dfrac{a(1-m)^{2}(\beta+\sigma y^{*})x^{*}y^{*}}{\{1+a(1-m)x^{*}\}^{2}} & -\dfrac{K\alpha x^{*}}{(1+Ky^{*})^{2}}-\dfrac{(1-m)(\beta+2\sigma y^{*})x^{*}}{1+a(1-m)x^{*}} \\ \dfrac{c(1-m)(\beta+\sigma y^{*})y^{*}}{\{1+a(1-m)x^{*}\}^{2}} & \dfrac{c\sigma(1-m)x^{*}y^{*}}{1+a(1-m)x^{*}} \end{pmatrix}$$

For $\sigma=\sigma^{[h]}$, the characteristic equation of J_2 becomes

$$\lambda^2-\lambda T+\Delta=0,$$

where T and Δ represent the trace and determinant of J_2, respectively. At the point of Hopf bifurcation, the above equation must have purely imaginary roots. Hence, the trace must be zero and the determinant must be positive, that is,

$$T=-bx^{*}+\frac{a(1-m)^{2}(\beta+\sigma y^{*})x^{*}y^{*}}{\{1+a(1-m)x^{*}\}^{2}}+\frac{c\sigma(1-m)x^{*}y^{*}}{1+a(1-m)x^{*}}=0,$$

and

$$\Delta=\frac{(1-m)y^{*}}{\{1+a(1-m)x^{*}\}^{2}}\left[\frac{K\alpha x^{*}(\beta+\sigma y^{*})}{(1+Ky^{*})^{2}}+\gamma\{\beta+(2+ax^{*})y^{*}\}-bc\sigma x^{*2}(1+a(1-m)x^{*})\right]>0 \text{ at}$$

$\sigma=\sigma^{[h]}$

$T=0$ provides

$\sigma^{[h]}=\dfrac{a\gamma^{2}y^{*}-bc^{2}\beta x^{*2}}{c^{2}x^{*}y^{*}(bx^{*}-\gamma)}$ and the characteristic equation becomes

$\lambda^2+\Delta=0.$

The roots of the equation are $\lambda_{1,2}=\pm i\omega_0$, where $\omega_0=\sqrt{\Delta}$ if $\Delta>0$.

Now, for the transversality condition, we choose any point σ in the small neighbourhood of $\sigma^{[h]}$, $\lambda_{1,2} = V(\sigma) \pm i\omega_0(\sigma)$, where $V(\sigma) = \frac{1}{2}T$ at $\sigma = \sigma^{[h]}$ and $\omega_0 = \sqrt{\Delta - \frac{1}{4}T^2}$. Hence, the transversality condition is satisfied if $\frac{d}{d\sigma}\text{Trace}(J_2)|_{\sigma=\sigma^{[h]}} \neq 0$ at $\sigma = \sigma^{[h]}$.

If the conditions hold, then the system experiences Hopf bifurcation around E_2 at $\sigma = \sigma^{[h]}$.

Note: Since the analytical proof of the transversality condition is quite difficult, we have numerically proved this for finding Hopf bifurcation point.

Direction of Hopf Bifurcation

The direction and stability of Hopf bifurcation of the system (2) around the coexistence equilibrium E_2 is discussed in this sub-section via calculating the first Lyapunov coefficient and applying the results described in (Wiggins, 2003).

For this purpose, to transform the equilibrium point E_2 into origin, we consider $z_1 = x - x^*$ and $z^2 = y - y^*$. The present system (2) transforms to

$$\frac{dz_1}{dt} = \frac{\alpha(z_1 + x^*)}{1 + K(z_2 + y^*)} - b(z_1 + x^*)^2 - \frac{(1-m)\{\beta + \sigma(z_2 + y^*)\}(z_1 + x^*)(z_2 + y^*)}{1 + a(1-m)(z_1 + x^*)},$$

$$\frac{dz_2}{dt} = -\gamma(z_2 + y^*) + \frac{c(1-m)\{\beta + \sigma(z_2 + y^*)\}(z_1 + x^*)(z_2 + y^*)}{1 + a(1-m)(z_1 + x^*)}.$$

Now, the expansion in Taylor series about (0, 0) upto order 3, provides the following system

$$\dot{z}_1 = a_1 z_1 + a_2 z_2 + a_3 z_1^2 + a_4 z_1 z_2 + a_5 z_2^2 + a_6 z_1^3 + a_7 z_1^2 z_2 + a_8 z_1 z_2^2 + a_9 z_2^3 + O(|z|^4)$$

$$\dot{z}_2 = b_1 z_1 + b_2 z_2 + b_3 z_1^2 + b_4 z_1 z_2 + b_5 z_2^2 + b_6 z_1^3 + b_7 z_1^2 z_2 + b_8 z_1 z_2^2 + b_9 z_2^3 + O(|z|^4) \quad (4.1)$$

where

$$a_1 = -bx^* + \frac{a(1-m)^2(\beta + \sigma y^*)x^* y^*}{\{1 + a(1-m)x^*\}^2} \qquad b_1 = \frac{c(1-m)(\beta + \sigma y^*)y^*}{\{1 + a(1-m)x^*\}^2}$$

$$a_2 = -\frac{K\alpha x^*}{(1 + Ky^*)^2} - \frac{(1-m)(\beta + 2\sigma y^*)x^*}{1 + a(1-m)x^*} \qquad b_2 = \frac{c\sigma(1-m)x^* y^*}{1 + a(1-m)x^*}$$

$$a_3 = -b + \frac{a(1-m)^2(\beta + \sigma y^*)y^*}{\{1 + a(1-m)x^*\}^3} \qquad b_3 = -\frac{ac(1-m)^2(\beta + \sigma y^*)y^*}{\{1 + a(1-m)x^*\}^3}$$

$$a_4 = -\frac{K\alpha}{(1 + Ky^*)^2} - \frac{(1-m)(\beta + 2\sigma y^*)}{\{1 + a(1-m)x^*\}^2} \qquad b_4 = \frac{c(1-m)(\beta + 2\sigma y^*)}{\{1 + a(1-m)x^*\}^2}$$

$$a_5 = \frac{K^2\alpha x^*}{(1 + Ky^*)^3} - \frac{\sigma(1-m)x^*}{1 + a(1-m)x^*} \qquad b_5 = \frac{c\sigma(1-m)x^*}{1 + a(1-m)x^*}$$

$$a_6 = -\frac{a^2(1-m)^3(\beta+\sigma y^*)y^*}{\{1+a(1-m)x^*\}^4} \qquad b_6 = \frac{a^2c(1-m)^3(\beta+\sigma y^*)y^*}{\{1+a(1-m)x^*\}^4}$$

$$a_7 = \frac{a(1-m)^2(\beta+2\sigma y^*)}{\{1+a(1-m)x^*\}^3} \qquad b_7 = -\frac{ac(1-m)^2(\beta+2\sigma y^*)}{\{1+a(1-m)x^*\}^3}$$

$$a_8 = \frac{K^2\alpha}{(1+Ky^*)^3} - \frac{\sigma(1-m)}{\{1+a(1-m)x^*\}^2} \qquad b_8 = \frac{c\sigma(1-m)}{\{1+a(1-m)x^*\}^2}$$

$$a_9 = -\frac{K^3\alpha x^*}{(1+Ky^*)^4} \qquad b_9 = 0$$

The truncation of higher-order terms (more than 3rd order) modifies the system as:

$$\dot{Z} = J_2 Z + Q(Z), \tag{4.2}$$

where $Z = \begin{pmatrix} z_1 \\ z_2 \end{pmatrix}$ and

$$Q = \begin{pmatrix} Q_1 \\ Q_2 \end{pmatrix} = \begin{pmatrix} a_3 z_1^2 + a_4 z_1 z_2 + a_5 z_2^2 + a_6 z_1^3 + a_7 z_1^2 z_2 + a_8 z_1 z_2^2 + a_9 z_2^3 \\ b_3 z_1^2 + b_4 z_1 z_2 + b_5 z_2^2 + b_6 z_1^3 + b_7 z_1^2 z_2 + b_8 z_1 z_2^2 \end{pmatrix}$$

If we consider $i\omega_0$ as one eigenvalue of J_2 at $\sigma = \sigma^{[h]}$, then the analogous eigenvector is $v = (a_2, i\omega_0 - a_1)^T$. Additionally, we define

$$Y = S^{-1}Z, \text{ where } Y = \begin{pmatrix} y_1 \\ y_2 \end{pmatrix} \text{ and } S = (Re(v), -Im((v)) = \begin{pmatrix} a_2 & 0 \\ -a_1 & -\omega_0 \end{pmatrix}$$

This transformation results the system (4.2) to $\dot{Y} = S^{-1}J_2S)Y + S^{-1}Q(SY)$. Normal form of the system is

$$\begin{pmatrix} \dot{y}_1 \\ \dot{y}_2 \end{pmatrix} = \begin{pmatrix} 0 & -\omega_0 \\ \omega_0 & 0 \end{pmatrix} \begin{pmatrix} y_1 \\ y_2 \end{pmatrix} + \begin{pmatrix} H^1(y_1, y_2 : \sigma = \sigma^{[h]}) \\ H^2(y_1, y_2 : \sigma = \sigma^{[h]}) \end{pmatrix} \tag{4.3}$$

where the non-linear functions H^1 and H^2 are given by, $H^1(y_1, y_2 : \sigma = \sigma^{[h]}) = \frac{1}{a_2} Q_1(SY)$, and $H^2(y_1, y_2 : \sigma = \sigma^{[h]}) = -\frac{1}{a_2\omega_0}\left[a_1 Q_1(SY) + a_2 Q_2(SY)\right]$, along with

$$\begin{aligned} Q_1(SY) = {} & (a_2^2 a_3 - a_1 a_2 a_4 + a_1^2 a_5 - a_1 a_2^2 a_7) y_1^2 - (a_2 a_4 - 2a_1 a_5 + a_2^2 a_7)\omega_0 y_1 y_2 + a_5 \omega_0^2 y_2^2 \\ & + (a_2^3 a_6 + a_1^2 a_2 a8 - a_1^3 a_9) y_1^3 + (2a_2 a_8 - 3a_1 a_9) a_1 \omega_0 y_1^2 y_2 + (a_2 a_8 - 3a_1 a_9)\omega_0^2 y_1 y_2^2 \\ & - a_9 \omega_0^3 y_2^3, \end{aligned}$$

$$\begin{aligned} Q_2(SY) = {} & (a_2^2 - a_1 a_2 b_4 + a_1 b_5) y_1^2 + (2a_1 b_5 - a_2 b_4)\omega_0 y_1 y_2 + b_5 \omega_0^2 y_2^2 + (a_2^2 b_6 - a_1 a_2 b_7 + a_0^2 b_8) a_2 y_1^3 \\ & + (2a_1 b_8 - a_2 b_7) a_2 \omega_0 y_1^2 y_2 + a_2 b_8 \omega_0^2 y_1 y_2^2. \end{aligned}$$

Now, the Lyapunov coefficient can be calculated from the normal form (4.3) using the following formula as:

$$L = \frac{1}{16}\left(H^1_{111} + H^1_{122} + H^2_{112} + H^2_{222}\right)$$
$$+\frac{1}{16\omega_0}\left\{H^1_{12}(H^1_{11} + H^1_{22}) - H^2_{12}(H^2_{11} + H^2_{22}) - H^1_{11}H^2_{11} + H^1_{22}H^2_{22}\right\},$$

where $H^l_{ij} = \frac{\partial^2 H^l}{\partial y_i \partial y_j}|_{(y_1,y_2;\sigma)=(0,0;\sigma^{[h]})}$ and

$H^l_{ijk} = \frac{\partial^2 H^l}{\partial y_i \partial y_j \partial y_k}|_{(y_1,y_2;\sigma)=(0,0;\sigma^{[h]})}, i, j, k, l \in \{1,2\}$. Hence, for the system (2), the expression of Lyapunov coefficient is given below:

$$L = \frac{\omega_0^2}{8a_2}[a_2a_8 - a_5(a_4 + 2b_5)] - \frac{a_1^2a_5}{4a_2}(a_4 + 2b_5) - \frac{a_1^4a_5}{8a_2\omega_0^2}(a_4 + 2b_5) + \frac{1}{8}[a_1(a_4^2 - 2b_5^2)$$
$$+ a_1(a_4b_5 + 2a_3a_5 + a_5b_4) + a_1(a_1a_8 - 2a_2b_8) - a_2(a_3a_4 - b_4b_5) - 2a_2(2a_1a_7 - a_2b_7) + 3a_2^2a_6]$$
$$+ \frac{1}{8\omega_0^2}[a_1^3(a_4^2 - 2b_5^2) + a_1^3(a_4b_5 + 2a_3a_5 + a_5b_4) - 3a_1^2a_2(a_3a_4 - b_4b_5) - a_1a_2^2b_3(a_4 + 2b_5)$$
$$+ 2a_2^2a_3(a_1a_3 + a_2b_3) - a_2^3b_4(a_1b_4 - a_2b_3 + a_1a_3)].$$

Results presented in Wiggins study (Wiggins, 2003) confirm that the Hopf bifurcation is stable, that is, supercritical if $L < 0$ and unstable, that is, subcritical if $L > 0$. For $L = 0$, the system experiences a generalised Hopf bifurcation.

Simulation Results

In this section, we numerically simulate (By using Matlab) the system considering various parameter values to explore the system behaviour explicitly. Also, here we numerically validate all the analytical findings. First, we fix the parameter values as stated in Table 21.1 and vary one or two parameters to demonstrate the parameter(s) effect on the system.

Global Behaviour Analysis

For the values expressed in Table 21.1, the trajectories initiating from various points ultimately converge to a particular point, that is, to $E_2 = (5.29828, 0.16762)$. This shows the global stability behaviour of the system. Global stability of the system (2) around $E_2 = (5.29828, 0.16762)$. Convergence of all the trajectories to E_2 after initiating from different points ensures this behaviour.

Parameter Variation Effect

In this section, we vary each parameter to record the overall behavioural changes of the system (2). We perturb each parameter up to 10-fold (only the refuge parameter is increased up to -fold as its highest value can be 1) and recorded the results in Table 21.1. In Fig. 21.2, we demonstrate the results described in Table 21.1. The critical roles of the three important parameters (σ, K, m) on the system behaviour disruption, are elaborated by the bifurcation diagram.

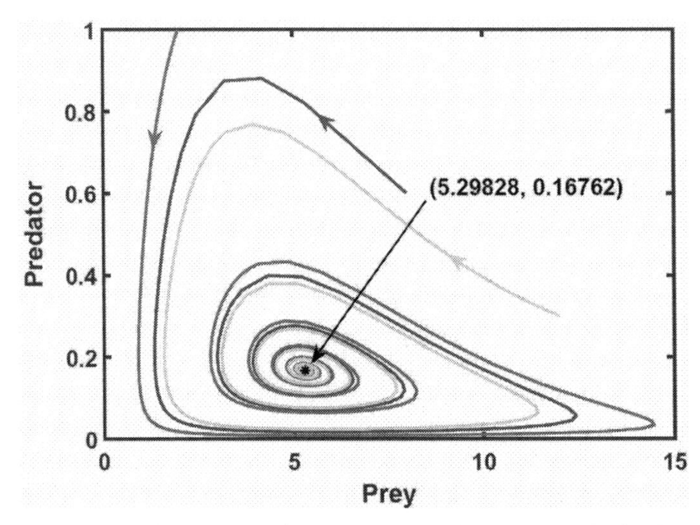

Fig. 21.1 Global stability of the system (II.1) around E2 = (5:29828;0:16762). Convergence of all the trajectories to E2 after initiating from different points ensure this behaviour

Table 21.1 Parameter values for the system (2) with references

Parameters	Values	References
K	10	(Wang, Zanette, & Zou, 2016)
α	0.2	(Zhang, Cai, Fu, &Wang, 2019)
b	0.01	(Zhang, Cai, Fu, &Wang, 2019)
a	0.5	(Zhang, Cai, Fu, &Wang, 2019)
σ	0.1	(S. Pal, Pal, Samanta, & Chattopadhyay, 2019)
m	0.25	(Zhang, Cai, Fu, &Wang, 2019)
γ	0.55	(Zhang, Cai, Fu, &Wang, 2019)
β	0.5	(Zhang, Cai, Fu, &Wang, 2019)
c	0.8	(Zhang, Cai, Fu, &Wang, 2019)

Effect of Hunting Cooperation

Predator cooperation in the prey hunting plays a very important role in system stability. The offset of the cooperation parameter (σ) value always drives the system towards coexisting steady state [Fig. 21.2(b)].

On the other hand, Fig. 21.2(a) shows that the rise of cooperation parameter (σ) from its default value upto -fold is capable of sustaining the CSS state. When σ increases ($\sigma > 1.8$), the system loses its stability and enters into an oscillatory coexistence state.

Effect of the Fear Parameter

Fear parameter positively affects the prey population and negatively on predator population. Figure 21.4 ensures that an increment of the value of K rises prey density and simultaneously

Table 21.2 Effect of the individual parameter on the overall system dynamics: By the variation of each parameter value individually the ranges of oscillatory coexistence (OC), co-existence steady state (CSS) and extinction of predator (EP) are recorded

Parameter	Range	System Dynamics
K	$K < 5.64$	OC
	$K > 5.64$	CSS
α	$\alpha < 0.291$	CSS
	$0.291 < \alpha < 1$	OC
b	$b < 0.0081$	OC
	$0.0081 < b < 0.0339$	CSS
	$b > 0.0339$	EP
a	$a < 0.45$	OC
	$0.45 < a < 0.66$	CSS
	$a > 0.66$	EP
σ	$\sigma < 0.183$	CSS
	$\sigma > 0.183$	OC
m	$m < 0.12$	OC
	$0.12 < m < 0.778$	CSS
	$0.778 < m < 1$	EP
γ	$\gamma < 0.54$	OC
	$0.54 < \gamma < 0.72$	CSS
	$\gamma > 0.72$	EP
β	$\beta < 0.39$	EP
	$0.39 < \beta < 0.854$	CSS
	$\beta > 0.54$	OC
c	$c < 0.62$	EP
	$0.62 < c < 0.85$	CSS
	$c > 0.85$	OC

decreases predator density. The numerical simulation presented in Fig. 21.2(a) portrays that the accession of fear from its default value permits the system to retain its coexisting steady state. The system switches from CSS to an oscillatory coexistence state when K decreases from the critical fold value 1.773-fold [Fig. 21.2(b)].

Effect of the Prey Refuge

Prey refuge has the most important role in the prey-predator interacting dynamics as it drives the system (2) through all three different states. The bifurcation diagram displayed by Fig. 21.5 demonstrates all the states as the initial values give the system an oscillatory coexistence state (green lines), followed by a coexistence steady state (blue line) and lastly extinction of the predators (red line). Also, the figure ensures that the increment of prey refuge gradually increases prey density and drives it to the maximum while decreasing the predator density and after a critical value of $m = 0.778$, it extinct. This behaviour justifies the biological

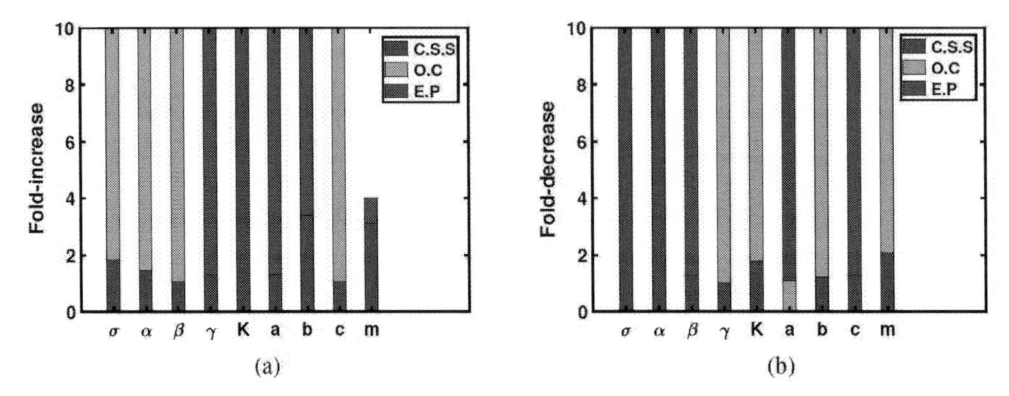

Fig. 21.2 Variation of system dynamics concerning parameter values perturbation: Parameters are perturbed over a range of ±10 folds from their defined values (except refuge parameter). Each parameter effect is represented by colour changing bars. Green colour indicates the parameter ranges. Blue represents the ranges associated with coexisting steady state (CSS), oscillatory coexistence (OC) and red represents the extinction of the predator (EP).

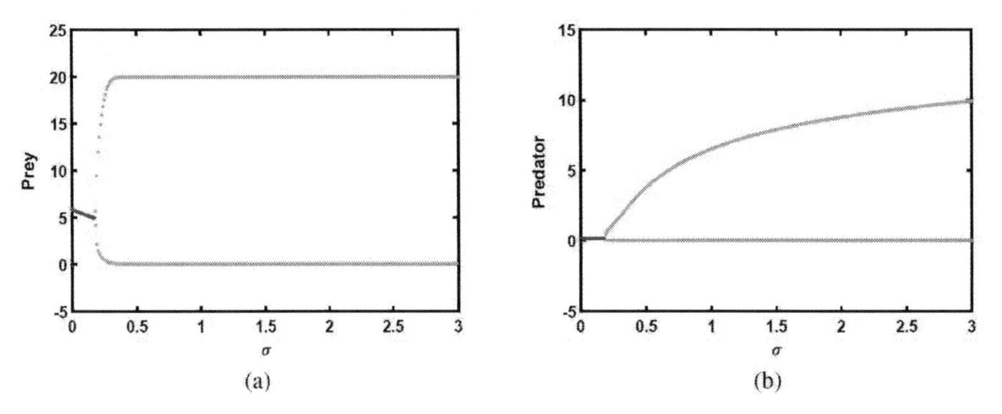

Fig. 21.3 Bifurcation diagram of the system (2) with respect to cooperation parameter σ. Here, the blue and green dotted lines represent the states CSS and OC, respectively

phenomena of prey refuge. The study suggests from Fig. 21.2(a) that the system shifts towards EP state from CSS with a rise to the prey refuge $m > 3.112$ folds. On the other hand, the system experiences a dynamical switch from CSS to OC state when prey refuge offsets towards the lower value $m < 2.083$ folds.

Simultaneous Effects of Parameters

Now, we present three-dimensional figures to analyse the effects of σ, K and m simultaneously on the system. The blue surface portrays the maximum population density whereas the red surface denotes the minimum population density. Both surfaces coincide at a stable state.

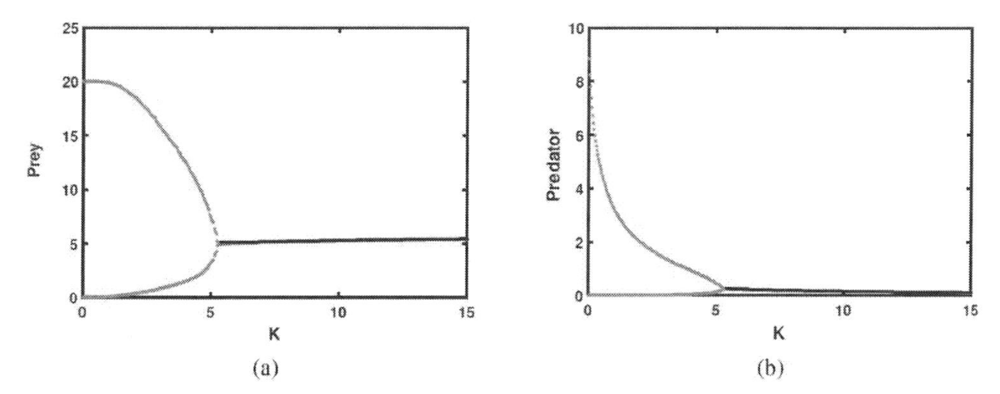

Fig. 21.4 Bifurcation diagram of the system (2) with respect to fear parameter K. Here, the blue and green dotted lines represent the states CSS and OC, respectively

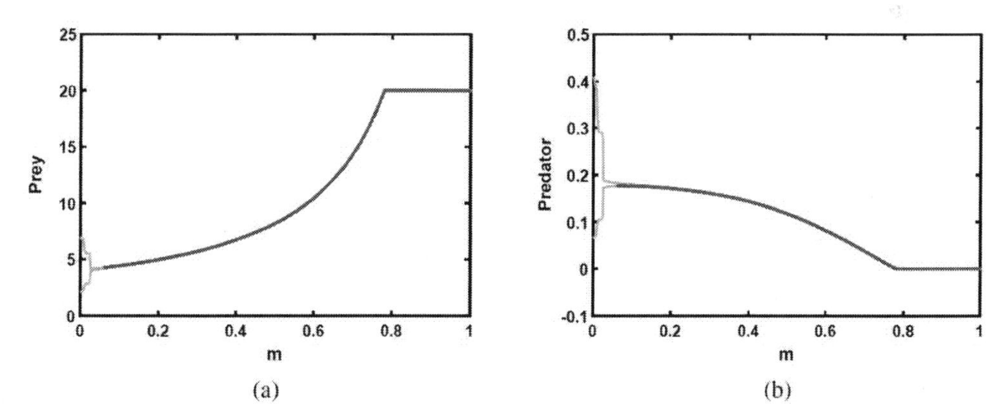

Fig. 21.5 Bifurcation diagram of the system (2) with respect to fear parameter m. Here the blue, green and red lines represent the states CSS, OC and EP, respectively

The simultaneous variation effect of cooperation (σ) and prey refuge (m) parameters is displayed in Fig. 21.6. Figures 21.6(a) and 21.6(b) show the dynamical changes of the prey and predator population, respectively. The three states of the system, that is, OC, CSS and EP are recorded on $m - \sigma$ surface for better understanding. The surfaces portray the oscillatory behaviour near the axial point (0.2, 0). It clearly represents that the lower value of prey refuge and higher value of predator cooperation supports system oscillation.

Figure 21.7 describes the system behaviour for the variation of fear and refuge level. Figure 7(a) and Figure 7(b) represent the effects on prey and predator population, respectively. The origin closed region, that is, lower value of both the variables force the system to oscillate. Moreover, the $K - m$ surface is presented in Fig. 21.7(c) that depicts all the regions showing different behaviours.

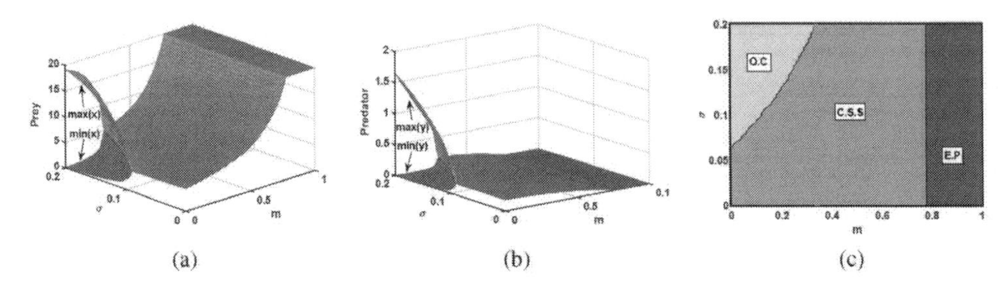

Fig. 21.6 Dynamical behaviour of system (2) depending on both the parameters σ and m: (a) Prey population dynamics, (b) predator population dynamics and (c) population behaviour on $\sigma - m$ surface

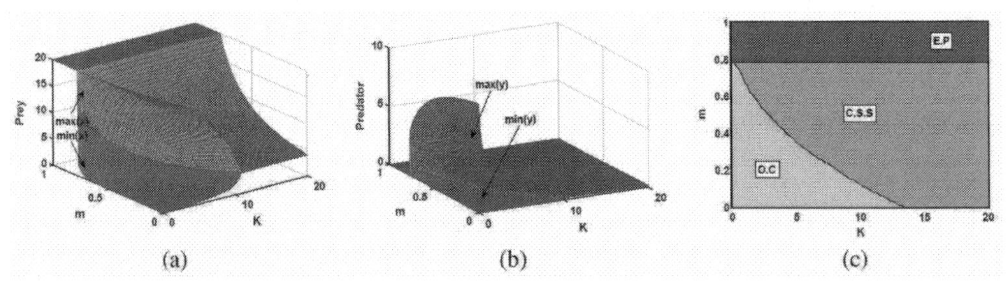

Fig. 21.7 Dynamical behaviour of system (2) depending on both the parameters m and K: (a) Prey population dynamics, (b) predator population dynamics and (c) population behaviour on $m - K$ surface

Occurrence of Hopf Bifurcation and its Direction

Considering two different fear effects on the system (2) we determined the Hopf bifurcation critical values of predator cooperation and presented the limit cycles and corresponding first Lyapunov coefficient in Fig. 21.9. For the default value $K = 10$, Trace(T) = 0 provides $\sigma^{[h]} = 0.183768$. At $\sigma^{[h]}$, $\Delta = 0.0123454 > 0$ and $\frac{dT}{d\sigma} = 0.333415 \neq 0$. For these values, the system oscillates in a bi-stability situation [Fig. 21.8(a)]. The corresponding first Lyapunov coefficient is positive at $\sigma^{[h]} = 0.183768$ [Fig. 21.8(b)], confirms that the Hopf bifurcation is subcritical. In this bi-stability situation, the system experiences two limit cycles simultaneously, depending on the initial values of the species. The blue circle represents the stable cycle, and the black dotted circle stands for unstable cycle. Species initiating from inside the unstable circle eventually stabilise to the interior equilibrium while initiation from outside provides periodic oscillation, that is, stable limit cycle.

On the other hand, for the value $K = 50$, Trace(T) = 0 provides $\sigma^{[h]} = 0.76268$. At $\sigma^{[h]}$, $\Delta = 0.00992416 > 0$ and $\frac{dT}{d\sigma} = 0.387909 \neq 0$. This scenario provides a negative value of the corresponding first Lyapunov coefficient, as shown in Fig. 21.8(d). Hence, the system experiences a supercritical Hopf bifurcation that produces a stable limit cycle around the coexisting equilibrium point [Fig. 21.8(c)].

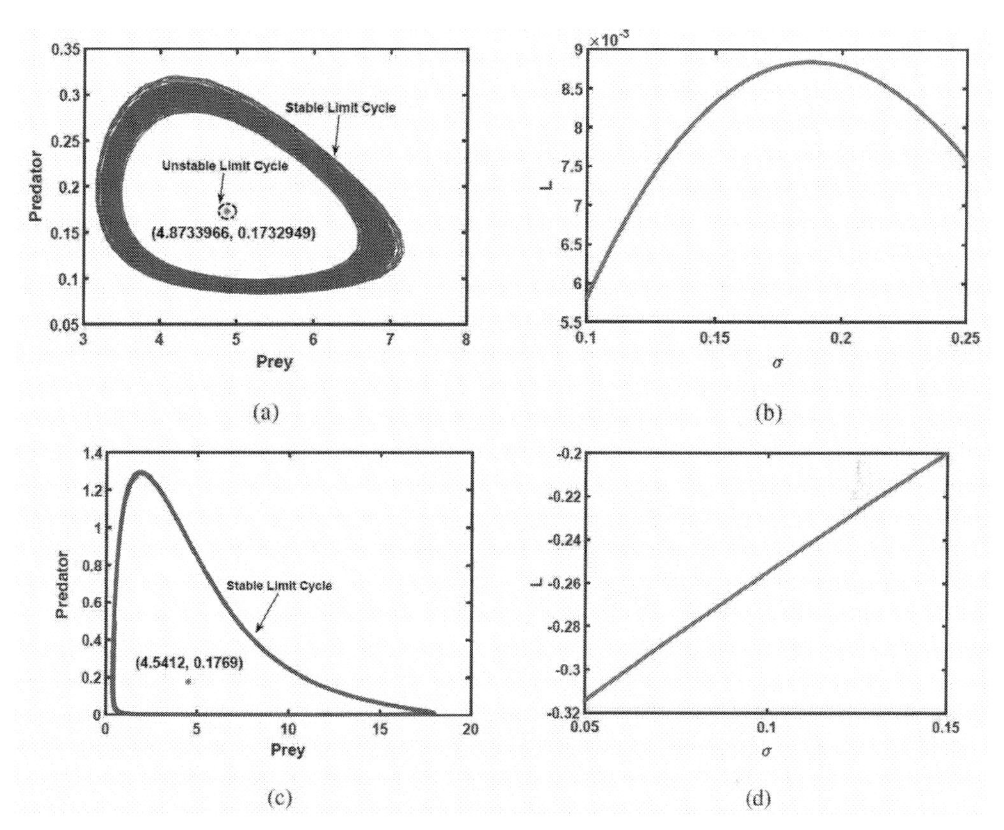

Fig. 21.8 Occurrence of limit cycle due to Hopf bifurcation: (a) Subcritical Hopf bifurcation/bi-stability, (b) First Lyapunov coefficient concerning Subcritical Hopf bifurcation ($K = 10$), (c) Supercritical or stable Hopf bifurcation and (d) First Lyapunov coefficient concerning supercritical Hopf bifurcation ($K = 50$)

A Case Study Analysis of Bow Valley, Banff National Park, Alberta

In this section, we analyse a case study held on Elk and Wolf populations of Bow valley in Banff National Park, Alberta. We have taken the Elk wolf interacting population data from the study of Mark Hebblewhite (Hebblewhite, 2013) and fit the model to 17 years' time course of interaction by using Matlab toolbox fminsearch. Hebblewhite presented the population densities from 1985 to 2007 but in this paper, we consider the data from 1991 to 2007 as the initial density of our model cannot be null. In the literature, a lot of models have been developed considering fear and cooperation factors, but none of them (to the best of our concern) shows data fitting to the real case study. In this case study, we try to emphasise our model validity.

The best-fitted parameter values are provided in Table 21.3. For the best-fit series, we have taken the data from 1991 as the initial value. Further, we calculate the RMSE (Root Mean

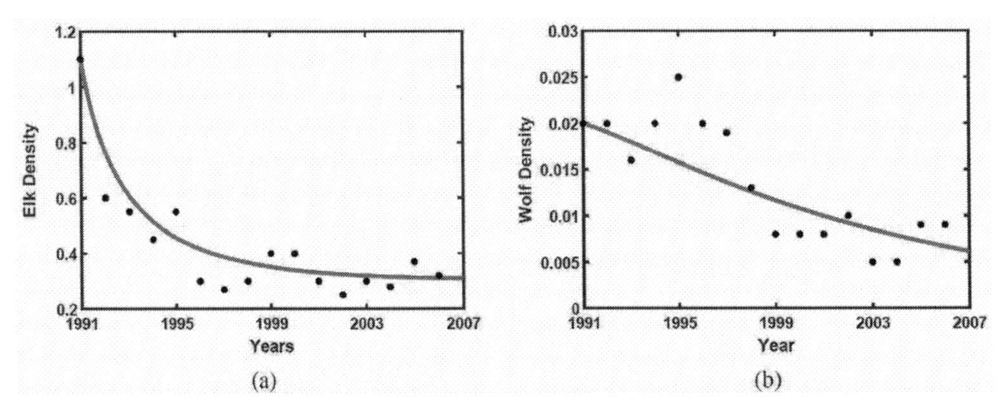

Fig. 21.9 16-year Bow Valley species data fitting to the proposed model. The black dots represent real species densities while the red lines represent the model density concerning the best-fitted parameter values

Table 21.3 Best-fitted parameter values of model (2)

Parameter	Values
K	0.1
α	0.15
b	0.5
a	0.2
σ	1
m	0.1
γ	0.1
β	0.127
c	0.5

square Error) that represents the deviation between the Bow valley population data and best-fitted data. Additionally, we calculate the corresponding p value. The RMSE value and p value are displayed in Table 21.4.

Table 21.4 RMSE value and p value for the data fitting of Bow valley case study and model (2)

Species	RMSE	p value
ELK	0.0939	0.000169
WOLF	0.0035	0.000105

The RMSE value and corresponding p value ensure that our model system (2) can efficiently elaborate the interacting population density and estimate future extinction threats on species.

Discussion and Conclusion

The foraging behaviour plays an important role in the dynamics of prey-predator system. Animals usually forage searching for food and mating. This phenomenon is very basic and natural for survival. Also, it exposes them towards predators. Naturally, prey population show some anti-predator behaviour concerning predator fear, for instance, less foraging, habitat switch and increased vigilance. This helps the prey population to increase their density. On the other hand, the predator population also undertakes some strategies to upgrade their density. For this purpose, predators usually attack in cooperation which helps them to chase prey more efficiently and also prevents carcass stealing. In the present study, we have considered a prey-predator interacting system concerning induced prey fear due to predators and predator cooperation hunting. Also, we incorporate the basic phenomenon of prey refuge and Holling type II functional response. First, we have shown the system is well-behaved and found three biologically relevant equilibrium points. The trivial equilibrium point always exists and unstable saddle. The predator-free equilibrium always exists and is locally stable if the predator death rate is too high or the attack rate is too low which is biologically significant. We also obtained the conditions for Hopf bifurcation and its direction. The analysis describes that the Hopf bifurcation can be supercritical or subcritical depending on fear and cooperation levels. This contrasts with the classical models that only experience supercritical Hopf bifurcation.

We perform numerical simulation to explore the potential roles of fear effect, cooperation level and prey refuge by releasing various parameters. We have portrayed the robustness of the system by varying each parameter over a range of ±10 folds from their default values. It provides the overall dynamics of the system. Also, we display the bifurcation diagram concerning the three major parameters cooperation level, fear level and prey refuge level. It satisfies the fact that the increment of fear and low cooperation provide global stability. Moreover, to reflect the effects of these three parameters simultaneously, we have taken two among them and represented the system behaviour by three-dimensional figures. Additionally, we portrayed the surface considering respective parameters to display different states on that surface. Limit cycles through Hopf bifurcation and respective first Lyapunov coefficient are presented for two different fear levels. For fear level $K = 10$, the system exhibits a subcritical Hopf bifurcation at cooperation level $\sigma = 0.183768$ and for fear level $K = 50$, a supercritical Hopf bifurcation at $\sigma = 0.76268$. Hence, cooperation along with fear can change the direction of Hopf bifurcation. Higher values support stable limit cycles.

Most importantly we have analysed a case study of Bow Valley, Banff National Park, Alberta. We have taken the real data of Elk and Wolf interaction from the study of Hebblewhite (Hebblewhite, 2013) and best fit our model to the densities from 1991 to 2007. The RMSE and p value for the data fitting ensures the validity of our model and enriched our work.

The present study has shed light on various behavioural dynamics in comparison to previous studies. The inclusion of cooperation and fear drags the system to a more natural scenario and describes various interesting behaviours, so that the system becomes more biologically significant. Hence, this study can help readers for better understanding and have enthusiasm toward ecology.

References

1. Beddington, J. R. (1975). Mutual interference between parasites or predators and its effect on searching efficiency. The Journal of Animal Ecology, 331–340.
2. Bednarz, J. C. (1988). Cooperative hunting harris' hawks (parabuteo unicinctus). Science 239, 1525–1527.
3. Berryman, A. (1992). The orgins and evolution of predator-prey theory. Ecology 73, 1530–1535.
4. Birkhoff, G., & Rota, G. C. (1978). Ordinary differential equations. John Wiley & Sons.
5. Brockmann, H. J., & Barnard, C. (1979). Kleptoparasitism in birds. Animal behaviour 27, 487–514.
6. Creel, S., & Christianson, D. (2008). Relationships between direct predation and risk effects. Trends in Ecology & Evolution 23, 194–201.
7. Creel, S., & Creel, N. M. (1995). Communal hunting and pack size in african wild dogs, lycaon pictus. Animal Behaviour 50, 1325–1339.
8. DeAngelis, D. L., Goldstein, R., & O'Neill, R. V. (1975). A model for tropic interaction. Ecology 56, 881–892.
9. Dill, L. M., & Fraser, A. H. (1997). The worm re-turns: hiding behavior of a tube-dwelling marine polychaete, serpula vermicularis. Behavioral Ecology 8, 186–193.
10. González-Olivares, E., & Ramos-Jiliberto, R. (2003). Dynamic consequences of prey refuges in a simple model system: more prey, fewer predators and enhanced stability. Ecological modelling 166, 135–146.
11. Hebblewhite, M. (2013). Consequences of ratio-dependent predation by wolves for elk population dynamics. Population Ecology 55, 511–522.
12. Huang, J., Ruan, S., & Song, J. (2014). Bifurcations in a predator–prey system of leslie type with generalized holling type iii functional response. Journal of Differential Equations 257, 1721– 1752.
13. Kooij, R. E., & Zegeling, A. (1997). Qualitative properties of two-dimensional predator-prey systems. Nonlinear Analysis: Theory, Methods & Applications 29, 693–715.
14. Kuang, Y., & Freedman, H. (1988). Uniqueness of limit cycles in gause-type models of predator-prey systems. Mathematical Biosciences 88, 67–84.
15. Lima, S. L. (2009). Predators and the breeding bird: behavioral and reproductive flexibility under the risk of predation. Biological reviews 84, 485–513.
16. Ma, Z., Li, W., Zhao, Y., Wang, W., Zhang, H., & Li, Z. (2009). Effects of prey refuges on a predator– prey model with a class of functional responses: the role of refuges. Mathematical biosciences 218, 73–79.
17. May, R. M. (1972). Limit cycles in predator-prey communities. Science 177, 900–902.
18. Pal, S., Pal, N., Samanta, S., & Chattopadhyay, J. (2019). Effect of hunting cooperation and fear in a predator-prey model. Ecological Complexity 39, 100770.
19. Pal, S., Pal, N., Samanta, S., & Chattopadhyay, J. (2019). Fear effect in prey and hunting cooperation among predators in a leslie-gower model *16*(5), 5146-5179.
20. Pitcher, T., Magurran, A., & Winfield, I. (1982). Fish in larger shoals find food faster. Behavioral Ecology and Sociobiology 10, 149–151.
21. Prasad, K. D., & Prasad, B. (2019). Qualitative analysis of additional food provided predator–prey system with anti-predator behaviour in prey. Nonlinear Dynamics 96, 1765–1793.
22. Ripple, W. J., & Beschta, R. L. (2004). Wolves and the ecology of fear: can predation risk structure cosystems?. BioScience 54, 755–766.
23. Samaddar, S., Dhar, M., & Bhattacharya, P. (2020). Effect of fear on prey–predator dynamics: Exploring the role of prey refuge and additional food. Chaos: An Interdisciplinary Journal of Nonlinear Science 30, 063129.

24. Schneider, D. C., & Harrington, B. A. (1981). Timing of shorebird migration in relation to prey depletion. The Auk 98, 801–811.
25. Sih, Schneider A., Bolnick, D. I., Luttbeg, B., Orrock, J. L., Peacor, S. D., Pintor, L. M., Preisser, E., Rehage, J. S., & Vonesh, J. R. (2010). Predator–prey naïveté, antipredator behavior, and the ecology of predator invasions. Oikos 119, 610–621.
26. Thieme, H. R. (2018). Mathematics in population biology. Princeton University Press, Vol. 12.
27. Vucetich, J. A., Peterson, R. O., & Waite, T. A. (2004). Raven scavenging favours group foraging in wolves. Animal behaviour 67, 1117–1126.
28. Wang, J., Cai, Y., Fu, S., & Wang, W. (2019). The effect of the fear factor on the dynamics of a predator prey model incorporating the prey refuge. Chaos: An Interdisciplinary Journal of Nonlinear Science 29, 083109.
29. Wang, X., Zanette, L., & Zou, X. (2016). Modelling the fear effect in predator–prey interactions. Journal of mathematical biology 73, 1179–1204.
30. Wiggins, S. (2003). Introduction to applied nonlinear dynamical systems and chaos. Springer Science & Business Media Vol. 2.
31. Zanette, L. Y., White, A. F., Allen, M. C., & Clinchy, M. (2011). Perceived predation risk reduces the number of offspring songbirds produce per year. Science 334, 1398–1401.
32. Zhang, H., Cai, Y., Fu, S., & Wang, W. (2019). Impact of the fear effect in a prey-predator model incorporating a prey refuge. Applied Mathematics and Computation 356, 328–337.